Crafting the Quantum

Transformations: Studies in the History of Science and Technology
Jed Z. Buchwald, general editor

For a list of the series, see page 379.

Crafting the Quantum

Arnold Sommerfeld and the Practice of Theory, 1890–1926

Suman Seth

The MIT Press
Cambridge, Massachusetts
London, England

© 2010 Massachusetts Institute of Technology

All rights reserved. No part of this book may be reproduced in any form by any electronic or mechanical means (including photocopying, recording, or information storage and retrieval) without permission in writing from the publisher.

For information on special quantity discounts, email special_sales@mitpress.mit.edu.

Set in Stone sans and Stone serif by Toppan Best-set Premedia Limited. Printed and bound in the United States of America.

Library of Congress Cataloging-in-Publication Data

Seth, Suman, 1974–
Crafting the quantum : Arnold Sommerfeld and the practice of theory, 1890–1926 / Suman Seth.
   p. cm. — (Transformations : studies in the history of science and technology)
Includes bibliographical references and index.
ISBN 978-0-262-01373-4 (hardcover : alk. paper)
1. Sommerfeld, Arnold, 1868–1951. 2. Quantum theory. 3. Physics. I. Title.
QC16.S76S48 2010
530.09'04—dc22

2009022212

10 9 8 7 6 5 4 3 2 1

Contents

Acknowledgments   vii

Introduction   1

I

1 The Physics of Problems: Elements of the Sommerfeld Style, 1890–1910   13

2 Pedagogical Economies: The "Sommerfeld School" and the Problems of Teaching   47

3 The Kaiser's Physicists: The Sommerfeld School Goes to War   71

II

4 The Practice of Principles: Planck, Experiment, and the "Thermodynamic Method"   95

5 The Dynamical and the Statistical: Sommerfeld, Planck, and the Quantum Hypothesis   139

III

6 *Prinzipienfuchser* and *Virtuosen*: Theoretical Physics after World War I   177

7 Crafting the Quantum: Sommerfeld, Bohr, and the Older Quantum Theory   201

Conclusion   247

Notes   271
Bibliography   347
Index   373

Acknowledgments

I began work on this book more than a decade ago and have accrued many debts since then. Funding to undertake and complete a doctoral degree was provided by a Sydney University Traveling Scholarship and by Princeton University. Grants for language study in Germany were supplied by Princeton's History Department and its Council for Regional Studies. Material support for research and writing has been provided by the Deutscher Akademischer Austausch Dienst (DAAD), the National Science Foundation (under award 0140149), the Deutsches Museum in Munich, the Max Planck Institute for the History of Science in Berlin, and the Department of Science and Technology Studies at Cornell University. Archivists at the Deutsches Museum in Munich have been extraordinarily helpful, and it is a pleasure to thank them. Publication of the book has been supported by a great from the Hull Mentorial Publication Fund of Cornell University.

I have been lucky in enjoying the support and guidance of a number of advisors and mentors. Ian Johnston and Michael Shortland set me on the path toward work in the history of physics by co-supervising an honours thesis at the University of Sydney. Alan Chalmers shaped my ongoing interest in blending the history and philosophy of science. I hope he can see his stamp in this book. At Princeton, I benefited enormously from discussions with the now sorely missed Gerry Geison and Mike Mahoney. Angela Creager taught me, by advice and example, more about being a scholar and a teacher than I can properly acknowledge here. Norton Wise, most simply, taught me how to think like a historian. He and Elaine remain ongoing resources for a life in academia and elsewhere. Otto Sibum has served as both a reader for my dissertation and as a post-doctoral supervisor. This book is immeasurably better for his intellectual openness and acumen. I could not ask for better colleagues than those in Cornell's Department of Science and Technology Studies. They have provided an intellectual environment at once stimulating, critical, and supportive. I am particularly indebted to Peter Dear, who read every page of my manuscript and whose insights, corrections, and requests for clarification have always made for better work.

Portions of the book have been presented in a number of different settings. For their comments, I am indebted to Eric Ash, David Aubin, Charlotte Bigg, David Bloor, Graham Burnett, David Cahan, Kristian Camilleri, Cathryn Carson, David Cassidy, Jamie Cohen-Cole, Jeroen van Dongen, Moritz Epple, Clayton Gearhart, Mike Gordin, Tony Grafton, John Heilbron, Mary Henninger-Voss, Michel Janssen, David Kaiser, Dan Kevles, Ursula Klein, Ron Kline, Kevin Lambert, Manfred Laubichler, Molly Loberg, Mike Lynch, Suzanne Marchand, Ole Molvig, Tania Munz, Cat Nisbett, Kalil Oldham, Trevor Pinch, Rebecca Press-Schwartz, Sarah Pritchard, Jürgen Renn, Simon Schaffer, Arne Schirrmacher, Skuli Sigurdsson, Arlan Smith, Helen Tilley, Helmuth Trischler, Jay Turner, Heidi Voskuhl, Matt Wisniowski, and Stefan Wolff. I am grateful to Margy Avery and Paul Bethge at the MIT Press. Thanks also to Robert Schombs for preparing the index.

I owe a special debt to Richard Staley for helping me think through many of the themes explored in this book, in particular the ways in which history functions simultaneously as an actor's and an analytic category. Andy Warwick's work has been a model for my analysis of Munich's pedagogical culture. His engagement with my project over the years has been crucial to its development. *Crafting the Quantum* would never have made it past the manuscript stage had it not been for Jed Buchwald's support and editorial eye. Without Michael Eckert's exemplary scholarship on Sommerfeld, the book could never have been attempted. For this, and for his unstinting personal and intellectual generosity, I am deeply in his debt.

Much of my gratitude is due to those who read comparatively little of the manuscript but whose friendship made its writing both possible and enjoyable. Thanks to Lisa Bailey, Celia Bloor, Scott Bruce, Holly Case, James Cunningham, Anne Lester, Vlad Micic, Ellie Pridham, Carol Smith, Jay and Darcy Turner, Donald Williams, and Chuck Wooldridge. My one sadness in finally completing this project is that my mother—the person who would be most proud to see it—cannot be here to share in the celebrations. My siblings, Sanjay and Vanita, have been, since childhood, my closest friends. Some debts cannot be articulated, let alone acknowledged, but my love for you both is boundless. Any place where I have visited Raju and Nishad has felt like home, complete with chaos, laughter, and long conversations. My father has encouraged me in all my endeavors. I hope that some of his profound skill as a writer has made it into these pages.

I dedicate this book to my wife, Aminda Smith. She has shared the entirety of my life in academia. More importantly, she has shown me how many lives there are beyond its walls.

Crafting the Quantum

Introduction

In early 1909, Arnold Sommerfeld, professor of theoretical physics at the Ludwig-Maximilians Universität in Munich, received a letter from a publisher in Leipzig. Samuel Hirzel told him of recent travels that had brought him into contact with "various physicists at universities and technische Hochschulen."[1] From all sides, Hirzel claimed, he had heard talk of "the desire and the need" for a "short, new, modern textbook of theoretical physics." Since producing such a text would be too time consuming for one man, the publisher suggested that Sommerfeld share the authorship with Max Planck, professor of theoretical physics at the Friedrich-Wilhelms-Universität in Berlin since 1889. Planck had earlier written to Hirzel, agreeing that a "new textbook written in a modern theoretical physical spirit" was thoroughly necessary. He had reacted with enthusiasm to the prospect of working with Sommerfeld, and he expressed his dismay when informed that Sommerfeld had declined the offer. "It is certainly a pity," wrote Planck, "that you do not want to collaborate on a textbook for theoretical physics. I had just been thinking that we would, in a certain sense, complement each other well."[2]

The letter reveals that theoretical physics—a relatively young sub-discipline at the time—was well defined enough in 1909 for a major publisher to seek a new textbook in the subject. One may also note that there was now the perception that a "modern" theoretical physics existed, and hence that an "older" theoretical physics had been superseded. Here Hirzel may have had in mind the *Compendium of Theoretical Physics* written in the mid 1890s by Woldemar Voigt, a theoretician at the Georg-August-Universität in Göttingen. Voigt bemoaned its timing, realizing that it was published at a moment of considerable change, as researchers began increasingly to take up problems in microphysics: "So the book was obsolete almost with its appearance! And the great work and joy I have invested in it almost make me sorry now."[3] The letter makes it evident that Sommerfeld was seen, at this point, only three years after taking up his position at Munich, as one of the major representatives of the "modern theoretical physical spirit," and as a counterpart to his older and more prominent colleague at Berlin.

Planck's textbooks on theoretical physics began appearing in 1916. Sommerfeld's would take much longer. Yet even after so much time had elapsed, Sommerfeld maintained the notion, suggested by Planck at the beginning of the twentieth century, that his approach and Sommerfeld's would "complement each other well." In his preface to the first edition of the *Mechanics* volume in 1943, Sommerfeld wrote:

> In print, as in my classes, I will not detain myself with the mathematical foundations, but proceed as rapidly as possible to the physical problems themselves. My aim is to give the reader a vivid picture of the vast and varied material that comes within the scope of theory from a suitably chosen mathematical and physical vantage point. With this in mind I shall not be too concerned if I have left some gaps in the systematic justification and axiomatic structure of the work.... Whereas the lectures of Planck are irreproachable in systematic formulation, I believe that I can claim for mine a greater variety of subject matter and a more flexible handling of the mathematical apparatus. Moreover, I gladly refer the reader to the more complete and often more thorough treatment of Planck, especially in thermodynamics and statistical mechanics.[4]

The present book offers a detailed treatment of Sommerfeld's approach to physics: its origins, its methods, and its pedagogical and disciplinary consequences.[5] As the quotation above may suggest, however, it is not possible to do this by treating his approach in isolation. Sommerfeld, and others, saw his vision of physics—concerned with "the physical problems themselves"—as qualitatively different from an alternative kind of physics, including Max Planck's or Albert Einstein's, that was concerned with subsuming all physical phenomena under a few abstracted, generalized axioms. Sommerfeld drew questions from a variety of sources, including and even emphasizing questions of economic, military, or technological benefit. The centrality of theoretical analyses of wireless telegraphy to the work of the Sommerfeld School may provide the most obvious example of such emphasis. The form of his solutions would be of a similar character, often attempting to get to a numerical answer that could be compared with real-world engineering problems, or working hand in hand—as with his work on the quantum in the 1920s—with extensive experimental data. In contrast, Planck promoted a practice of theoretical physics devoted to abstract, de-anthropomorphized, de-historicized, "pure" principles, with vastly fewer references either to experimental data or to the application of his work.

Thus, in contrast with the "physics of principles" espoused by Planck, Einstein, and (to an extent) Niels Bohr, Sommerfeld's was a "physics of problems."[6] Peter Debye, the first of Sommerfeld's many students and later a Nobel laureate, described one conversation as follows:

> [Sommerfeld] said, "Well, Maxwell's equations are no good if you are going to talk about"—I can still hear him say that—"about general solutions." He always said: "I have learned from Fourier that you have to make a solution and develop it so far that you can really put numbers in it!" And apply it, you see. That was his main purpose, not fundamental equations themselves, but after you have equations, to make a mathematical solution which is really practical.[7]

Introduction

The Oxford University physicist Frederick Lindemann made a similar point—although it was directed at Sommerfeld's approach to pedagogy—in a 1933 letter to Einstein:

> I have the impression that anyone trained by Sommerfeld is the sort of man who can work out a problem and get an answer, which is what we really need at Oxford, rather than the more abstract type who would spend his time disputing with the philosophers.[8]

In a sense the present work is a study of contrasting if mutually illuminating styles, although by "style" I mean more than what is connoted by "professional style" and less than is included in the broad-scale notion of "styles of scientific thought."[9] I discuss the physics of principles and the physics of problems in terms of their contrasting methodologies and ontologies, as well as in terms of their differing attitudes toward pedagogy and the place of experimental thinking within theoretical practice. Methodologically, the distinction was between the search for general principles and the focus on specific problems. What difference this would make to the detailed technical practices of physics is a fundamental question that I seek to answer here.[10] Ontologically, one sees in the twentieth-century writings of both Planck and Einstein an attempt to found physics on principles that transcended any particular substantivist basis. In place of a physics founded on mechanics, energetics, or electromagnetic theory, Planck and Einstein offered a physics based on de-materialized principles. "The energy view [*Naturanschauung*]," Planck wrote in 1910, "rests on the principle of energy, while the mechanical view rests on the principle of momentum."[11] Sommerfeld, on the other hand, remained a committed proponent of the (substantivist, if not materialist in the strict sense) electromagnetic view of nature, criticizing both the quantum and relativity theories on the basis of the programmatic aims of that worldview and then (after he came to accept and even further both) insisting on its more limited place within the structure of modern physics.

Pedagogically, the physics of principles had limited success. Neither Planck nor Einstein would ever have a school associated with him. The Sommerfeld School, on the other hand, could make a strong claim to have been the most successful educational institution for theoretical physics in the twentieth century. At least eight Nobel laureates were associated with it. Einstein once commented that Sommerfeld seemed to stamp good students out of the ground, and in 1928 Max Born spoke of Sommerfeld as the founder of a school that was spreading his "spirit" to positions in theoretical physics throughout the world.[12] Finally, as a detailed contrast between Planck and Sommerfeld will demonstrate, the difference between a principled thermodynamics and a problem-based electrodynamics or mechanics carried with it a difference in attitude toward the relationship between theory and experiment. Whereas for Planck the place of experiment within theoretical practice was limited to the testing of conclusions, for Sommerfeld experiment was a constitutive element at multiple stages in the production of theoretical work. In the 1920s, in fact, Sommerfeld would abandon the quest for theoretical explanation altogether, seeking instead to produce rules for the understanding and organization of atomic spectra from the direct study of empirical regularities in spectroscopic data.

The contrast of styles sketched above also illuminates the multifaceted nature of theoretical physics as a discipline in the early twentieth century. Earlier studies of the subject tended to define it either too narrowly or too broadly. On the one hand, one must avoid the temptation to equate physical theory with theoretical physics. Whereas physical theory has been an integral part of analyses in natural philosophy and physics since at least the scientific revolution, theoretical physics (as an independent research sub-discipline) is a much more recent creation.[13] Only a few chairs for the subject existed in the 1890s, and several of those were, according to Wilhelm Wien, moribund. "Theoretical physics in Germany," Wien wrote to Sommerfeld in 1898, "lies as good as completely uncultivated." One could find "pure" theoretical physics only at the universities in Berlin and Göttingen. Munich's chair was empty. "Theoretical physics today," Wien concluded grimly, "finds no takers."

The paucity of full professorships for theoretical physics in the late nineteenth century, though it would not last long, implied that a few individuals (Planck and Sommerfeld among them) would play disproportionate roles in shaping the development of the field[14] and that the path was relatively open for these practitioners to define the field as they saw fit. This latter point is crucial, for to the modern eye many of the subjects examined by Sommerfeld (including ballistics, hydrodynamics, the theory of wireless telegraphy, and the analysis of gyroscopic motion) belong more obviously to mathematical physics or applied mathematics. Sommerfeld and his contemporaries, however, considered them perfectly appropriate topics for dissertations within theoretical physics. One can thus see how an excessively narrow focus by historians on the "revolutions" in microphysics, quantum theory, and relativity has led to analyses of these subjects and their methods of solution at the expense of a range of topics that made up theoretical physics more generally.[15] This book thus aims to recover both the wide field of possibilities open to theoretical physicists in the first decades of the twentieth century and the way this field would gradually narrow to such an extent that Max Born would declare in 1952 (with a hindsight that ignored his own earlier statements to the contrary) that Sommerfeld had never been a theoretical physicist.[16]

It should be clear that what is offered here is a history of the social construction of theoretical physics. The discussion thus far, however, has been concerned largely with explicating the latter part of this term, describing the modes of practice and meaning making through which theoretical physicists made (or constructed) and propagated their forms of knowledge and their socio-disciplinary environment.[17] Understanding what is involved in these processes, however, also requires close attention to the multiple areas of intellectual, political, and social life from which science draws resources and to which it contributes. As several scholars have shown, for Planck and Einstein a physics based on absolute principles was an answer to questions posed in fields as diverse as history, philosophy, and political theory, as well as in the physical sciences broadly construed.[18] For the physics of problems, the need to situate research papers, textbooks, reports to the faculty on students' dissertations, lecture notes, and the texts of public speeches within a specific historical context is equally necessary. Recent scholarship has emphasized how, after 1871, the sciences joined industrial and military concerns to support the new German empire. David Cahan, for example, has argued that the "institutional revolution" in German physics that occurred between 1865 and 1914 was "due principally to the social needs of modernizing German society."[19] Certainly contemporary voices noted the connection between the success of certain disciplines and those disciplines' social, political, and (especially) economic utility. "In the last few decades," Friedrich Kohlrausch argued before the Prussian Academy of Sciences in 1896, "physics has grown out of its formerly outwardly very modest place into a recognized position of the first rank. It owes this, on the one hand, to its teaching, whose manifold power of cultivating thought and creativity no longer can be denied by anyone. However, if one were to analyze the motives which have led governments and legislatures to approve many millions for physics institutes, then the effective motive would prove to be the connection of physical research with life, with technology.... Physics has come to its rich resources through its interaction with cultural development, according to the modern principle, that one must risk capital in an enterprise that promises to be of use."[20]

Yet it was not merely new *institutes* that signaled a changing relationship between the sciences and the state. Perhaps the true revolution was signaled by the creation of an essentially new educational *institution*: the *technische Hochschule* (technical college). Especially after the 1870s, these institutions formed the training ground for the German engineering profession, which gained significantly in social power as the new nation rapidly industrialized. It is crucial to an understanding of Sommerfeld's physics of problems that he spent considerable time teaching at one of these technical colleges. "The time is gone once and for all," he wrote while there, "when the physicist and mathematician could superciliously hold himself aloof from the pursuits of engineering [Technik] because he perceived a smaller degree of scientific rigor in these branches than in his own particular sphere. The technical sciences [*technischen Wissenschaften*], at least here in Germany, have developed from their own innate power a confident and self-sufficient position. We the theoretical enquirers record it to our honor, if we can assist in the building up of the technical sciences [*technischen Wissenschaften*], and we appreciate our good fortune whenever it brings us into active contact with the problems of engineering [*Technik*]."[21]

For six years before 1906, when he took up his professorship at Munich, Sommerfeld was at the technische Hochschule in Aachen. While there he pursued questions about gyroscopic theory, continuing his co-authorship of a four-volume work that looked at the mathematical theory but also at applications to gyroscopic compasses, torpedoes, and spinning shells. He also conducted research into the hydrodynamic theory of lubrication and the electrodynamic theory of wireless telegraphy. Many of these topics would be carried over to the university at Munich, where Sommerfeld taught them in theoretical physics courses, assigning related problems as dissertation topics. In 1922 he would even speak of "the craft of the quantum" [*die Technik der Quanten*]. It is, in part, the aim of the present book to understand precisely what such a phrase might mean.

Part I of the book describes the roots of Sommerfeld's style, its transmission to his students, and its application and adaptation during World War I. Part II explores the methodological, ontological, and philosophical implications of the physics of principles as it was deployed by Planck, then turns to an explicit comparison of "principles" and "problems" at the 1911 Solvay conference. Part III covers the postwar years, continuing the story of Sommerfeld's methods while expanding the range of "physicists of principle" under discussion to include the work of Einstein and Bohr for the years up to and just after the development of quantum mechanics.

Chapter 1 paints a picture of Sommerfeld's physics of problems as a fusion of mathematics, physics, and engineering—the three phases of Sommerfeld's professional and pedagogical life, drawn together by the pursuit of solutions to specific questions. After 1906, having held positions teaching first mathematics and then technical mechanics, Sommerfeld refashioned himself into a theoretical physicist. Treating problems that he previously had referred to as mathematical, Sommerfeld now consciously approached these problems from a more physical perspective. Similarly, he sought to meld technological applications with his mathematical and physical methods. Yet, at the same time, the construction of theoretical physics was not merely about the appropriation of parts of fields beyond the disciplinary borders of physics, for the process was also one of reconceptualizing physics itself. If Sommerfeld selected aspects of mathematics and technical mechanics, then he did the same with physics, emphasizing those parts of the field that were in accord with the electromagnetic view of nature. It was the fusion of these selections from three different fields that made up the theoretical physics of the Sommerfeld School.

Chapter 2 seeks to account for Sommerfeld's prodigious talents as a teacher—talents that produced two generations of theoretical physicists. The first section describes the day-to-day practice of pedagogy in Munich. My aim there is to reconstruct, for the Sommerfeld School, what Andrew Warwick has called a "pedagogical economy"—the structural, institutional, and personal elements that made successful training possible.[22] The second section offers a comparative perspective, contrasting pedagogical economies in two of the most successful schools of the late nineteenth and early twentieth centuries: Cambridge under the tutelage of Edward Routh, and

Sommerfeld's Munich school. In particular, I contrast the "discipline" that has been deemed characteristic of the former with the institutionalization of Sommerfeld's more flexible and topically wide-ranging teaching practices. Sommerfeld, I argue, taught creativity.

In chapter 3 I examine the tasks undertaken by members of the Sommerfeld School during World War I, pointing to the strong continuity between the war work of Sommerfeld's students and the dissertation topics that they had previously undertaken: between the physics of problems and the problems of war. Ludwig Hopf, whose dissertation had dealt with the hydrodynamic problem of turbulence, was reclaimed by the aircraft command and was assigned the task of determining the stability of aircraft wings. Fritz Noether, who had edited the fourth volume of Sommerfeld's and Felix Klein's text on gyroscopes, worked (as did Sommerfeld) on the stability of spinning shells. And in 1911 Wilhelm Lenz completed a dissertation that dealt with the electromagnetic field around a spool of current carrying wire, and with its resistance, self-induction, and capacitance. During the war, Lenz worked as a *Funker* (wireless operator). He sent Sommerfeld regular reports on his studies of the physical functioning and directional operation of antennas, a topic regularly considered by members of the Sommerfeld School.

As well as illustrating the profoundly practical aspects of Munich theoretical physics and the impact of Sommerfeld's teaching practices, the experiences of these "Kaiser's physicists" make it possible to re-imagine the war years not in terms of the perversion of a "pure" discipline such as theoretical physics, nor even as an interruption of real work in the field, but as a militarily inflected continuation of that work.

Chapter 4 supplies a direct parallel to chapter 1. Just as the earlier chapter discusses the practices and methodologies involved in Sommerfeld's physics of problems, chapter 4 offers an analysis of Planck's "practice of principles." Planck's anti-positivism, his emphasis on the transcendental truth value of general principles, and his distaste for micro-physical modeling have all been well studied.[23] Less closely analyzed has been the means by which Planck could move, as he did, from the generality of abstracted thermodynamic laws to the details of atomic and molecular processes. As Planck himself noted in a lecture describing his thermochemical studies in 1891, it seemed paradoxical that a method that eschewed all "special molecular representations" should provide such insight into "the world of the molecule."[24] The means of resolving the paradox were Planck's "ideal processes." Chapter 4 tracks the changes in Planck's attitudes toward the place of the ideal within theoretical physics during three distinct periods of his working life. The first section examines his comparatively unstudied work on thermochemistry from the 1880s to the first years of the twentieth century. In the early 1890s, Planck accorded processes that were ideal in the sense that they could not be experimentally realized (even approximately or momentarily) only a limited, heuristic role within theoretical reasoning. Ten years later, when Planck defended his theory of solutions in the pages of the *Annalen der Physik*, much of this timidity was gone. Now he merely deemed it a matter of personal preference if a physicist wished to avoid the use of ideal processes, "a tool of research that is just as fruitful as it is well tried and well tested."[25]

The ten-year period during which Planck expanded the role of ideal processes beyond that of mere "pathfinders" was also, of course, the period in which he completed the analysis of black-body radiation that led to his postulation of the quantum hypothesis. The connection is more than merely temporal. The idealist logic that was essential to his theory of dilute solutions may also be discerned in Planck's idiosyncratic deployment of a Hertzian resonator in his work on irreversible radiative processes, the topic of the second section of chapter 4. The third section pursues the theme into what have usually (and problematically) been understood as Planck's philosophical rather than scientific works, interrogating the logic behind the vision of the "physics of the future" that he articulated in a public lecture in Leiden in 1908. His speech concerned the need to "de-anthropomorphize" physics, in part by removing the question of experimental realizability from the definition of its principles. Keeping Planck's responses to critiques of his ideal processes in mind allows us to comprehend what he called "the unity of the physical world picture" as not merely the articulation of a long-held philosophical position, but a temporally specific reaction to objections to his mode of theoretical analysis.[26]

In chapter 5, Planck's and Sommerfeld's visions are compared explicitly. Their two papers at the 1911 Solvay conference in Brussels—the first major conference devoted to the problem of the quantum—are near-emblematic examples of their broader positions. That these positions may be understood as oppositional, both in actors' categorical terms and in analytic terms, is evident from Paul Langevin's characterization of the Solvay conference as laying out two antithetical ways to proceed toward a clarification of the quantum concept: Planck's "statistical" method and Sommerfeld's "dynamical" method. Chapter 5 concludes with a look at the successive attempts by Planck and Sommerfeld to adjust to, modify, and make use of each other's visions and techniques. In Planck's adoption and adaptation of Langevin's distinction between the "dynamical" and the "statistical," and in Sommerfeld's utilization of Planck's "statistical" method in his reformulation of Niels Bohr's quantization conditions, one can see the flexibility and (ultimately) the resilience of the not quite dialectically opposed physics of principles and problems.

Chapter 6 begins with a general description of the state of theoretical physics in the aftermath of Germany's defeat in the World War. It continues by following the spread of the language of the physics of principles in the 1920s, seeking to describe and explain the success of this discourse in sites as varied as Berlin, Cambridge, and Copenhagen. The strength and robustness of the physics of principles, I contend, did not arise from a consensus over the meaning or content of such a physics, but rather from the capacity of the word 'principle' to embrace at times dramatically different—indeed diametrically opposed—philosophical positions.

The significance of Sommerfeld's work to the practice of theoretical physics remained unchanged in the Weimar era, and the 1920s saw the production of a new generation of students, including Wolfgang Pauli, Werner Heisenberg, Linus Pauling, and Hans Bethe. Chapter 7 explores the continuation of the Sommerfeld School, after World War I, as once again the representative problem-based opposition to a physics (or several different kinds of physics) based on generalized principles. Yet much was also new in the postwar years. Sommerfeld and others increasingly contrasted his distinctive method not to that produced in Planck's Berlin, but rather to that produced in Bohr's Copenhagen. That method, too, underwent a striking alteration. In 1916, as he completed his extension of Bohr's earlier analyses, Sommerfeld affirmed his strong commitment to a realist understanding of Bohr's model: periodic orbits within the atom possessed a "real existence." Hopes were high that theoretical physics could reveal the mechanisms behind both atomic structure and even the most complex spectral lines. Within five years, however, in the face of increasing difficulties in achieving a *modellmässig* (model-based) account of either set of phenomena, Sommerfeld began to emphasize an alternative. Models would be eschewed in favor of a phenomenological investigation of spectral data, one that sought to draw information concerning *Gesetzmässigkeiten* (lawful regularities) directly from the vast array of empirical materials then available. Combining phenomenological analysis with what he termed *Mystik* (an aesthetic sense of the harmony that ruled within the atom), Sommerfeld contrasted his approach with the "more indirect" methods involved with Bohr's correspondence principle. In the chapter's detailed reconstruction of Sommerfeld's oft-noted yet rarely analyzed study of "number mysteries" within the atom, the physics of problems re-emerges as an essential element in the craft of the quantum.

In the conclusion, I discuss the historical and historiographical implications of my study of Sommerfeld's practice of theory. For the former, I suggest that the historians' neglect of Sommerfeld's contributions to the development of quantum mechanics has been unfortunate. Pauli and Heisenberg were active members of the Sommerfeld School during precisely the period in which their teacher and mentor shifted his intellectual focus from an analysis of models to one of *Gesetzmässigkeiten*. Sommerfeld's craft served as a resource for theirs, and the origins of Pauli's exclusion principle may be explicated by understanding Pauli's problem as central to the work of the Sommerfeld School, and his solution as drawing heavily on Sommerfeld's phenomenological methods. This new understanding of the history of Pauli's principle illuminates, in turn, Heisenberg's quantum-mechanical *Umdeutung* (re-interpretation) of quantum theory in 1925. From the wording of the title of Heisenberg's paper to its discursive insistence that any new approach must only use quantities that were (in principle) observable, the birth of what would come to be known as matrix mechanics was infused with both Pauli's and Sommerfeld's skepticism toward atomic modeling.

The book ends with a description of the reception of Schrödinger's quantum wave mechanics within Sommerfeld's Munich school after 1926. Utilizing familiar mathematical techniques, and seemingly well suited for problem solving, Schrödinger's approach trumped Heisenberg's as the calculational method of choice. For many outside of Munich, however, the significance of quantum mechanics lay less in the ease with which seemingly intractable puzzles could now be solved than in the apparently necessary and radical changes in epistemology the new theories appeared to herald. Historians have tended to share this reading of events and have echoed the claims of Bohr, Born, and others in quickly declaring the existence of a "revolution" in science. Perhaps most famously, Thomas Kuhn, in *The Structure of Scientific Revolutions*, saw the birth of quantum mechanics as the epitome of a paradigm-altering revolution, replete with anomalies, crises, and epistemic breaks. That it was seen as such by "physicists of principle" is not in question; however, for those working within the context of a physics of problems, neither crises nor revolutions came to pass in the mid 1920s. Understandings of the processes and meanings of historical change in physics—by both historians and contemporary historical actors—have been strongly tied to particular modes of theoretical analysis. Theories of history are themselves part of the practice of theory.

I

1 The Physics of Problems: Elements of the Sommerfeld Style, 1890–1910

In 1906, Sommerfeld was called to Munich to fill the chair in theoretical physics. The position had been vacant for a dozen years, ever since Ludwig Boltzmann had left it to return to Vienna. The high standards required by the Munich faculty and the paucity of practitioners in theoretical physics led to an almost comical situation in the intervening years, as the job was repeatedly offered to the Austrian in an attempt to lure him back. Failing both in this and a further attempt to win Hendrik Antoon Lorentz for the position, the search moved on to younger men. Of the three candidates (Emil Wiechert, Emil Cohn, and Sommerfeld), only Cohn held a position in physics. When Wiechert declined the position, the ministry offered it to Sommerfeld, who had come highly recommended by both Lorentz and Boltzmann.[1] Sommerfeld's work on Röntgen-ray (i.e., X-ray) diffraction and the electron theory had probably also attracted the attention of Wilhelm Röntgen, Munich's professor of experimental physics. Röntgen signaled his approval, and the 38-year-old Sommerfeld jumped at the opportunity to occupy a full professorship at the prestigious university.

The opportunity, however, brought with it a major challenge. "In Munich I had for the first time," wrote Sommerfeld in an autobiographical sketch, "to give lectures on the different areas of theoretical physics and special lectures about current questions. From the beginning I plugged away at—and wouldn't let any trouble divert me from—the founding through Seminar and Colloquia activities of a nursery [*Pflanzstätte*] for theoretical physics in Munich." These lectures,[2] written in Sommerfeld's hand and delivered during a critically formative period in the development of theoretical physics, provided a means for Sommerfeld to educate a new generation of students and researchers in his methods. They also offered a means for him to develop these methods and to master the relevant material himself. These early lectures (1906–1910), Sommerfeld's published writings, and his reports on his students' dissertations to the Munich philosophical faculty constitute the basis for this chapter, which is an attempt to describe the "Sommerfeld style" of theoretical physics. This is the first of three chapters devoted to an examination of the Munich school in the years before 1918. Chapter 2 takes up the question of the transference of this style to a generation of students and researchers via an examination of the day-to-day practices of pedagogy. In chapter 3, Sommerfeld's correspondence, particularly letters written to him from his students during the First World War, is used to delineate the subsequent applications of Sommerfeld's teachings during the years of the conflict.

Sommerfeld worked closely with his students and was known as an excellent teacher. He fused teaching and research, incorporating his most recent work into lectures. Concentrating on Sommerfeld's lectures thus provides a means of bridging the gap between pedagogy and research practice and a means of better understanding Sommerfeld's approach to the problems of physics and the state of the field as he saw it.[3] The casual prose of the lectures is in stark contrast to more guarded comments in Sommerfeld's published writings. For example, in his lectures on heat radiation, first delivered in 1907, one can clearly see an attempt to master Max Planck's *Vorlesungen über die Theorie der Wärmestrahlung*, published the year before.[4] At the same time, Sommerfeld was (as he was not in his publications) openly skeptical toward Planck's black-body work, preferring the theories of Lorentz and James Jeans.[5] The lectures thus provide insight into both the early years of one of the most important sites for theoretical physics in the early twentieth century and Sommerfeld's own approach to the problems of contemporary physics. Sommerfeld's case also, and more generally, provides a particularly telling example of one of the central arguments of this book: that theoretical physics at the turn of the twentieth century cannot be understood as a "distillation" of theory from physics, but rather must be seen as having been actively constructed from multiple and varied parts. Far from being merely a subset of an existing discipline, the subject that emerged in Munich was a blend of at least three components: mathematics, physics, and engineering. Drawing from his experience in each of these three fields, Sommerfeld selected and modified components that would make up the theoretical physics of the Sommerfeld School.[6]

After 1906, having previously held positions teaching first mathematics and then technical mechanics, Sommerfeld quite consciously "refashioned" himself into a theoretical physicist. Problems previously deemed mathematical were now reformulated to emphasize a new, more physical perspective; technical applications were blended with mathematical and physical methods that may well have seemed alien to Sommerfeld's former colleagues in engineering. Not merely an incorporation of fields distinct from physics, however, the process also involved the selection and emphasis of specific areas within physics itself—in particular, those parts of the field that were in accord with the electromagnetic view of nature, a worldview of which Sommerfeld was an ardent supporter.

Central, indeed perhaps essential, to Sommerfeld's work in these eclectic fields was his and his students' emphasis on the solution of specific problems in areas such as wireless telegraphy, the wearing on ball bearings, the turbulent flow of water in the channels of the Isar, and black-body radiation. It would be remarkably fitting that the

Festschrift prepared for him by his students on the occasion of his sixtieth birthday should be titled simply *Problems of Modern Physics*.[7] Eschewing an axiomatic and generalized approach, Sommerfeld sought out, both in his teaching and his research, issues of contemporary interest that he would then attempt to understand in theoretical detail. And it would be these problems—both in terms of topic and their forms of solution—that would provide coherence—"technical unity"—for the wide-ranging work of the Sommerfeld School.

Mathematics: The Kind of Notion We Call Heat

Der Verstand schöpft seine Gesetze (*a priori*) nicht aus der Natur, sondern schreibt sie dieser vor.
—Kant

The words above—meaning "The understanding draws its laws (*a priori*) not from Nature, but rather prescribes them to her"—are to be found at the beginning of a draft of what is probably Sommerfeld's first scientific work.[8] They should not, perhaps, be taken as evidence of any particular philosophical commitment. Sommerfeld was born and grew up in Königsberg, where his father was a practicing physician.[9] At the local university, he had walked the same halls as Germany's most famous philosopher. Yet, insofar as Kant was important to Sommerfeld's development in this period, it was probably only through his dictum that "in every specific natural theory only so much actual science can be found as there is mathematics within it."[10]

The seven-page paper never takes up philosophy again, but a parenthetical remark at the beginning lays out Sommerfeld's main aim: "The leading thought in my work is to simplify the problem of heat conduction by establishing a characteristic function."[11] While his approach here was principally mathematical, the inspiration for the work could be found in an existing physical problem, one set as a prize question, worth 300 Marks, by the *Physikalisch-Ökonomische Gesellschaft* (Königsberg Physical-Economical Society), of which Sommerfeld's father was a member.[12] In the local Botanic Garden, Franz Neumann, a co-founder of the mathematical-physical seminar at Königsberg, had established a meteorological station that measured the temperature below the surface. The task—set by three of Neumann's students and four other members of a commission established by the society in 1889—was to analyze the data the station produced in its measurements of temperature at different depths. "The Society," read the question, "would like as comprehensive as possible a theoretical evaluation of the geothermal measurements made at Königsberg, especially to understand the thermal conductivity of the earth and the causes of it...."[13]

Sommerfeld set to work on solving the problem in a long essay, clearly intended to be presented to the society, and approached the institute for theoretical physics and Neumann's eventual successor there, Paul Volkmann (also a member of the prize commission), for aid in evaluating the terms that arose in his efforts to reduce an arbitrary curve to the sum of a trigonometric series.[14] Sommerfeld and Wiechert, Volkmann's assistant, built an integrating machine to deal with the calculations, although this met with limited success in its operation as a result of what Sommerfeld described only as "an insufficient practical understanding of the apparatus."[15] In the end, owing to a significant error made in assumptions about appropriate boundary conditions, Sommerfeld was forced to withdraw the solution. As it stood, however, the paper he had prepared was a good example of what Olesko has described as the theoretical physics of the Königsberg School.[16] Sommerfeld, who had attended Volkmann's lectures, had clearly noted his teacher's enthusiasm for the problem set forth by the Physical-Economic Society, and provided a solution that would conform to his expectations. In the first of two sections, Sommerfeld set out the mathematical theory for the ideal case of heat conduction in terms of Fourier series and then in terms of Fourier integrals. A short discussion of the operation of the integrating machine followed. In the second section, Sommerfeld dealt with the modifications to the theory that had to be considered in the "real world": corrections for nonperiodic temperature functions, for inhomogeneous surfaces, and for non-level ground (the station stood at the foot of a small hill). Finally, he considered the non-ideal character of the measuring instruments, offering a theoretical treatment of the air thermometer. The project shows a striking similarity to part of the Königsberg paradigm for theoretical physics: essentially the mathematical analysis of experiment. The only important element missing was a numerical error analysis of the results.

Yet, while he could do the problem "Königsberg-style," Sommerfeld did not revel in it. His dissertation on "arbitrary functions in mathematical physics," which he later claimed to have conceived and written out in a few weeks, made use of his earlier work on Fourier series and integrals, but largely without mention of the physics of heat conduction.[17] Consistently, over the course of his life, Sommerfeld would refer to himself in this period as a mathematician. His papers, even when dealing with possible topics in physics, would often emphasize that they were mathematical treatments, such as his 1894 work *Zur mathematischen Theorie der Beugungserscheinungen* (On the Mathematical Theory of Diffraction Phenomena) or his Habilitationsschrift, *Die Mathematische Theorie der Diffraction* (1896) (The Mathematical Theory of Diffraction), which he bragged would wake physicists up to the flaws in their analysis. In a letter to his mother in 1894, Sommerfeld referred to all that physicists had done for the mathematical theory of optics as "humbug and meaningless words."[18] The same year he refused the offer of an assistantship with the theoretical physicist Woldemar Voigt on the ground that he would have to work there on matters "which I do not wholeheartedly consider as my mission." That mission, especially given that Sommerfeld was happy to take up an assistantship with the Göttingen mathematician Felix Klein, was clearly mathematics.[19]

Yet even more than a preference for mathematics over other fields, the young Sommerfeld seems to have had a distaste for the physics of Volkmann and others on the prize commission. Upon hearing of the death of Heinrich Hertz, for example, he wrote to his mother, asking her whether she had read about it yet: "It is Awful! The man began his brilliant experimental investigations five years ago. Half of all physicists at the moment are following in his footsteps and are working on Hertzian oscillations. There are few discoveries that can stand next to his electromagnetic light waves. If it had to be a physicist that died, why couldn't it have been one of the useless Papes, Volkmanns etc."[20] In 1908, however, when he took up the problem of heat conduction again, Sommerfeld's approach to physics (if not to particular physicists) had changed completely. The title of his summer lectures—"Heat Conduction, Diffusion and Conduction of Electricity, together with their Molecular and Electron-Theoretical Connections"—already attested to an involvement with matters of profound interest to physicists at the time. The electron theory in particular was one of the main areas of research of Lorentz, one of the most respected members of the physical community, and the faculty at Munich had, in their choice of Boltzmann's successor, made clear their desire for someone expert in what Sommerfeld would term "the burning questions of electrons."[21] In fact, if any one topic could be said to have been the center of path-breaking theoretical research it was this one, and Sommerfeld himself had already contributed three significant articles to the field, published in the reports of the Göttingen Science Society between 1904 and 1905.[22]

If the topic was explicitly physical, the approach was even more so. Commenting on the topic of his Königsberg lecture, heat conduction, Sommerfeld noted that it was "the source of the methods of mathematical physics": "[T]he book by Fourier is the original Organon of these methods. At the beginning of the 19th century the problems of heat conduction were the order of the day. Apart from Fourier: Poisson, Lamé, Kelvin. Nowadays it's totally out of fashion, because its physical result [is] not great."[23]

Yet these mathematical methods would not be the sum total of the course. "Because the mathematical approach [Gesicht] is too one-sided," Sommerfeld wrote "we will give the lecture in addition a more physical orientation. This then has a more current interest."[24] Thus, Sommerfeld took up the topic in a way that Paul Ewald, one of his students, described as characteristic. He "penetrated quickly through the classical parts of the subject and, after having laid this foundation, dwelt on the topical problems requiring research—ship waves and turbulence in Hydrodynamics, theory of relativity in Electrodynamics, radiation theory, specific heat and energy quanta in Thermodynamics."[25] Recent research topics discussed in the lectures included not only electron theory in general but also Nernst's osmotic theory and the Wiedemann-Franz law, which postulated a proportionality between coefficients of heat and electrical conduction.

The most fundamental equations of early-nineteenth-century mathematical physics were consciously redirected toward a more physical—albeit *theoretical* physical—end. Under the German academic principle of *Lehrfreiheit* (academic freedom), Sommerfeld had no requirements other than to teach something called "theoretical physics." No particular subjects were prescribed, and no specific curriculum had to be worked through. His mode of structuring his lectures can thus be seen as representative of his own vision of the shape of the field. It was, clearly, a vision that made use of the methods of mathematicians, but did not necessarily accept their mindset. From his origins as a mathematician in Königsberg, working on Fourier series and arbitrary functions, Sommerfeld had begun, after the turn of the century and certainly after the move to Munich, to refashion himself.[26]

Perhaps most telling of Sommerfeld's explicit attempts to make such a change in persona was his insistence on being provided with experimental facilities. Experimental ability was still considered a fundamental prerequisite for a fully trained physicist, and Sommerfeld clearly felt that he lacked such ability. His previous chair had been in "technical mechanics" (a fusion of mathematics and engineering), and it seems reasonable to read his desire for an adequate laboratory as part of an attempt to shift fields, and, just as importantly, to be seen as doing so.[27] This latter aspect comes to seem even more important in light of the fact that very little experimental work was actually done at the institute. Sommerfeld himself did none, and while he began at Munich emphasizing the importance of laboratory work to his students, comparatively few of them actually followed that path. When Ewald arrived, in 1908, only one student was to be found in the fourth room of the building, that devoted to experiment: Ludwig Hopf was attempting, without a great deal of success, to observe and measure the onset of turbulence in an open trough as he varied the velocity of flow and the viscosity.[28] If a detailed engagement with experimental and observational data would remain a lasting characteristic of the Sommerfeld School, a dual experimental-theoretical strategy would slowly drop away.[29] By 1910, Sommerfeld had given more than a dozen courses and had published a series of well-received papers in physics. One might also imagine that his anxiety—the worry expressed to his mother in 1894 that he knew nothing about experiment and feared making a fool of himself in physics—had receded.[30] Sommerfeld's shift from a mathematician to a theoretical physicist was underway.

Technical Mechanics: Gyroscopes and Ship Waves

The year after completing his thesis in 1891, Sommerfeld sat for the state exam that would qualify him as a teacher. After satisfying the requirements of military service, he moved in 1893 to Göttingen, the site of (as he put it) "mathematical high culture." Family connections initially won him a position as assistant to Theodor Liebisch in the mineralogical institute, but his interests continued to lie in mathematics. In 1894 he became an assistant to Felix Klein, with whom he completed his Habilitation thesis two years later. It was during this period that Sommerfeld came to an appreciation of the value of connections between mathematics, physics, and engineering. He later credited Klein with "giving to my mathematical outlook that sense which is best suited to applications."[31]

In 1897 Sommerfeld became professor of mathematics at the mining academy in Clausthal, a position that seems to have involved teaching basic mathematics to largely uninterested students. Klein's maneuverings eventually resulted in a better offer, and in 1900 he took up the professorship for technical mechanics at the technische Hochschule at Aachen.[32] As a student of Klein's, however, he was initially greeted with suspicion: "In 1900 I was called as a professor of technical mechanics to the Aachen Hochschule. As a result, I was compelled for several years to apply the main focus of my works to engineering problems. I had there the satisfaction, that my Aachen colleagues and students, who first regarded the 'pure mathematician' with mistrust, soon recognized me as a useful member, not only in education, but also in practical matters of engineering, so that I was consulted for expert reports, for collaboration for the engineering society etc."[33]

That Sommerfeld was not exaggerating the degree of hostility which commonly met non-engineers when hired into a technical college may be seen from the welcome that greeted Otto Krigar-Menzel, a student of Helmholtz's, when he was hired to the professorship of theoretical physics at the technische Hochschule at Berlin in 1904.[34] Alois Riedler, who had previously held a chair at Aachen and who numbered among the most prominent of the engineering professors on the faculty in Berlin, serving as Rector in 1899, reacted rapidly and with aggression. There had been an "urgent need," he wrote to the Kultusminister (minister of education and arts) in July, for a teacher "who knows our needs and who gives to physics, which at the moment has proven to be a scarcely fruitful activity, a new style and content."[35] That need, Riedler made clear, had not been satisfied through the hire of Krigar-Menzel. "The teacher and the teaching task stand in the most glaring opposition to the urgent needs of our Hochschule." The Hochschule, he continued, "is no playground for areas of science that belong exclusively in universities, and can find with us as many listeners as engineering lectures among theologians."[36]

A tension, apparent and even acknowledged, thus existed between Sommerfeld's early training as a mathematician and his need to fit into his new role in an engineering school. One can begin to see that there would be two reasons that Sommerfeld's Aachen years would be such a formative element of his "physics of problems." Not only did he arrive at the school at an early, impressionable stage of his career, but also—and more importantly—he was compelled while there to adapt himself to his (at times antagonistic) environment.

Things had changed dramatically since 1870, when the Rheinisch-Westphälisch Polytechnical School in Aachen celebrated a long-awaited opening as Prussian and French forces faced one another across battlefields not far away. Then technical colleges had sought to imitate universities as closely as possible, and a mathematician like Sommerfeld would have been welcomed as one who could be counted on to provide the requisite *Bildung* (cultivation) for engineering students. Indeed, it was in this period that Klein himself took up a position at the technische Hochschule in Munich, his classes on analytic geometry drawing over two hundred students.[37] By the turn of the century, however, an increasingly muscular and praxis-minded engineering community could set its own terms, consciously contrasting the aims of technische Hochschulen with the classical university.

Around the middle of the nineteenth century, when the Verein Deutscher Ingenieure (Society of German Engineers) was founded, an enormous distinction still divided the "artisanal" engineer from the Hochschule engineers who led the society. Without the social status that in England, for example, had followed from an obvious economic utility, and in the absence of an effective German bourgeoisie, the society's leadership had sought a means of raising the public status of engineers by allying them with an intellectual aristocracy, the so-called German Mandarins.[38] The espoused ideals of engineering became those of the mandarinate: pure over applied science, general cultivation over specific training. Engineers would pursue the abstract ethics of German Kultur in an institution for higher learning rather than grub after money in a workshop or factory.

In the 1880s, however, "entrepreneurial" visions of engineering replaced "professorial" ones. Under the leadership of Theodor Peters, who became the business manager of the Verein Deutscher Ingenieure in 1882, the society emphasized far more "practical" issues, attracting to its Zeitschrift authors who "wrote about concrete matters and emphasized criteria such as application, cost effectiveness, fuel efficiency, material strength and ease of construction."[39] In terms of education, this change in outlook meant radical alteration. At the level of secondary schooling, moving away from its former support for the classical Gymnasium, the association began a vehement push for the modern Realschulen.[40] At the tertiary level, new community leaders argued for a vision of technische Hochschulen as institutions equal to universities, but nonetheless completely independent of, and fundamentally different from them (see figures 1.1–1.3).[41] This effort to keep the two institutions separate but equal would constitute one of the principal differences between these engineers and Klein, who sought to bring them ever closer.

After the late 1870s, the emphasis was on increased shop and laboratory training for engineering students at the technische Hochschulen. In terms of detailed curricular change, the 1890s and later years saw the "de-emphasis of calculus in favor of less precise but much more pragmatic graphic methods."[42] More generally, and

Figure 1.1
The four faculties of the university (theology, philosophy, medicine, law), represented by four spheres, balanced by the six sections of the technische Hochschule (architecture, civil engineering, mechanical engineering, shipbuilding, chemistry and mining, general). An imperial eagle perches above the scale. Source: *Die Hundertjahrfeier der kgl. T. H. zu Berlin, 1899* (1900).

controversially, prominent engineers, including Alois Riedler, argued against what they portrayed as an excessive reliance on theory over practice—on *Wissen* (knowledge) over *Können* (Capability)—and began to agitate for a "demotion" of mathematics and other theoretical subjects to the status of "auxiliary sciences" [*Hilfswissenschaften*].[43] Just how vehement this agitation was can be seen from the lines below, from a tract written in 1895. Having acknowledged that engineering requires the aid of sciences to climb the "heights" of the knowledge of reality, Riedler asserted that

the ruling "Theory" remains below in the comfortable valley: The terrible preparatory education forces it to do so. Down there in the valley, it pursues all kinds of Gymnastics, [yet] knows not itself the efforts and dangers of the mountains and also deceives its disciples about them. The disciples probably storm the heights sometimes, far away from the creating world, with neither aim nor purpose. But the learned, unfruitful theory, when it raises itself to bold flight, then flies from the sight of the real world, up over the clouds to Abel and Riemann, where the Theta functions disappear, where the "special" concept "Dimension" is replaced by the general concept "manifold" and then can be gymnastically performed [*geturnt*] in a world of four and more manifolds.[44]

Figure 1.2
The university (represented by four texts, labeled according to the four faculties) clasps hands with the technische Hochschule (represented by a hand using a compass). Note, further, that the V shape subtended by the quills in the ink-pot on the left opens to the sky, whereas the V formed by the compass offers an inverse image, opening to the Earth. Source: *Die Hundertjahrfeier der kgl. T. H. zu Berlin, 1899* (1900).

In view of how closely this mocked mathematics resembled some of the work of the Göttingen school, it should be taken as a mark of Klein's power that he was able to place Sommerfeld in a technische Hochschule at all.

A great deal was at stake for young theoreticians in the debate over the importance of the mathematical sciences to engineers. Riedler and others were arguing for a substantial reduction—in schools that had long been seen as stepping stones to better jobs in universities—in both the numbers and significance of such young scholars. In this environment, Sommerfeld's 1903 speech "The Scientific Aims and Results of Modern Technical Mechanics" takes on a particular significance. One can read it both as a defense of the technical sciences, now that "they have developed from their own innate power a confident and self-sufficient position," and as a claim for the relevance of theory to such sciences.[45] Sommerfeld, in other words—as befitted a man in his position—was seeking to play two roles, both that of theoretician and of engineer.

Figure 1.3
The six sections of the *technische Hochschule*. Note that the cherub representing the "general" section is the only one not using his hands. Instead, he stares (rather perplexed) at an integral.
Source: *Die Hundertjahrfeier der kgl. T. H. zu Berlin, 1899* (1900).

Sommerfeld and Klein

Sommerfeld presented his paper on modern technical mechanics as part of a panel organized by Klein, and there are strong similarities between the positions of the two men. Klein's program sought to bring Hochschulen and universities closer together. This program, however, did not advocate equality for engineers; rather, it sought to subordinate all fields to the central control of a mathematics that would function as "a specific power pervading the whole."[46] Thus, his plan for a Göttingen institute for physical technical research—posited as an educational counterpart to the physikalisch-technische Reichsanstalt—would also have a complementary relationship to the technische Hochschulen. The latter would provide an engineering education for the "Front-Line Officers," while Klein's smaller, more exclusive institute would supply a specific education in the exact sciences for future "General Staff Officers."[47] It was not difficult to work out the implied power relations between the two, and one can see why engineers might fight so strongly to keep the two institutions separate while guaranteeing their equal status.[48]

It is perhaps not surprising that Sommerfeld, Klein's student and supporter, was met with distrust when he took up his position at Aachen. And although Sommerfeld claimed to win over many of his doubters, there is a strong sense in which he remained far more of a member of the general staff than a front-line officer. He certainly worked on engineering topics, examining the problems involved in railway braking, studying the operation of dynamos, and, in what he saw as his most important project, working on the hydrodynamic theory of lubrication.[49] Yet even in the latter the aim of the task was less that of providing an effective solution to a practical problem—Sommerfeld admitted that his theory "doesn't correctly treat in all parts the real occurrences of bearing friction"—and more that of demonstrating the general applicability of the mathematics. His purpose, he claimed, was merely "to show how far one can come with the pure hydrodynamic theory."[50] The enjoyment of the project, he would write later, came from "helping the power of mathematical-physical thought to a victory in its exact treatment of an apparently inapproachable subject."[51] The papers on this and other "technical" topics should be seen as displays of mathematical virtuosity and proofs that engineering was part of the whole that mathematics pervaded.

And yet it is also necessary to distinguish between Klein and Sommerfeld. Klein, it has been argued, was driven to his attempts to bring universities and technische Hochschulen together by the fear that an increasingly powerful engineering movement would sideline university mathematics.[52] He thus pushed for a mathematics that would be more useful for the technical sciences, irritating both engineers (who were increasingly independent) and neohumanist mathematicians (who resented the importation of "Americanismus" and the loss of their pure ideals).[53] But, as the comments about the Göttingen institute imply, Klein's position was not completely centered between these two poles. While his rhetoric was as much about mathematicians learning from engineers and vice versa, this was not the way that things actually played out. Thus, in spite of his support for the efforts of those who worked on technical mechanics, it was not such a compromise between engineering and mathematics that emerged in 1898 as a result of Klein's maneuverings over teaching reform, but the far less industrial "applied mathematics." An alliance with the physical sciences was pursued largely in order to strengthen the position of mathematicians, and a separation between mathematicians and natural scientists could still occur, Klein noted to Paul Gordon in 1890, if "we are strong enough or feel overly restricted."[54] The ultimate aim was the salvation of mathematics, and Klein's own neohumanist predilections meant that he leaned more strongly toward a solution that would keep mathematics strong and independent. Quite simply, he advocated mathematical engineering rather than engineering mathematics.[55]

For Sommerfeld, on the other hand, with his chair in technical mechanics, the applications of mathematics were avowedly industrial. In his 1903 paper, Sommerfeld spoke on his own field, discussing problems of machine construction and of gyroscopic compasses on ships. The audience, of course, was different here, and Klein had organized the session, so one can assume he was in full accord with the idea that mathematicians should be told to wake up to the need to enter the industrial age. Yet Sommerfeld's "active contact with the problems of technology" was, as the topics of his papers demonstrate, more than an occasional tactic. His aim may have been, like Klein's, to sell mathematics to the world, but Sommerfeld also clearly believed that mathematicians and physicists had something equivalent to buy. Since Sommerfeld was the outsider at Aachen, it was largely a one-way trade. But once he had (much to Klein's chagrin) given up that post in favor of Munich, the situation changed. Like his interest in mathematical physics, Sommerfeld's enthusiasm for technical mechanics became an integral part of his theoretical physics.

One need not portray Klein as a Machiavellian figure, politicking in support of his own field, in order to acknowledge the local and contextual differences between him and Sommerfeld. Those differences are clear in a book they wrote together: the four-volume, 966-page *Über die Theorie des Kreisels* (On the Theory of the Gyroscope).[56] The idea of working on such a topic came to Klein during the annual meeting of the Verein Deutscher Ingenieure in 1895. "Here the thought came to me," he wrote later, "of connecting theoretical reflections that were familiar to me with the needs of physical and technical understanding via a detailed lecture over a specific mechanical problem, like gyroscopic theory."[57] A two-hour lecture followed, after which Klein approached Sommerfeld about writing up and publishing a more detailed but still small treatment of the subject, much as had been done with the lectures that same semester on number theory.[58]

The timing of the book, with the first volumes appearing at the height of Klein's battles, is crucial for understanding its purpose. Whereas the second volume was a more collaborative effort, the first volume was almost entirely based on Klein's ideas, and we can use it as a means of better understanding his position.[59] Klein had long argued for a more intuitive means of teaching mathematics, one that would have an appeal beyond the ranks of pure mathematicians and would "establish not only a knowledge of mechanics, but so to say, a feeling for it."[60] At the same time, he wanted to push the idea that mathematics was useful beyond its own boundaries and that it had a role to play in practice as well as in theory. *Über die Theorie des Kreisels* would thus have two aims—one for each of its audiences. Gyroscopic motion was a well-known but rather poorly understood problem in mechanics—one, moreover, that had applications in many neighboring fields, including astronomy and physics. The gyroscope thus offered a bridge to the natural sciences. At the same time, an appropriate analysis of gyroscopic motion could provide an example of a general approach to mechanics for pure mathematicians. Klein deplored what he referred to as the too abstract and formal direction that mechanics had taken in Germany, a tack that hindered a direct understanding: "The student who probably learns general mechanical principles analytically by heart, for that reason never grasps their actual mechanical meaning in a lively enough way and appears, when positioned before a specific problem, clumsy in its solution."[61]

The solutions the text offered, which it drew largely from the pattern of English textbooks, would be manifold. First, it eschewed the approach of Lagrange and those who followed him, and emphasized the importance of geometrical representation. Second, it emphasized the importance of understanding the *causes* of motion and made these comprehensible through the introduction of vector notation. Finally, emphasis was placed not on understanding the mechanics of the problem through the equations, but rather on allowing "the analytical formulation, as a final consequence, to appear of its own accord from a fundamental understanding of the mechanical relationships."[62]

The book that Klein had planned was thus principally one that would elucidate complicated mathematical concepts by applying them to a more easily grasped, more intuitive mechanical object. Through a work originally intended to span at most two volumes, the reader would be introduced to some of the mathematics of Euler, Lagrange, and Jacobi, and to elliptical functions and integrals. The place of these topics in areas outside of mathematics would be emphasized to appeal to a wider audience. And yet, although Klein was to talk in 1922 of "the needs of physical and technical understanding" that the gyroscope book was supposed to satisfy, the book that he planned in 1895 did not deal with any technical problems. The "applications" mentioned were to astronomy and physics, and even these were largely dealt with in the volume that Sommerfeld completed while in Aachen. The technical applications of *Über die Theorie des Kreisels* were all undertaken by Sommerfeld.[63]

The Physics of Problems

The opening words of Klein and Sommerfeld's joint introduction to the fourth volume (1910) laid out the division of labor that had developed between the inception and the completion of their work:

When Felix Klein held a two-hour lecture in the winter semester of 1895/96 "On the Gyroscope" he intended it in the first instance to emphasize the immediate grasp of mechanical problems that was extended particularly in England in opposition to the more abstract coloring [*Färbung*] of the German schools and, on the other hand, to render fruitful the methods of Riemannian function theory that were principally built up in Germany. The consideration of applications and physical reality was at the time certainly indicated and heartily supported, but was not yet carried out to its full extent.

In the extensive publication stemming from the pen of A. Sommerfeld the interest in applications took over more and more, particularly after his taking up of chairs in technical mechanics and later in physics.[64]

The result of the increasing emphasis on astronomical, geophysical, and technical applications, the authors continued, was a forced reworking of the kind of mathematics being utilized within the text. Thus, although the original aim of the text under Klein had been to produce intuitive instruction methods for particular mathematical techniques, Sommerfeld had introduced a change in mathematical content to suit an emphasis on scientific and engineering applications.[65]

Hydrodynamics: Theory in Practice

One can track, through the volumes of *Über die Theorie des Kreisels*, the changes in Sommerfeld's intellectual interests: mathematics when he arrived in Göttingen, applied mathematics after his time as assistant to Klein, increasingly industrial problems while a professor for technical mechanics at Aachen. In the same period, however, Sommerfeld had been engaged on another extensive project: the editorship of the physics section of *Die Enzyklopädie der mathematischen Wissenschaften mit Einschluss ihrer Anwendungen*.[66] This latter, after his failed attempt to pass the job on to his colleagues, became Sommerfeld's letter of introduction to leading members of the physics community, including, in particular, Boltzmann and Lorentz. Between these and the contacts made during the work on gyroscopic theory (including Carl Cranz, the "Pope of Ballistics"), Sommerfeld came to be connected to a majority of mathematics-minded professors in the German-speaking world.[67]

This multitude of interests came together after Sommerfeld's call to Munich in 1906, as can be seen in his work on hydrodynamics. A subject Sommerfeld had studied as a mathematician in 1900, and had called upon in his examination of the theory of lubrication while at Aachen, now became the topic of a lecture series in the summer semester of 1908.[68] In 1909, after a five-year hiatus, he returned to the question of hydrodynamics in his published work. His paper "A Contribution to the Hydrodynamical Explanation of Turbulent Fluid Motion" displayed the multiple resources upon which its author was able to draw: published as part of the proceedings of the fourth international mathematical congress in Rome, its second reference was to a book on theoretical physics, and its first section offered a literature review of works that dealt with the relationship between theory and praxis.[69] The paper began as follows: "The cleft between technical hydraulics on the one side and theoretical hydrodynamics on the other, between the investigations of large-scale properties on the one hand and laboratory investigations of capillarity on the other, has often been noted and bemoaned."[70] The gap was first bridged, Sommerfeld claimed, by Osborne Reynolds, who introduced a theoretical means of examining the phenomenon of turbulence. He did not, however, supply an exact method of calculating the value of the parameter that bears his name (the Reynolds number). This was the role that Lorentz played, simplifying the case of Couette's flow (that between two cylinders) to the geometrically easier case of the flow between two parallel plates, one moving at a fixed velocity with the velocity gradient approximated as linear.[71]

Sommerfeld's work, which took up the problem at this point, restricted itself to examining the mathematically easier case of the *onset* of turbulence—the border between turbulent and non-turbulent flow—and did not consider the character of the flow above the critical value of Reynolds's parameter. This problem, as well as a complete discussion of the equation he arrived at "which appears to me to form the actual content of the problem of turbulence," he reserved for another time.[72] The problem chosen, that of turbulence, was portrayed as standing on the border between technical and theoretical studies, one foot on either side. Technical hydraulics and theoretical hydrodynamics come together in a subject that would form a part of the lectures on theoretical physics. The fusion would pervade the work of the Sommerfeld School more generally. Sommerfeld's research topics often became dissertation projects for his students; thus he passed on the problem of turbulence to Ludwig Hopf in an attempt yet again to unite the worlds of theory and engineering.

Hopf's partially experimental work was touched upon above, but little mention was made there of the importance of technical applications within it. At first glance, studies of sugar water in troughs may seem obscure, but the point was to extend Sommerfeld's own studies. The open channel in which the water flowed would be a different case—and one more often seen in the real-world systems, such as rivers or canals—from that usually studied, where the fluid was enclosed on all sides. Sommerfeld hoped that the different set-up would allow the observation of different phenomena, such as the effect of capillarity, as the water shifted from laminar to turbulent flow.[73] Hopf's project would thus, quite literally, combine what Sommerfeld saw as "large-scale properties," and phenomena studied in the laboratory, the sets of observations that Sommerfeld associated, respectively, with technical hydraulics and theoretical hydrodynamics.

In almost perfect mimicry of his supervisor's paper, Hopf's dissertation began with the placement of hydrodynamics within physics, and continued with a discussion of the theory/praxis divide:

In perhaps no other branch of physics do theory and praxis remain as foreign to each other as in Hydrodynamics; and it is particularly peculiar and regrettable, given that both have grown into very extensive and important areas. Alone, the hydrodynamic theory on the one hand faced insurmountable mathematical difficulties in its treatment of phenomena in real fluids; on the other hand it opened up an easily accessible and praiseworthy [*dankbar*] area by going around these difficulties via idealized frictionless and eddyless fluids. Engineering [*Technik*] naturally developed completely independently of these theories and created its own mathematical fundamentals, with all the pros and cons of empirical formulas. Right at the moment the need for a metamorphosis of these conditions has become many times more keen, and a great deal of attention has been turned to real fluid [dynamical] problems.[74]

Of these latter, that of fluid flow, Hopf claimed, "is in fact the most urgent and most interesting of hydrodynamical problems, precisely because it can be completely and satisfactorily solved."[75] The laboratory examination of these "real-world" flows would constitute half of Hopf's dissertation.

Sommerfeld described the second half, "On Ship Waves," as "purely theoretical" when he reported on the project to the philosophical faculty.[76] That was certainly true in the sense that there were no *experimental* observations in that section. Yet even here, Hopf made clear the dual role that his investigations served. "The shape of waves which accompany a ship moving on tranquil water, and the corresponding resistance, against which the ship exerts continual effort, are not only of great significance practically, but also offer the theoretician several interesting problems."[77] As Sommerfeld had advised, the theoretician was coming into active contact with the problems of technology—so active, in fact, that Hopf's investigation later found outside financial support from the Gesellschaft Mittlere Isar, a group that wanted to regulate the flow of Munich's largest river.[78] "Purely theoretical" it may have been in one sense, but Hopf's dissertation found a significant role to play in practice.

Physics: The Electromagnetic Worldview[79]

It should by now be clear that the theoretical physics espoused by Sommerfeld was considerably more than simply a subset of physics. Sommerfeld drew heavily from his knowledge of both mathematics and technical mechanics in constructing his own brand of theory. Theoretical physics in Munich was an interdisciplinary creation. Yet it should also be clear that Sommerfeld's was a selective borrowing. Although he collaborated on two texts with the mathematician Felix Klein, one on number theory and the other on gyroscopes, it was only the latter that became a resource for dissertation topics in Munich. Although he published papers during his Aachen period on both railroad brakes and lubrication (in fact related topics), it was the hydrodynamic theory utilized in the latter that passed into his lectures on theoretical physics. Along with this selectivity came a process of modification. If the (traditionally mathematical) problem of the diffusion of heat was to become a topic in theoretical physics, then it was to be altered to fit into its new surrounding. If technical hydraulics was to be part of the Munich curriculum then this would be in combination with more overtly theoretical considerations—the analysis of wear on ball bearings would be replaced by the still "practical" but less entirely industrial problem of turbulence in open channels.

Sommerfeld's papers and lectures dealt, not with general principles, but with specific problems. Mathematical resources and theoretical underpinning were marshaled to get to a point where these problems could be dealt with by Sommerfeld in his research and by his students in their dissertations. The aim was not, that is, to create a unified, general physics and a community that corresponded to it, but to study—and to create students who would study—particular problems, drawing topics from a wide range of sources. The wideness of this range contrasts with historians' concentrations on what have come to be called the "twin pillars" of modern theoretical physics—the quantum and relativity theories. Yet Sommerfeld studied these subjects too. With the nature of theoretical physics still under construction, it was not merely the "outside" disciplines that were selected from and modified in the building up of the Sommerfeld style of theoretical physics. Sommerfeld's approach to the third strand—physics—was also one of partial inclusion.

By 1911, the year he presented a paper on the "quantum of action" at the Solvay Conference, Sommerfeld vocally espoused the necessity of some form of a quantum hypothesis. In his earlier lectures, however, his reservations concerning the validity of Planck's position were far more apparent. While Sommerfeld's resistance to the theory has been noted before, usually in passing, there has been no detailed analysis of why this should have been so. What his lectures make clear is that Sommerfeld's attachment to the electromagnetic worldview led him to favor the so-called Lorentz-Jeans formula for black-body radiation, despite its known failure to accord with experiment as well as Planck's did. This conclusion, as well as illustrating elements of the Sommerfeld style, has deep implications for our understanding of the "conversion" of several leading physicists to the quantum theory. In *Black-Body Theory and the Quantum Discontinuity*, Thomas Kuhn claimed that a lecture that Lorentz gave in Rome in 1908 marked a turning point in the history of the early quantum theory and led to a growing acceptance of the idea of a quantum discontinuity. While I grant the importance of the Rome lecture, it is argued here that the acceptance of discontinuity followed what was, in fact, a more profound realization. In 1906, Sommerfeld still assumed that electromagnetic theory was untroubled by the problems that plagued mechanics. It was to be expected, he implied, that a mechanical description of the electromagnetic ether should produce inconsistency in the form of the incorrect blackbody curve. Lorentz's lecture of 1908 was the first statement by one of its leading proponents that the electromagnetic worldview must fail for the case of radiation. What was at stake for a significant part of the theoretical physics community was, far more critically than the question of discontinuity, the question of whether the electromagnetic worldview could incorporate Planck's results, or whether the universalizing dream of the electromagneticists had to be abandoned. As Sommerfeld's Solvay paper would show, it would be the latter view that prevailed.

Relativity and the Electromagnetic Worldview
In the winter of his first year at Munich, Sommerfeld began a series of lectures on "Maxwell's Theory and Electron Theory," a topic described in a letter that December to Lorentz as the "burning questions of electrons."[80] Sommerfeld introduced his students almost immediately to the current problems plaguing the subject area at hand. After a short historical overview of the topic, he noted:

Of course all this is only valid for the negative electron and the apparent mass bound to it. About the positive electron and the matter apparently inseparably bound to it, we know nothing. Also there are still serious difficulties to overcome regarding electro-optical phenomena, which should, according to the electron theory, show the influence of the earth's movement. Lorentz recently said, in reply to my question as to how the electrons were doing: badly. The difficulties are not overcome in the face of Kaufmann's newest experiments. Therefore Planck was also pessimistic.[81]

The reference was to the experiments of Walter Kaufmann, who had attempted to distinguish between the two most prominent electron theories at the time: that of Max Abraham, a former student of Max Planck's, who assumed a rigid, spherical electron, and the so-called Lorentz-Einstein theory, which assumed a deformable electron.[82] Planck, in his report on Kaufmann's results at the 78th meeting of the Gesellschaft Deutscher Naturforscher und Ärzte (GDNA, Society of German scientists and physicians) in September 1906, called the two possibilities respectively the "sphere" and the "relative" theories and concluded that one still couldn't work out which of them was right. "Therefore," he wrote,

no option remains but to assume that some essential gap is still contained in the theoretical interpretation of the measured values, which first has to be filled before the measurements can be utilized for a definitive decision between the sphere theory and the relative theory. One could think here of various possibilities, but I don't want to discuss these further, because to me the physical foundations [*Grundlagen*] of each theory appear too uncertain.[83]

Sommerfeld commented on Planck's "pessimism" in the discussion that followed. A strong supporter of Abraham's theory, Sommerfeld disliked in equal measure, as he wrote in a letter to Lorentz, both the latter's deformable electron and Einstein's "deformed" time.[84] The 38-year-old Sommerfeld was clearly one of the younger contributors to the conversation in Stuttgart, and his suggested explanation for the difference in opinions caused some merriment:

Sommerfeld (München): I would not, for the time being, like to ally myself with the pessimistic standpoint of Mr. Planck. In the extraordinary difficulties of measurement the deviations could perhaps yet have their ground in unknown sources of error. In the question of principles formulated by Mr. Planck, I would suspect that the men under forty years of age will prefer the electrodynamic postulate, those over forty the mechanical-relativistic postulate. I give preference to the electrodynamic. (laughter)[85]

In an article written in 1970, the historian Russell McCormmach explained Sommerfeld's hostility to the "mechanical-relativistic postulate" as deriving from the younger physicist's devotion to an "electromagnetic view of nature."[86] This worldview encompassed three related positions: a distaste for, and mistrust of, mechanical modeling, especially as applied to microscopic phenomena; a belief that the only physical realities were electromagnetic in nature; and a programmatic commitment toward a "concentration of effort on problems whose solution promised to secure a universal physics based solely on electromagnetic laws and concepts."[87] This last notion of an electromagnetic *program* is crucial. Many, if not most physicists at the turn of the century made some use of electromagnetic concepts in their work. Yet, as Sommerfeld's comments make clear, even when working on the subject of the electron itself, not all methods and approaches were equally palatable. Preference, from the proponent of the electromagnetic worldview, would go to those theories that used only electromagnetic properties (or those assumed electromagnetic in nature, such as the electron's mass),[88] eschewing mechanical concepts like deformability.[89] It was thus not merely the problems chosen, but also the modes of solution deemed acceptable, that marked Sommerfeld out from his older colleagues.

In common with many physicists of the generation that completed their university studies in the last two decades of the nineteenth century, Sommerfeld had seen what was portrayed as the gradual failure of the mechanical worldview, which held that all physical phenomena could be explained in terms of the equations and concepts of mechanics. His generation had also witnessed both Hertz's discovery of electromagnetic waves and the later successes of Lorentz's electron theory—crowned with the discovery of the electron in the last years of the century. In 1901, Wilhelm Wien (37 years old at the time) discussed the possibility of reducing all of mechanics to electromagnetic theory. A return to mechanical explanation thus seemed to be a "throwback." "For the younger physicists," wrote McCormmach, "the electromagnetic concepts clearly pointed to the future of physics."[90]

One should not, however, confuse support of the electromagnetic program with a form of theoretical determinism. Many of the problems studied in Munich can certainly be understood as part of the furtherance of the program of the electromagnetic worldview, understood not merely as support for microphysical studies, such as those of the electron theory, but also macrophysical ones, like those of wireless telegraphy. Nonetheless, the aim of Sommerfeld's lectures does not seem to have been to establish the electromagnetic worldview as the sole philosophy of the Munich school, a "hegemony" like that supposedly put forward by Niels Bohr in Copenhagen in the 1920s with respect to the correspondence and complementarity principles.[91] This would have been almost impossible in any case, for as a program, the electromagnetic worldview was as much a promise for the future as a claim about the present. As Sommerfeld noted in his lectures: "Certainly, the electromagn[etic] foundation of mechanics is the music of the future [*Zukunftsmusik*]. But I am convinced that matters will proceed here just as with the music that received the name *Zukunftsmusik* thirty years ago."[92]

The worldview functioned as one standpoint from which Sommerfeld and his students could understand and critique the work of others, and the lectures are not (even where it would have been possible) written entirely from a single perspective. The electromagnetic worldview provided the glasses through which Sommerfeld examined the world at this time, but one does not get the sense that he insisted that everyone else use the same prescription. He was selective, but neither exclusive nor dogmatic. With this proviso regarding the flexibility of the electromagnetic program in mind—at least in Sommerfeld's hands—the next section explores the program in action, for it was not merely the relativity theory that was viewed through the lenses of the electromagnetic worldview. The quantum theory—more particularly, the theory of black-body radiation—was judged according to its fit with the requirements of the electromagnetic program as well.

Teaching Planck's *Lectures*

In 1906 Planck published his now-famous book *Vorlesungen über die Theorie der Wärmestrahlung*, both a summary of his earlier work and a continuation and re-examination of it.[93] Sommerfeld appears to have gone over the text with a fine-toothed comb, adopting both Planck's summary of previous approaches to the problem of radiation and, for the most part, his units.[94] After beginning with the rapidly written claim that "radiation is a focus of modern research," Sommerfeld divided previous approaches, much as Planck had, into three types: thermodynamic, electrodynamic, and statistical methods.[95] In the thermodynamic category he placed the work of Kirchhoff, Stefan and Boltzmann, and Wien; in the electrodynamic, that of Helmholtz, Maxwell, Rayleigh and Jeans, and Lorentz; whereas the statistical principally dealt with the methods of Boltzmann, Gibbs, and Planck. An outline of the structure of Sommerfeld's lectures is reproduced here as table 1.[96] The column on the left represents the proposed structure of the course, as laid out in Sommerfeld's first lecture. The column on the right provides the topic headings for the course actually delivered. The last three sections of the proposed course were compressed into a single one.

Table 1: Structure of Sommerfeld's Lectures

Sommerfeld's Lecture First Outline	**Sommerfeld's Lectures on *Theorie der Strahlung* Sommer 1907**
§1. Kirchoff 1859 §2. Stefan-Boltzmann 1879 u. 1884 §3. W. Wien. 1893 Das Wien'sche Verschiebungsgesetz Elektrodyn. Methoden §4. Helmholtz 1856 Reciprocitätssatz. Umkehrbarkeit des Strahlenganges. §5. Maxwell 1873 Strahlungsdruck §6. Rayleigh-Jeans 1905 §7. Lorentz 1903 Statistische Methoden §8. Verteilungssatz der Energie §9. Wahrscheinlichkeit und Entropie, nach Boltzmann §10. Der Planck'sche Oscillator §11. Das Planck'sche Strahlungsgesetz §12. Das Planck'sche Elementarquantum h der Energie. Folgerungen von Einstein	Einleitung u. Übersicht (Introduction and Overview) §1. Kirchhoff §2. Stefan-Boltzmann'sches Gesetz (Stefan-Boltzmann Law) §3. Wien'sches Verschiebungsgesetz (Wien's Law of Displacement) Elektrodynamisches Teil §4. Maxwell'sches Strahlungsdruck (Maxwell's Radiation Pressure) §5. Bewegter Spiegel (Moving Mirror) §6. Jeans Ableitung eines Grenzfalles des Strahlungsgesetzes (Jeans Derivation of a Limiting-Case of the Radiation Law) §7. Lorentz Ableitung derselben Grenzformel aus der Elektronenth (Lorentz's Derivation of the Same Limit-Formula from the Electron Theory) Dritter Abschnitt. Statistisches. §8. Beispiel aus der Gastheorie §9. Planck'sche Theorie

The difference between Sommerfeld's discussion of previous treatments and his analysis of Planck's own contribution jumps to the eye. Whereas his summary of earlier research (sections 1–8) was in some cases as detailed as Planck's, his discussion of Planck's theory in section 9 is remarkable concise. Sommerfeld achieved this by excising almost entirely the discussion of the production of radiation by Hertzian resonators, a topic that took up nearly one-third of Planck's text. Instead, within half a page of expressing the energy of a Hertzian dipole in terms of its total energy U, the electromagnetic moment f, and constants K and L,

$$U = Kf^2 + \tfrac{1}{2}L\dot{f}^2, \tag{1}$$

Sommerfeld merely restated the relation Planck derived between the total energy of a resonator and its average energy $\bar{u}$ at frequency ν:

$$U = \frac{c^3}{8\pi\nu^2}\bar{u}. \tag{2}$$

A parenthetical note after the equation promised that a proof would follow, perhaps as an exercise, since no such proof appeared in the lecture notes themselves. Following Planck closely from this point on, Sommerfeld eventually arrived at Planck's equation, relating energy to frequency for a black body at a particular temperature T:

$$U = \frac{h\nu}{(e^{h\nu/kT} - 1)}. \tag{3}$$

Having obtained Planck's formula, Sommerfeld immediately launched into a section labeled "critical remarks." Sommerfeld appears to have paid close attention to comments made by Paul Ehrenfest in the *Physikalische Zeitschrift,* the year Planck's book appeared.[97] Planck had introduced the resonators into radiation theory in part in order to obtain a parallel to the Maxwell-Boltzmann distribution in kinetic theory. Just as interaction between molecules brought about the Maxwell-Boltzmann distribution as an equilibrium distribution of velocities, the interaction of resonators was supposed to ensure that an initially arbitrary distribution of energies in a black body would result in an equilibrated radiation. Ehrenfest quashed that possibility by showing that the resonators could not do what was required. Since they emitted and absorbed energy at characteristic frequencies, only resonators at the same frequency interacted, producing an equilibrium distribution of intensity and polarization for each color. For resonators at different frequencies, however, no interaction was possible, so any arbitrary frequency distribution would persist. Ehrenfest wrote:

1) The frequency distribution of the radiation introduced into the model [described by Planck] will not be influenced by the presence of arbitrarily many Planck resonators, but will be permanently preserved.

2) A stationary radiation state will [nevertheless] result from emission and absorption by the oscillators in that the intensity and polarization of all rays of each color will be simultaneously equilibrated in magnitude and direction.

In short: radiation enclosed in Planck's model may in the course of time become arbitrarily disordered, but it certainly does not become blacker.—For the discussion to come the following formulation is especially suitable: Resonators within the reflecting cavity produce the same effect as an empty reflecting cavity with a single diffusely reflecting spot on its wall.[98]

Planck had made similar remarks at the end of his lectures, realizing, in his book's conclusion, that much of his analysis to this end had been fruitless.[99]

It is clear that Sommerfeld drew his inspiration from Ehrenfest's critique. His first objection, under the heading "The Role of the Resonators," reads:

The resonators only operate like a Reagent, strips of blotting paper, not like a catalyst [*Ferment*], coal dust. The non-black radiation remains non-black. The resonators can only increase the disorder of directions, not the color distribution. Because the resonator only works in the region (ν, $d\nu$) to which it is allotted [*abgestimmt*]. The resonator does nothing more than a diffusing mirror. (Cf. § 6 Jeans).[100]

Another comment referred to the dissimilarity between the methods of Boltzmann and Planck. Whereas Boltzmann had proven that the entropy, S, was a maximum for the Maxwell-Boltzmann distribution (that is to say, that the equilibrium distribution was the most probable one), Planck had skipped this step. Sommerfeld noted, apparently again following Ehrenfest, that the "substitution for this unfortunately missing consideration" was the "auxiliary assumption" [*Hilfsannahme*] that we now know as Planck's hypothesis, $\varepsilon = h\nu$. It was only with this hypothesis that Planck was able to get to a result that provided the requisite dependence of the total energy on both temperature and frequency.[101]

Yet, while Ehrenfest was prepared to take the close accordance of Planck's formula with experimental data as proof that there was some validity to his analysis, Sommerfeld was less enthusiastic. In fact, although Ehrenfest rejected the resonator approach, he did not reject the recourse to combinatorics. Rather, he explained the fundamentally different assumptions that led to the different results of Boltzmann (his former teacher) and Planck. For Ehrenfest, Planck's hypothesis was an additional (if peculiar) constraint that led to an experimentally verifiable result. Ehrenfest was willing to accept a version of Planck's thermodynamical/statistical approach as long as the appeal to resonators was abandoned. For Sommerfeld, on the other hand, the failure of Planck's resonators seems to have appeared emblematic of the problem with Planck's method in general, and Sommerfeld treated the "auxiliary assumption" as little more than something that allowed Planck to get to the desired result. "I think it is very possible," he wrote in the lecture, "that Planck's formula is only a good approximation."[102]

As an approximation, Planck's equation was not alone. Sommerfeld described the result of the Englishman James Jeans as an "approximation" as well. Jeans had assumed that energy could be distributed equally among the eigenvalues of vibrations within a cube of side length L. Doing so, however, resulted in a curve that was not in accordance with the experimental data of researchers like Planck's friend at the Berlin *technische Hochschule*, Heinrich Rubens.[103] Sommerfeld explicitly compared the assumptions implicit in Jeans's derivation to those of Planck in section 6 of his lectures (the section to which he pointed at the end of his first "critical remark"):

The most interesting question is now this: Why do we only obtain an approximate formula?

1. The assumption of the equipartition of energy is not generally valid for the Aether, *it is derived mechanically*. It is, so to speak, [mere] chance that it is still valid for long waves. Long thereby means nothing: Size depends on L, L drops out.

2. Standp. of Planck. The quantity h is the quantum of action of energy. The energy can not be divided arbitrarily. If the smallest amount of energy were $h = 0$, then Planck's formula would also reduce to that of Jeans.[104]

And yet, although both approaches were seen as only partially successful, and although Planck's formula seemed to fit the data better, Sommerfeld did not accept Planck's derivation. In deciding which theory to reject, the impotence of Planck's resonators outweighed the failure of the Rayleigh-Jeans equation to match available experimental results. In effect, Sommerfeld bracketed off the question of which expression was more correct in terms of its relation to experimental data. The choice between Planck's and Jeans's formulas was, rather, reframed as a choice between two distinct *methods*. In fact, Jeans's result, as Jeans himself had derived it, did not receive Sommerfeld's support. The italicized lines in the quotation above suggest that Sommerfeld was skeptical of the very basis of Jeans's derivation: the Englishman had assigned a *mechanical* property (the equipartition of energy) to what was—for a proponent of the electromagnetic worldview—a fundamentally non-mechanical ether. Section 7 of Sommerfeld's lectures, however, was titled "Lorentz's derivation of the same [i.e. Jeans's] limit formula from the electron theory." Methodologically, then, Jeans's expression was (or could be shown to be) a result following from the electron theory, and that spoke strongly in its favor.

Lorentz's derivation thus provided, in Sommerfeld's eyes, a positive endorsement of Jeans's formula. On the other hand, Sommerfeld saw significant difficulties in accepting Planck's approach to the theory of radiation. He laid these out in a series of "General Comments" toward the beginning of the lectures, offering a critique of the three possible methods for approaching the problem of radiation. Of the first, thermodynamics, he noted that it was at once "the most secure but the least satisfying" possibility. "In opposition to Energetics," he wrote, "one demands an understanding of Mechanism or Electrodynamism." The kinetic theory, Sommerfeld claimed, had eliminated thermodynamics by explaining its laws in terms of statistical mechanics. Along similar lines, "The program offered by Planck of radiation th[eory] should offer: to explain thermod[ynamics] electro-statistically."[105]

Planck, however, while utilizing the statistical techniques of the kinetic theory, had come out firmly on the side of thermodynamics in his analysis of heat radiation. Discussing the calculation of radiation intensity in his lectures, he noted that it was "in no way determined," so that "in a case where according to the laws of thermodynamics and according to all experience a single valued result is to be expected, pure electrodynamics leaves [one] completely in the lurch," and one in fact ends up with infinitely many solutions. Mechanics served no better: "The temporal course of a thermodynamic process cannot be calculated on the mechanical heat theory or the electrodynamic theory of heat radiation under the [same] initial and boundary conditions that completely suffice in thermodynamics for the single-valued determination of the process."[106]

For Sommerfeld, the fact that Planck did not seek to explain radiation solely in electro-statistical terms spoke against his methods: "Planck's theory is therefore not ideal; the theories of Jeans and Lorentz are better in principle." [107] Here, then, was the programmatic aim of the electromagnetic view of nature in operation—programmatic because, as noted earlier, Sommerfeld had specific objections to Lorentz's particular version of the electron theory, preferring Abraham's. Nonetheless, he clearly deemed either better than one that did not seek to reduce all other explanatory means to electrodynamics. Jeans's result, as derived through the electron theory, was to be preferred over any result following from a system of thought that might seek to deny the unificatory capacities of electromagnetism. No doubt, like Lorentz himself, Sommerfeld hoped that a more complete electromagnetic theory would result in an expression in better accordance with experience and experiment. Until then, however, an "approximation" derived along correct programmatic lines trumped one derived in a manner deemed "not ideal."

The continuation of Sommerfeld's "General Remarks" shows him waxing lyrical over the total explanatory possibilities of electrodynamics, which "creates here as well the highest unity":

Heat (radiated) is light, therefore electr[icity?]; but heat is, on the other hand, molecular motion. ~~How it should~~ It must convert electr[ical] action into inertial action; as it does so, the theory shows the apparent degree to which kinetic energy actually should be electromagn[etic] energy of the charged matter. Therefore in short: From the ident[ity] of light $^{\text{Leslie Prevost Rumford 18th Cent.}}$ and heat, the id[entity] of light and electr[icity] $^{\text{Maxwell Hertz end of the 19th Cent.}}$ and the id[entity] of heat and molecular mechanics $^{\text{Clausius Maxwell Boltzmann 19th Cent}}$ follows necessarily the id[entity] of molecular mech[anics] and electrodynamics (20th Century).[108]

If Boltzmann had shown that thermodynamics reduced to mechanics, this last identity showed that both thermodynamics and mechanics could be reduced to electrodynamics. This conclusion, in turn, suggests a pointwise refutation of Planck's "introductory theses" on the basis of the electrodynamic worldview. The responsive sentences below indicate Sommerfeld's position:

1) Heat diffuses [fortpflanzt sich] in two different ways, conduction and radiation.

1a) Heat diffuses in only one way, electrod[ynamic], in conduction the electr[ic] fields of the charges are bound to the molecule, in radiation they spread out freely in the Aether.

2) Heat rad[iation] is much more compl[icated] than heat conduction, because in that case the state cannot be characterized through a vector.

2a) Heat rad[iation] is much easier than heat conduction, because the particularities of the charge distribution (matter) don't play a part. In the Aether only the direction and intensity of the radiation, in heat conduction the directions of movement of the molecule as well.[109]

One can read these responses as the principle (and the "in principle") reasons that Sommerfeld approached *Wärmestrahlung* with skepticism. Although Planck's approach allowed a simpler calculation of certain fundamental constants (and here Sommerfeld was thinking much more of *k*, Boltzmann's constant, than *h*), it went against the worldview that Sommerfeld had adopted. Certainly, Planck's resonator approach had its own, intrinsic difficulties, but, more generally, it suffered from its adherence to a viewpoint that Sommerfeld and others were seeking to supersede with the physics of the future. As with his response to relativity theory, Sommerfeld considered quantum theory a step backward, presumably also the domain of men over 40, not the youngbloods in whose camp he placed himself.

Black Bodies in an Electromagnetic World

Kuhn's argument that Lorentz's lecture in Rome in 1908 marked the beginning of the acceptance of the "quantum discontinuity" runs as follows:

During 1908 Lorentz produced a new and especially convincing derivation of the Rayleigh-Jeans law. Shortly thereafter he was persuaded that his results required him embracing Planck's theory, including discontinuity or some equivalent departure from tradition. Wien and Planck quickly adopted similar positions, the former probably and the latter surely under Lorentz's influence. By 1910 even Jeans's position on the subject had been shaken, and he publicly prepared the way for retreat. These are the central events through which the energy quantum and discontinuity came to challenge the physics profession.[110]

In the Rome paper Lorentz proved that the electron theory *must* lead to Jeans's result. That is to say, there could no longer be any suggestion of his electromagnetic approach avoiding the problems that followed from the equipartition theorem. Lorentz stated that such had been his hope, after reading Jeans's papers. Now that hope was officially dashed.[111] Without at this point making a choice between them, Lorentz then stated the difference between the Rayleigh-Jeans and the Planck case as baldly as possible. Accepting Planck would bring theory in line with experiment, but "we can adopt it only by altering profoundly our fundamental conceptions of electromagnetic phenomena." Accepting Jeans on the other hand, would "oblige us to attribute to chance the presently inexplicable agreement between observation and the laws of Boltzmann and Wien."[112] For experimentalists, the issue was now clear: Jeans's equation did not work at all. If the choice was between it and Planck's, then the latter had to be accepted. In a paper published a few months after the Rome lecture, Lorentz acknowledged that he had been convinced in the interim by the arguments of experimentalists (including Wilhelm Wien, Otto Lummer, and Ernst Pringsheim**)** and had abandoned any support for Jeans's equation.[113] For the final step, Kuhn claimed, Lorentz's "great personal authority" was responsible, to a great extent, for spreading the gospel to the rest of the physics community.[114]

But exactly what gospel was being spread? Participants in the discussion referred, variously, to the "Rayleigh-Jeans," the "Jeans," and the "Jeans-Lorentz" formula. While the two former do not necessarily carry with them the association of Jeans's result with the electron theory, the latter definitely does. Kuhn's inconsistent attention to this fact elides the difference.[115] Proponents of the electromagnetic worldview (including Sommerfeld, Wien, and to a lesser extent Lorentz) may not have regarded the choice between continuity and discontinuity as the central issue. Rather, the question that "came to challenge" them, the question over which they struggled, was whether the electron theory could produce a Planck-like formula. Once it was accepted that this was impossible, discontinuity was adopted quite readily by this group.

Lorentz wrote to Wien early in June 1908, noting that he had been "ceaselessly racking his brains over the last few years" over the question of deriving Planck's formula (or something similar) from the electron theory. Contrasted to this language of constant struggle, Lorentz's description of Planck's alternative solution, the introduction of elementary quanta of energy, seems almost casual: "In and of itself, I have nothing against it; I concede at once that much speaks in its favor and that it is precisely with such novel views that one makes progress. I would, therefore, be prepared to adopt the hypothesis without reservation if I had not encountered a difficulty."[116] Kuhn highlighted this difficulty to explain Lorentz's hesitancy in accepting discontinuity, but the problem Lorentz outlined was not that of discontinuity *per se* but rather that of an asymmetry between the (continuous) absorption and emission of energy by resonators in interaction with the ether, and discontinuous emission and absorption otherwise. This specific question would continue to bother those who had accepted the idea of a quantum discontinuity for some time, and would eventually lead Planck to his so-called second and third theories, each of which posited different mechanisms (one continuous, one discontinuous) for resonator emission and absorption. Lorentz did not have a difficulty with discontinuity "in and of itself." What counted was whether the electromagnetic worldview could include it. "I can only conclude," he wrote in the *Physikalische Zeitschrift*, "that a derivation of the radiation law from electron theory is scarcely possible without profound changes in its foundation. I must therefore regard Planck's theory as the only tenable one."[117]

The first radical move for proponents of the electromagnetic worldview was not the adoption of a new theoretical position—that would come later—but the forced abandonment of their old one. For Wien, it was not immediately obvious after the Rome lecture that such an abandonment was even being posited, and his route toward Planck's theory can be understood as the inverse of Lorentz's. If Lorentz tried to obtain Planck's result by beginning with the electron theory, Wien—after dismissing Jeans's result on experimental grounds early on—began with Planck's energy elements and then sought to understand them in electromagnetic terms. His original reaction to the Rome lecture evinced a certain irritation with what he saw as Lorentz's rather "poor" rederivation of Jeans's result in Rome. "The lecture which Lorentz gave in Rome," he wrote to Sommerfeld:, "has disappointed me greatly. That he presented nothing more than the old Jeans theory without bringing in any sort of new viewpoint I find a little poor. Besides, the question of whether one should regard the Jeans theory as discussable lies in the region of experiment. His opinion is not discussable here because observations show enormous deviations from the Jeans formula in a range in which one can easily control how far the radiation source deviates from a black body. What's the point in presenting these questions to the mathematicians, who can make no judgment on precisely this point? It seems, in addition, a little peculiar to seek the advantage of the Jeans formula, in spite of the fact that it corresponds with nothing, in the fact that it can preserve the whole unlimited multiplicity of electron oscillations. And the spectral lines? Lorentz has not shown himself to be a leader of science this time."[118]

Ditching Jeans's result on experimental grounds, however, was relatively unproblematic compared with doing so for methodological reasons. It was not until he read Lorentz's second paper that Wien realized, with some dismay, what giving up Jeans's result meant in relation to electromagnetic theory. He wrote to Sommerfeld: "Lorentz has recognized his error over radiation theory and that Jeans's hypothesis is untenable. Now, however, the situation is not so simple, since in fact it appears as if Maxwell's theory must be abandoned for the atom. Hence I have a problem to pose you again. Namely, to check how far Lorentz's statistical mechanics and proof is founded on the fact that a system obeying Maxwell's equations (including electron theory) must also obey the supposition of the 'equipartition of energy,' from which Jeans's law is deduced. Namely a restriction of the degrees of freedom, as required by Planck's energy element, must also require an electromagnetic interpretation. Now it seems to me almost as if such [an interpretation] would be impossible, as if precisely this restriction requires additional forces (fixed connections and the like) that don't fit in with a Maxwellian system. If that's really the case, one doesn't need to rack one's brains any more about an interpretation of the energy element and a representation of spectral series on an electromagnetic basis, but rather must seek to find an extension of Maxwell's equations within the atom."[119]

Standing almost as bookends, outlining first the problem and then the proposed solution, are the statements "it appears as if Maxwell's theory must be abandoned for the atom" and "we rather must seek to find an extension of Maxwell's equations within the atom." Between the two is an interpretation of *both* Jeans's and Planck's derivations in electromagnetic terms. That is, Wien translated the question of equipartition and the question of the meaning of Planck's energy elements into the language of electromagnetic theory. The contradictions that arose in so doing led him to both echo and reject a comment written to him less than two weeks earlier by Lorentz: "One doesn't need to rack one's brains any more." The effort to save the electron theory and the electromagnetic worldview in its entirety now seemed fruitless, and

Wien pointed quite calmly to the need for an intra-atomic extension of Maxwell's equations.

Sommerfeld's reply, dated 20 June, was less pessimistic. He claimed that he did not find Lorentz's electrostatistical derivation of Jeans's result conclusive.[120] He promised Wien that he would communicate his objections to Lorentz, and indeed he did so the same day. Rather than accept Lorentz's calculations as a proof that the electromagnetic worldview must fail in the face of Planck's result, Sommerfeld merely used the opportunity to emphasize what was at stake in such a question. "At one time," he wrote, "when I lectured on the theory of radiation, I believed Jeans's paradox could be overcome by saying that electrodynamics is not subject to mechanical laws. Your present remarks seem to me to be an excellent foundation for the resolution of this question."[121]

Fixing a date for Sommerfeld's acceptance of the necessity of discontinuity is not easy.[122] In November 1908 he wrote to Lorentz urging him to ignore his earlier criticisms, but did not explicitly retract his objections to Lorentz's theory in general.[123] It was, however, in the latter part of 1908 that Sommerfeld attended Minkowski's lectures on relativity and was "converted" by them.[124] This is critical, since Abraham's rigid spherical electron theory, which Sommerfeld had originally favored over Lorentz's, was not relativistically invariant. If Sommerfeld applied the relativity theory consistently to the choice between competing electron theories, that is, he would have been induced to accept Lorentz's some time after 1908. By late 1909, Sommerfeld would make this point explicitly, in lectures that mark the first classes taught anywhere in the world on relativity theory. In introductory comments, Sommerfeld noted that the hypothesis of the rigid electron "was dropped because it includes the hypothesis of absolute space" and that "the deformable electron follows from the concept of relative space-time, which experience demands."[125] This, in conjunction with the removal of his specific reservations about Lorentz's derivation, would imply that Sommerfeld accepted Lorentz's conclusion that the electron theory and the electromagnetic worldview were incapable of dealing alone with the theory of radiation.

Three quite different responses to the questions raised by Planck's black-body theorem are sketched above. Yet it should also be clear that Lorentz, Sommerfeld, and Wien held much in common.[126] All three men conceived the problem of radiation as one to be cast at first solely in electromagnetic terms. If, after repeated efforts, that should prove impossible, the answer was not to abandon electrodynamics in favor of some other extant approach, but to find a new way of extending it. That is, electrodynamics provided the only standpoint from which one could begin to construct the future steps required to come to a comprehension of the puzzles introduced by black-body theory. And the question at hand was not "the problem of the quantum"—such a problem did not yet exist in such terms for the majority of physicists. For electromagneticists, Planck's result was a problem for and of the electromagnetic worldview in general and Lorentz's electron theory in particular. Only after they had acknowledged the reality and insurmountability of the problem within present electromagnetic theory—after June 1908—did they focus on discontinuity.

On the other hand, for those not committed to the electromagnetic worldview, the issue of discontinuity was an important means of understanding Planck's result. Einstein and Ehrenfest, who approached the issue from the perspective of Boltzmannian statistical mechanics, were the first, Kuhn argues, to "discover" the quantum discontinuity, some years before the Rome lecture. Jeans, on the other hand, initially denied the force of experimentalists' arguments, not conceding their validity until 1910. His description of the choice on offer at the time does not include discussion of electron theory, but does place the issue of discontinuity—expressed in terms of differential equations—front and center:

Planck's treatment of the radiation problem, introducing as it does the conception of an indivisible atom of energy, and consequent discontinuity of motion, has led to the consideration of types of physical processes which were until recently unthought of, and are to many still unthinkable. The theory put forward by Planck would probably become acceptable to many if it could be stated physically in terms of continuous motion, or mathematically in terms of differential equations.[127]

For proponents of the electromagnetic worldview, the most important issue introduced by black-body theory was the apparent failure of electron theory to incorporate or duplicate Planck's more experimentally verified result. The acceptance of discontinuity followed with comparatively little struggle after that blow to their shared worldview had been assimilated. For those who were not wedded to the electromagnetic picture, however, discontinuity became the most troubling thing about Planck's energy elements. Thus, perhaps one should, if one is to adapt Kuhn's religious language, speak not only of "converts" to discontinuity, but also of "lapsed" or at least disillusioned electromagneticists.

Conclusion

All those who have written on Arnold Sommerfeld in any detail have noted the number and eclecticism of both the problems he studied and the methods of their solution. This emphasis on specific questions and their specific solutions, the search for a mechanism or a process rather than a generalizing postulate is what distinguishes Sommerfeld's "physics of problems" from Planck's "physics of principles." Thermodynamics, which provided the model for Planck's unifying methodology, was to Sommerfeld "the most secure, but the least satisfying" approach to physics, for it failed to provide the specificities of mechanism. Historians have, perhaps naturally, tended to fragment Sommerfeld's various projects, attributing some to theoretical physics, others to mathematics or technical mechanics. Doing so is, in some ways, an obvious way of understanding a "physics of problems," for the specificity of problem solving can suggest a lack a coherence, an inability to be unified. Heretofore the discussion in this work, has also considered—separately—the three elements that went into making up theoretical physics in Munich: mathematics, technical mechanics, and physics. It remains to be considered how these three elements formed a recognizable single style. What, to phrase the question in its starkest form, was Sommerfeld's theoretical physics other than a single name given to a collection of disparate interests?

Perhaps not surprisingly, the problems themselves provide the answer. The problems themselves would often cross and hence blur the disciplinary boundaries that composed theoretical physics in Munich, producing what Andrew Warwick, in his discussion of mathematics training in nineteenth-century Cambridge, has termed a "technical unity."[128] The problems accorded with Sommerfeld's physical worldview and thus dealt on a majority of occasions—in the early years—with electromagnetic theory. At the same time, they were genuine problems of current technological interest, solved with mathematical prowess turned to physical ends. It was this quality of interdisciplinary fusion within the problems studied in the Sommerfeld School that brought a commonality of approach.

One may clearly discern this emphasis on both interdisciplinarity and technical unity in the selection and solution of problems in Sommerfeld's reports to the Munich Philosophical Faculty on his students' dissertation and habilitation projects. As noted earlier, most of these topics flowed from his own research, and the range of titles provides a good insight into the problems that Sommerfeld deemed significant. In addition, in the short commentaries describing the work, he would pick out those elements he deemed most important, so that even within the context of a given problem, one can discern those aspects representative of Sommerfeld's own interests.

The extent of Sommerfeld's pursuit of the electromagnetic view of nature appears in the number of his students' projects that deal with problems related to electromagnetic theory. Of the ten theses supervised or co-supervised by Sommerfeld in Munich between 1908 and 1911, eight discussed some aspect of electromagnetism, such as wireless telegraphy, electrical conduction in gases, measurements of capacitance, or the calculation of light pressure on spheres of arbitrary material. In many of the theses the question was not merely one of sheer theoretical analysis, but was derived from a practical problem. Hence, the project conducted by Hermann von Hoerschelmann took up the (very topical) question of the "Mode of Operation of the Bent Marconi Sender in Wireless Telegraphy." The problem, which Sommerfeld called "rather mysterious," lay in the discrepancy between the theoretical and the actual operation of a Marconi radio station. A bent sender should provide a signal in a preferred direction, that given by the antenna wire. Some of those who used such senders, however, had failed to detect such a preferred direction. Sommerfeld's discussion reveals his close interaction with those at the forefront of technological use and production, as he cites Count Arco (one of the doyens of German telegraphy) and an unnamed marine officer:

> Even though this [theoretical] effect is called into doubt by several practical men—Count Arco told me that in Marconi's opinion the antenna only conducted horizontally in this way because otherwise he couldn't accommodate the large length of the wire, and a marine officer in the radio commandos [*Funkencommandos*] wrote to me that he had experienced no directional effect in the vicinity of such a Marconi station—nonetheless Marconi's data has, given all previous experiences, the greatest right to attention. Therefore, because Marconi at his distance station [*Fernstation*] now uses the bent sender throughout and has invested significant capital in it, a clarification of its mode of operation is an important theoretical task.[129]

Rather than merely a question of electromagnetic theory, it is the practical issue of the operation of an existing radio station—an issue considered by those who made use of such stations—that provides the impetus for a theoretical investigation. Armin Hermann has noted that Sommerfeld would often pursue "physical questions that he examined up to their technical application." Here the situation is reversed, as *Praxis* provides the problem for *Theorie*. In a similar fashion, commenting on a project that dealt with the spreading out of wireless telegraphic waves on the Earth's surface, Sommerfeld effectively chided previous, more mathematically inclined researchers (Poincaré and Nicholson) for their failure to explain clearly the success of practitioners. Wireless telegraphers had succeeded in overcoming the problem of the curvature of the Earth in their attempts to send long-distance signals, and the project of Hermann March was devoted to explaining this practical success theoretically.[130] At the same time, this problem-focused fusion of practical technology and electromagnetic theory required the development of a sophisticated mathematical apparatus, and Sommerfeld lauded his student's work in developing the means of representing the electromagnetic fields in terms of the integrals of spherical functions. This, he claimed, was "important for several problems of mathematical physics and also appears noteworthy to me from a pure mathematical standpoint."[131] The one problem, in other words, would fuse all three elements of the Sommerfeld Style.

Specific problems could, on occasion, not merely require all three aspects of Sommerfeld's theoretical physics, they could recur in areas that corresponded to different disciplinary contexts. In Hopf's project, for example, Sommerfeld noted the similarity of one part of the solution with another well-known phenomenon. For ship waves in water of finite depth, the angle subtended by the wake is a constant, a result similar to that arrived at by Ernst Mach through his studies of the shock waves that were produced by an object moving at supersonic speeds (e.g., a bullet) that underwent a rapid deceleration (by hitting a wall, for example). Both of these, of course, were problems in mechanics (with technical applications), but Sommerfeld, following Stokes, Lenard, and Wiechert, had already used such a model in the case of X-ray production through the braking radiation of an electron. "According to this theory," Paul Ewald wrote, "X-rays are the electrical analogue to the sound cracks which travel forth in air from a target hit by shot."[132] Yet later the model would be used as Sommerfeld's point of entry into studies on quantum theory.

The problems that characterized Sommerfeld's theoretical physics were thus among the agents that provided a form of unity for his eclecticism, both through the fact that multiple elements were mobilized toward their solution and through the recurrence of particular problems and modes of modeling and solution in different disciplinary contexts. Technology, mathematics, and physics were planted together in Sommerfeld's nursery for theoretical physics. In spite of the diversity of its subject matter, the eclectic physics that emerged there shared common roots, and grew to bind its tendrils together.

2 Pedagogical Economies: The "Sommerfeld School" and the Problems of Teaching

"Munich shone." So begins "Gladius Dei," Thomas Mann's sketch of the Bavarian capital and the near-inevitable fin-de-siècle clash between the city's long-held conservative Catholicism and the licentiousness of its modern artists. The severity of the cowled Hieronymus, damning the owner of a gallery displaying a sensual Madonna, stands in stark opposition to the gorgeous relaxation of his urban surroundings, where "[a] shining vault of silky blue sky stood above the festive squares, the white colonnades, the classicist monuments and baroque churches, the leaping fountains, palaces and parks of the capital city, and its broad bright vistas, tree-lined and beautifully proportioned, basked in the shimmering haze of a fine early June day." And where "across the squares, and past the rows of houses, the droll unhurried life of this beautiful leisurely town dawdled and trundled and rumbled along."[1]

It was among the younger artists, as Paul Ewald noted in his reminiscences of his student days, rather than in the older courtly circles, that aspiring theoretical physicists, drawn as to a beacon by the resplendent city, found their comrades.[2] Yet in the early decades of the twentieth century another community developed in Munich. Here it was not Art that "held smiling sway" with its "rose-entwined scepter," nor was "diligent work and propaganda" devoted to a "pious cult of line, of ornament, of form, of the senses, of beauty."[3] It was, rather, a new kind of physics, one relaxed in its boundaries, eschewing the principled stance found under Berlin's gray skies in favor of the eclectic, and often collective, problems of the Sommerfeld School. For Ewald, the city itself, as much as Sommerfeld's teaching methods and his "warm-hearted and fundamentally decent attitude," was responsible for the prodigious success of this key site for the development of theoretical physics:

Munich as city, cultural milieu, and way of life exerted a great appeal on the students and offered Sommerfeld a large selection of young people who were intellectually moved and not merely bent on money making. Almost all his students were physicists "with body and soul," who shouldered a long and difficult course of study enthusiastically, without attending too much to their future prospects. That they grew up in a time of growing need for theoretical physics was their good fortune, as was the fact that their teacher had, from the beginning, almost a monopoly for the satisfaction of this need.[4]

The first of Sommerfeld's students came with him from the technische Hochschule in Aachen as his assistant. Peter Debye was an important fixture in the early years of the Sommerfeld School, aiding Sommerfeld in the preparation of lectures and the supervision of his students. The first dissertation candidates were drawn by Sommerfeld's skill as a lecturer from experimental physics, mathematics, and other fields. His reputation spread rapidly. Less than two years after Sommerfeld's arrival in Munich in 1906, he received a letter from Einstein, then a patent clerk in Bern. "I assure you," wrote the man who many would consider the most talented theoretical physicist of the century, "that if I were in Munich and had the time, I would sit in on your lectures in order to complete my mathematical-physical knowledge."[5] In 1922, Einstein was still marveling at Sommerfeld's pedagogical capabilities: "What I particularly wonder at about you," he wrote, "is that you have stamped so great a number of young talents as if they came out of the ground. That is something quite remarkable. You must have a gift for activating and engaging the talents of your listeners."[6] On the occasion of Sommerfeld's sixtieth birthday in 1928, Max Born would count up some of these "young talents," claiming expansively in an article titled "Sommerfeld as a Founder of a School": "If we look for the visible successes of Sommerfeld's teaching activity, we find—if I did not count wrongly—ten professorships of theoretical physics in universities in which German is spoken, which are occupied by his direct pupils. How many are spreading Sommerfeld's spirit as assistants, school teachers, or in industry, is impossible to quantify."[7] Of the caliber of the students, one commentator noted in 1948 that one could come up with a list more than twenty strong of those who had been trained in whole or in part by Sommerfeld and whose surnames alone would be enough for them to be identified.[8]

If the fact of the success and importance of the Sommerfeld School is somewhat obvious, the reasons for this success are far less so. How can we account for Sommerfeld's talent for producing two generations of theoretical physicists? What made the ground in Munich so fertile that Sommerfeld could stamp exceptional talent out of it with such regularity? The answer may be broken into two parts. The first attempts as complete a description of the practice of pedagogy in Munich as is possible, aiming to reconstruct the "pedagogical economy" of the Sommerfeld School, those structural, institutional and personal elements that made successful training possible.[9] The second offers a comparative perspective, contrasting pedagogical economies in general and the role of "disciplining"—in the Foucauldian sense—in particular, in two of the most important centers for physics education in the late nineteenth and early twentieth centuries: Sommerfeld's Munich and Cambridge, especially under the tutelage of Edward Routh. I conclude this chapter with a brief discussion of the limits of Foucault's

Pedagogical Economies

notion of discipline as an explanatory tool for the history of pedagogical practice in modern science.

The Munich Economy

Arriving in Munich in 1906, Sommerfeld could have been forgiven a trace of disappointment upon seeing the space granted his "institute" for theoretical physics. He had been promised room in a new main building of the Ludwig-Maximilians University, but it would not be until 1910 that construction was complete and he could make the move to Amalienstrasse. Until then he was housed in a building (once owned by the Society of Jesus) that served as the meeting place for the Bavarian Academy of Sciences, as well as housing the university's zoological, mineralogical, and geological institutes. Arbitrary at first glance, the temporary location of the institute for theoretical physics here made sense given one of the responsibilities Sommerfeld had inherited from Boltzmann. As a supplement to his professorial salary, Sommerfeld's predecessor was paid by the state to act as conservator for the "mathematical physical state collection of the Bavarian Academy," a collection of models, instruments, and teaching aides, containing, by 1894, more than 1,200 different instruments.[10] The collection occupied, as Ewald remembered, about eight rooms on the second floor of the building known as "the old academy," which one reached by passing the zoological collection, where the smell of naphthalene mingled with that of the sand and soap used to scrub the stairs. Useful for Boltzmann's experimental researches, the mathematical-physical collection had clearly fallen into neglect and disrepair by the early twentieth century, leading Ewald to describe it as "a cabinet of dusty junk," which owed its continued existence to the salary attached to the conservator's position, and that of a mechanic—Herr Sinz—who Sommerfeld inherited as well.[11] It would be Sinz who in 1908, with a shuffling tread and with a great rustling of keys, admitted the young Ewald to his first meeting with Sommerfeld, leading him down a dimly lit corridor to the four rooms that received direct sunlight. Of these, one contained Ludwig Hopf as he carried out his experiments on turbulence. Sommerfeld occupied a large office, while Debye sat nearby in a smaller one. The final room contained benches, desks and a large table, a lecture room capable of seating about 20–25 students. For larger lectures (including those on hydrodynamics which Ewald and Demetrios Hondros attended at the beginning of the summer semester of 1908), students walked to a small lecture theatre in Röntgen's institute on Ludwigstrasse.[12]

After the first decade of the century, Sommerfeld's institute occupied a far more spacious set of rooms. The ground floor contained a lecture room capable of seating about 60 students as well as four rooms for Sommerfeld and those working closely with him. Those students admitted to the mathematical-physical seminar had access "at any time and without formalities" to a small library, containing texts and journals.

The floor below contained four additional rooms for experiments as well as a workshop and darkroom. The set-up was unusual: few, if any, institutes for theoretical physics were so well provided for, especially when Sommerfeld was granted a second assistant in 1911.[13]

If the curatorium of the state's collection of instruments provided one ongoing connection to the Boltzmann years, even after the move to new premises, the presence of an additional professor of theoretical physics provided another. The subject had a long history at Munich, being taught by the Privatdozent Friedrich Narr since 1870.[14] Max Planck completed his dissertation at Munich in 1879 and habilitated the next year, proceeding to offer classes in theoretical physics and co-organizing what was termed the "physics colloquium" with another Privatdozent, Leo Graetz, until Planck's call to Kiel in 1885. The following year, Narr was appointed as Ausserordinarius professor, with duties including lectures on "the discipline of theoretical physics."[15] Yet frequent bouts of illness made Narr essentially incapable of teaching for the years between Boltzmann's arrival in 1890 and Narr's death in 1893. That year, Graetz was appointed to the Ausserordinarius professorship.

Graetz's story was a sad one, as even Wilhelm Röntgen, who arrived in 1899 and who had no great respect for Graetz's abilities, acknowledged.[16] Fifty when Sommerfeld arrived, Graetz had filled in for Röntgen's predecessor, Eugen Lommel, when the latter had been ill. Quarreling with Röntgen publicly after his hire, Graetz withdrew as director of the student laboratory, thus forfeiting half of his teaching responsibilities. The other half, of course, involved teaching theoretical physics, a task that became largely redundant after Sommerfeld's hire in 1906. The faculty responded by granting Graetz an honorary Ordinarius professorship, which provided him with a measure of dignity and a certain reward after so many years of service, but did little to resolve the question of his teaching responsibilities. As it turned out, Graetz—a popular teacher—offered classes that ran in parallel to both Röntgen's and Sommerfeld's, and continued to do so well into his seventies. A student who sat in on Graetz's lectures on electricity in the winter semester of 1915–16 noted that the man whose textbook *Die Elektrizität und ihre Anwendungen* (*Electricity and Its Applications*) was known as "the great Graetz" was as faultless in his teaching as he was in his texts. A nearly full lecture theatre—even in the middle of the lean war years—testified to the clarity and comprehensibility of his courses. Röntgen's classes barely filled the third row.[17]

Although Graetz's ongoing teaching presence clearly irritated Röntgen, who had been quite explicit in his desire not to hire him in the position Sommerfeld eventually accepted, Sommerfeld himself seems to have been well, if obliquely, served by Graetz's courses. Graetz's lectures (pitched at a level lower even than that of Röntgen's, which, one must assume, were neither as mathematically rigorous nor as up to date as Sommerfeld's) were presumably attended by a large number of those students for whom the physical sciences were a requirement rather than a calling.[18] Better students, or at least those who intended to pursue further studies in physics, seem to have gravitated toward Röntgen and Sommerfeld, the latter of whom was quite explicit in the case of the young Werner Heisenberg in recommending that he stop attending Graetz's classes now that he would be admitted into the theoretical physics seminar. "Well, these courses of Graetz are just a kind of general survey," Heisenberg recalled Sommerfeld saying. "You better let that go."[19] Graetz's presence meant that Sommerfeld did not have to teach such a survey, allowing the focus on more specific and contemporary problems that many students later remarked upon.

Sommerfeld's own description of the lectures makes clear the rather high standard at which they were delivered: "This was an introductory course, and was attended not only by the physics majors of the University and the Polytechnic Institute (Technische Hochschule), but also by candidates for teacher's degrees in mathematics and physics, by students of astronomy and some few of physical chemistry—all usually in their third or fourth years."[20] Heisenberg elaborated on the rationale for this occurrence, noting that Sommerfeld usually accepted students into his class who were in their fourth or fifth semester, having already taken at least three terms of differential and integral calculus. "[O]nly when the student really knows these things and can solve simple differential equations," Heisenberg remembered Sommerfeld saying, "only then shall he attend the course on mechanics."[21]

In the early years of the school, Sommerfeld's lectures attracted roughly 30–40 students.[22] In the postwar years this number had essentially doubled; by Heisenberg's estimate, the hall regularly contained 60–100 students.[23] Sommerfeld, small in stature, nonetheless dominated the room, combining clear, carefully prepared lectures with a slow, almost halting diction, which gave students the impression that he was working through difficulties in the theory even as he spoke, rather than offering well-worn physical truths. In common with his mentor, Felix Klein, Sommerfeld had no time for Jacobi's boast that he had been able to produce a text on analytic mechanics without including a single diagram: his lectures were well illustrated with figures, surrounded by his large, firm handwriting.

The content of some of the lectures, in particular that on radiation, was discussed in the last chapter. Here the intent is to capture something of the experience of being in Sommerfeld's classes.[24] The lectures' relevance was a matter commented upon by many. Linus Pauling remembered that "Sommerfeld would point out the places where the theory was still uncertain, in order that the student would know that his failure to understand was due to deficiency in the state of the science and not in his reasoning ability."[25] In 1903, Sommerfeld himself emphasized this element of his pedagogical practice in a critique of older technical mechanics textbooks. These, he noted, "have a strongly deductive, almost dogmatic character." The student reading them could easily gain the impression that the subject could be understood after the fashion of Euclidean geometry, with seamless statements and proofs, and little more to be added than the occasional empirical coefficient. "I do not believe," he stated, "that this is the spirit of modern treatments of nature [*Naturbetrachtung*] in which we should educate our students; I believe, in fact, that it is just as informative to have the attention brought to the deficiencies of theory as to have to continually marvel at its putative completeness."[26]

Not content merely with flagging lacunae, Sommerfeld marked these "deficient" areas for future exploration, either by students or by Sommerfeld himself. As Otto Scherzer put it: "It is characteristic of his way of teaching that many of the problems he discussed in his lectures for advanced students, and in his seminar, were those which he was just going to solve himself."[27] The rapidity with which both Planck's work on radiation and Einstein's relativity theory became topics of lectures and seminars provides a ready demonstration of this claim, but even in more traditional fields (e.g. hydrodynamics or mechanics) lectures were sites for the description of future research problems as well as for the transmission of fundamental concepts and techniques. Heisenberg's dissertation problem, to take only one example, came up in a discussion of murky areas in the study of turbulence held in Sommerfeld's room after his lecture.[28] Results from dissertations were then folded back into the lectures, so that Sommerfeld's texts are littered with references to students' work.

Unfortunately, many of the original manuscripts for the lectures have been lost, but the textbooks published in the 1940s and the 1950s provide an indication of structure and of Sommerfeld's general approach, even if we cannot be sure of their exact faithfulness to the lectures actually delivered in Munich. We have seen already that his introduction to the first volume on mechanics delineated the way in which Sommerfeld saw his own teaching style as one that emphasized problems rather than axiomatics, as well as against whom he defined himself: Planck.[29] A comparison between Sommerfeld and Planck's textbooks on mechanics indeed proves insightful. One notices immediately, for example, the use of illustrations. Planck's text has about 40, Sommerfeld's one-third again as many. And while Planck's are, for the most part, representative of generalized angles, positions, or forces, Sommerfeld's display a characteristic specificity, one devoted to mechanical and engineering applications. There are no equivalents in Planck's book of Sommerfeld's schematic diagrams of the differential of an automobile (the topic of one of the problems) or of a "Gyroscope in Cardan's suspension" (to accompany section 27, "Demonstration Experiments Illustrating the Theory of the Spinning Top; Practical Applications"), or of a "Schilk mass balance of a vertically arranged four-cylinder piston engine" (figures 2.1–2.3). Just as in Thermodynamics and Statistical Mechanics Sommerfeld vaunted the Carnot-Clausius proof of the second law because it "makes use of concepts derived from engineering" and "after all, thermodynamics did originate from the needs of steam engine builders,"[30] so too—throughout the lectures—Sommerfeld emphasized engineering applications.[31]

VI.5 Problems 255

The axle of the rear wheels of an automobile is cut at the center (Fig. 58, right). Fixed to the left end of its right half is the bevel gear (ω_1), to the right end of its left half, the bevel gear (ω_2). The two halves of the rear

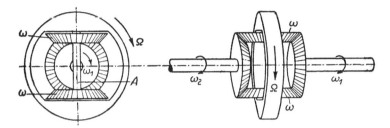

FIG. 58. The differential of an automobile, at the same time a model (after Boltzmann) for the induction effect of two coupled circuits. Left: view along rear axle of vehicle. Right: side view of this axle.

axle are therefore coupled by the differential in such a way that they can turn with different angular velocities.

Set up the kinematic relations between angular velocities Ω, ω, ω_1 and ω_2. Next make use of the principle of virtual work to derive the condition of equilibrium between the driving torque L acting on (Ω) and the torques L_1 and L_2 acting on (ω_1) and (ω_2).

What is the equation of motion of the system? Let I_1 and I_2 be the moments of inertia of (ω_1) and (ω_2), I that of the pair of gears (ω) about the axis of A, I' that of (ω) about the axis of the driving wheel. Neglect the contribution of (Ω) to I'.

If one rear wheel is accelerated, for instance by decreasing friction, the other wheel is retarded, even if driving torque and frictional torque remain equal there.

Figure 2.1

Problem involving the differential of an automobile in Sommerfeld's *Mechanics*.

§ 27. Demonstration Experiments Illustrating the Theory of the Spinning Top; Practical Applications

We begin by describing the well-known device known as *Cardan's suspension*, which affords an unusually effective means of demonstrating the properties of tops and gyroscopes.

FIG. 47. Gyroscope in Cardan's suspension. Axis of rotation of outer ring = vertical, axis of rotation of inner ring = horizontal perpendicular to paper, axis of rotation of gyroscope = horizontal in plane of paper.

The suspension consists of an outer and an inner ring. The outer ring has a vertical axle borne by the outer frame or cage; the inner ring has an horizontal axle with bearings in the outer ring. The flywheel-shaped top revolves with its axis perpendicular to the axis of rotation of the inner ring. Fig. 47 shows the flywheel axle pointing normal to that of the outer ring, which causes the inner ring to lie in a horizontal plane. We shall designate this arrangement of the apparatus as its normal position.

On the axle of the flywheel provision is made for a means by which angular momentum can be imparted to the wheel while in its normal position, with the gimbals at rest. This angular momentum must be so great that all phenomena are essentially dominated by it and the effect of the mass of the gimbals becomes negligible.

In the following experiments a considerable angular momentum and the initial normal position are presupposed.

Figure 2.2
Practical applications of gyroscopic theory.

The lectures were eventually offered in a six-semester cycle, beginning with Mechanics and moving through Mechanics of Deformable Bodies, Electrodynamics, Optics, Thermodynamics and Statistical Mechanics, and Partial Differential Equations in Physics.[32] Younger colleagues taught additional classes on Hydrodynamics, Electrodynamics, and Thermodynamics, while Vector Analysis was offered in a separate course, according to Sommerfeld, "so that its systematic development could be omitted from my lectures."[33] Classes met in the morning four times a week for lectures and were supplemented by a two-hour period devoted to the solution of problems. More advanced topics, often connected with Sommerfeld's current research, were considered in special two-hour seminars held weekly.[34]

The nature of the problems in Sommerfeld's lectures requires some elucidation, for they were not of the same character of those delineated in speaking of a "physics of problems" in the preceding chapter. The problems solved for the award of a doctorate

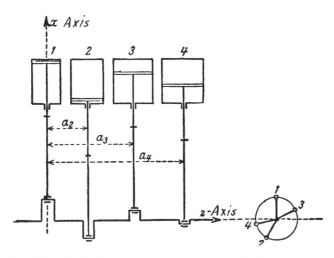

FIG. 17. Schlick mass balance of a vertically arranged four-cylinder piston engine. Diagram at lower right shows the position of the four crank pins relative to each other.

Figure 2.3
Diagram of a piston engine in Sommerfeld's *Mechanics*.

were necessarily far more open-ended than the training problems of the introductory lectures. Nonetheless, as will be seen below, a homology existed, and problem solving as an exercise was an integral part of this level of teaching. While Planck's *Mechanics* contains no additional problems for the student to answer, Sommerfeld's texts includes 15 pages of them, followed by 26 pages of hints for their solution, making the problem section roughly 15 percent of the total. Sommerfeld's introduction notes explicitly that the problems should be regarded as a valuable supplement to the text, and that his own students would hand in solutions every week, later working them through in problem sessions.[35]

Sommerfeld had introduced the two-hour problem-solving period in his Aachen years. He continued the tradition in Munich by enrolling Debye to set the questions, but he reserved the handling of the problems for himself.[36] Sommerfeld shared the task, but was no less involved, calling on students to step to the blackboard to demonstrate their method of solution and then interrupting to suggest improvements, so "it was a kind of permanent discussion between Sommerfeld, the assistant, and the student at the blackboard." Each problem took about 15 minutes to solve, and two or three were discussed in detail each week. "These problems were then solved, and the solution was fixed, and then the students knew what it was all about." The session ended with the distribution of the problems for the next week, to be solved at home before the next session.[37] The focus in these sessions was less on mathematical virtuosity—although Sommerfeld did demand "clean" mathematics—than it was on the appropriate mathematical formulation of the problem. Any mathematical expression, for example, had to be related to the physical problem term by term, while the more mathematical questions of the existence and uniqueness of solutions were largely left by the wayside.[38] Beginning with students who had taken several semesters of calculus already, Sommerfeld could assume a certain mathematical facility. Friedrich described Sommerfeld's attitude as follows: "You know, it was that way in his seminars: making the initial equation was the ticket. Because he demanded from the students that had certainly heard mathematical lectures before, that could differentiate and integrate. But formulating the initial equation for a physical problem, that they learned primarily with him...."[39]

We are now so familiar with the notion of training physics students through exemplary problems that it is hard to imagine a time when such pedagogical practices were new. Nonetheless, the absence of problems from Planck's lectures requires that their presence in Sommerfeld's lectures be explained. Karl Pearson's memories of his time in Germany in the 1880s suggests that Planck was not the exception. "Every bit of mathematical research is really a 'problem,' or can be thrown into the form of one, and in post-Cambridge days in Heidelberg and Berlin I found this power of problem solving gave one advantages in research over German students, who had been taught mathematics in theory, but not by 'problems.'"[40] The riddle is only made more perplexing when one considers the reason for the presence of such problems in England due to the Cambridge Tripos system. The introduction of timed written examinations provided a logic for training students in the rapid solution of carefully crafted examination problems. The problem and the examination, that is, went hand in hand. Germany, however, had no parallel system of examinations.[41] Students were awarded degrees on the basis of dissertations, and the questions or problems explored in such dissertations bore little resemblance to the kind of problem that a skilled practitioner could solve in a few hours.

There were people, however, who saw benefits in the English model, Felix Klein chief among them. In his introduction to a translation of Edward Routh's *Mechanics*, Klein argued that the book "must be of the deepest significance for every man that does not want to restrict himself to abstract principles, but who would like to comprehend the application of the principles to concrete problems." He continued:

In fact, Routh's work occupies a very specific position even within the English textbook literature. It is, one might say, the logical consequence of the educational methods developed at and dominant within Cambridge University, [methods] of which the author has been the acknowledged master for many years. Without doubt, this method, by laying the greatest emphasis on the working through of individual applications, brings the abilities of the students in a given direction to extraordinary development. In exchange something else takes a back seat, that to which we in our German academic lectures are accustomed primarily to strive for: the systematic construction of a general overview and the encouragement toward one's own independent conceptual education [*Ideenbildung*]. One can fully acknowledge the benefits that our German method may in this respect possess without overlooking the fact that the English methods possess alongside [ours], as a supplement, so to speak, their extraordinary significance. This in any case is the view from which we most particularly wish to recommend the work of Routh to the German public.[42]

In the absence of a culture of examinations, it is the connection, in Klein's explanation, between problems and *applications* that should be emphasized.[43] Klein's foreword was written the year after the first volume of Klein and Sommerfeld's *Über die Theorie des Kreisels* was published. That text, with its emphasis on the gyroscope as a means to teach students mathematics through its applications, seemed to both authors a fine application of a more English than German approach to education, one that found its highest form in the works of Thomson and Tait, and of Routh.[44] *Kreisels* was also written, as I noted in the preceding chapter, at a time when Klein (and with him Sommerfeld) was involved in a determined effort to prove the relevance of mathematics to fields like engineering and physics. Routh's approach, therefore, which Klein depicted as by its very nature focused on applications, would have seemed a natural way to teach a new kind of applied mathematics. As for Sommerfeld, it was Klein whom he credited with his own emphasis on a mathematics "adapted to applications." Moreover, Sommerfeld wrote, it was through Klein's "mastery of the art of lecturing" that he "exerted a strong indirect influence on my own teaching."[45] Klein's dichotomy between an English emphasis on the working through of individual applications and a Germanic systematic construction of a general overview—between the "application of principles to concrete problems" and a focus on "abstract principles" alone—maps neatly onto Sommerfeld's distinction between his textbooks and Planck's.[46]

The nature of Sommerfeld's solutions to the training problems he assigned was itself characteristic of his approach to physics. Just as he eschewed the search for general principles in his papers, so too Sommerfeld taught his students specific techniques adapted for specific kinds of problems. In place of the underlying logic of a set of physical phenomena, his students learned "tricks" (as Heisenberg phrased it) to aid in their solution:

He would never teach general rules, general principles, according to which you could always do that kind of thing. He would rather teach them that there is always some trick with which you can do the problem.... He did not treat the group theoretical side of the problems, which nowadays everybody would start with. He rather just told them mathematical tricks—how to solve the problem. Such a trick was that you have to take the equation of motion and multiply it by the velocity and then integrate it and then you can find the law of energy conservation. So he would teach them these kinds of tricks, but not teach the general way of solving differential equations or the underlying group theoretical principles.... He would rather teach them that there is always some trick with which you can do the problem and the group theory came only as a trick. He would say, "Now here we have a problem which has rotational symmetry." But he would not say, "Now consider the group of rotations;" he would say, "Since we have rotational symmetry, then, of course, it's a nice trick to introduce polar coordinates; then you will see that things work out."[47]

For Debye, Sommerfeld's focus on "problems and pencil-and-paper work" rather than understanding "the whole business" was "just the thing that I did not like so much."[48] Certainly in practical terms such a method has clear potential drawbacks. As at least one student of the Italian theoretical physicist Enrico Fermi noted, students taught at such a level of specificity may well have difficulties generalizing the method to solve unfamiliar problems. "If there was a difficulty," Marvin Goldberger remembered, "Fermi usually had invented at some time, perhaps in connection with an entirely different work, a trick to get the answer. But the unfortunate student confronted with a hard problem of a similar variety couldn't invent for himself a corresponding trick to save the day."[49] Heisenberg drew a contrast with training in Göttingen under Born:

In Sommerfeld's institute one learned to solve special problems; one learned the tricks, you know. Born took it much more fundamentally, from a very general axiomatic point of view. So only in Göttingen did I really learn the techniques well. Also in this way Born's seminar was very helpful for me. I think from this Born seminar on I was able really to do perturbation calculations with all the rigor which was necessary to solve such problems.[50]

Nonetheless, Sommerfeld seems to have been able to fashion a method that functioned between principled generalization and idiosyncratic one-off solutions. If students were not given an explanation at the most fundamental level as to why a particular trick would work, they were nonetheless informed of a class of problems for which such tricks were suitable, and with enough consistency so that they could tacitly acquire an understanding of the underlying logic. One understood that polar coordinates are useful in problems with cylindrical symmetry, in Heisenberg's example, even if the group-theoretical explanation for this result was left unexamined. The result was that one acquired a "feeling" for problem solving without necessarily being able to articulate the logic of particular methods of solution. "You know he was in some way always fond of these tricks, but he did not try to look behind the tricks and see why does this trick work and why doesn't any other trick work. And so in this way you might say it is not a good education; but the funny thing is, that just instinctively one did acquire a feeling why one had to use this trick here and the other trick there."[51]

For most students, these lectures and the accompanying problem sessions constituted the totality of their interaction with Sommerfeld. For a small number, however, usually between 5 and 10, the lectures were merely the beginning of a much more sustained training. As was traditional, while lectures were open to all those with sufficient preparation, admittance to Sommerfeld's seminar required his personal permission, granted on the basis either of his own knowledge of a student's capabilities, gleaned from participation in lectures and problem-solving sessions, or via recommendations from a colleague, or, occasionally, merely after an interview, as was the case with Heisenberg.[52] The vetting process was necessary, for the standards of the seminars were high; Paul Epstein later compared them to American graduate-level classes.[53] While Sommerfeld picked the general topic, students were expected to provide much of the content, giving prepared talks on original research or on papers in the field. In the summers of 1910 and 1911, for example, seminars were explicitly listed as "lectures by participants on statistical methods in physics" and "lectures by participants on relativity."[54] Topics to be reported on were handed out every two to three weeks. Often Sommerfeld merely passed a paper to a student and said, for example: "Now here you have the paper of Mr. Kramers. You give a talk at the seminar next week and explain to us what Kramers actually means by his paper and what you think about it."[55] Before Schrödinger's papers on his wave equation were published, Sommerfeld distributed galley proofs of them to his class and required each member to give a talk on them.[56]

If occasionally casual in distributing problems, Sommerfeld was less so in judging them, not hesitating in telling students that they had clearly failed to understand the central point of the paper or at least to articulate that point in a meaningful manner. In view of this element of correction and control, it is legitimate to see the seminars as exercises in intermediate problem solving. Unlike the training problems of the lectures, the tasks distributed in seminars were not well-known, exemplary problems, aiming to teach the student particular concepts, techniques, or "tricks." Neither, however, were they the open-ended problems required for the dissertation, where a student may well have had to invent a new mathematical technique or physical interpretation in order to proceed. Seminar problems were part of the training to become a researcher. Understanding a paper, or working through a small problem where the means of proceeding was suggested in advance, the purpose of such seminars was pitched between didactic pedagogy and free research. Students were trained, tacitly, in how to read and understand papers in the field—a necessary skill for a publishing researcher. At the same time, they were forced to begin to learn the basics of a field in the making. And, in contrast to the problems that followed the lectures, students were assigned the tasks individually: no two students were assigned the same task.

Where seminar problems and "training problems" overlapped, however, was in the *form* of their solution. Just as problems in lectures were solved collectively, with the solution fixed so that all students understood the method, so too in seminars all calculations were shown explicitly. "One of the things in the seminars," Bethe noted, "was that everything was done explicitly including all the algebra, so that you could just follow the seminar and then you knew all there was to know."[57] A means for both Sommerfeld and his students to keep up with advances in rapidly changing fields and a place where students could begin their training as researchers, seminars were also pedagogical spaces where a collective and shared understanding was the final aim.

Lectures and seminars constituted the principal elements of the structured learning environment at Munich. However, for advanced students—those who participated in seminars and were expected to continue on to complete a dissertation under Sommerfeld—a considerable informal economy of learning operated. At Ewald's instigation, for example, Debye agreed to organize a students-only colloquium—an institution peculiar to Munich—as a means of bringing everyone up to speed on issues that might come up in the daily discussions of more senior students. "We wanted a seminar," Debye wrote, "but a seminar in which there were no professors, because we wanted to be free to be as stupid as we wanted to be."[58] Debye consulted Sommerfeld, who allowed the use of the lecture room for the purpose, offering a carton of cigars "to sharpen the thinking capabilities."[59]

Students admitted to the seminar had free access to a small library near Sommerfeld's office. In Heisenberg's recollection, the gathering of students who intended to proceed to the dissertation was so regular that he gave it a name: "Sommerfeld's seminary." Regularity, in fact, seems to have been a motif, since Sommerfeld's schedule was remarkably consistent, with lectures earlier in the morning and then the hours until lunch devoted to meetings with his students, calling them into his office every one to two days for an hour or two at a time:

…it was a rule that those few students who really were interested in the game would sit out in the seminary in the morning—say 9:00 till at least 1:00 or half past one. During that time there were these discussions and also during this time people were called in by Sommerfeld to his room to discuss things with him. So I should say that almost every morning, but at least every second morning, I was called in to Sommerfeld. I had to tell him about what I had tried in my own work, and what I thought about Landé's paper, and so on. In this way, Sommerfeld has really made an enormous effort to get his students into the game—to get them interested, to get them to take part in the scientific life, and, of course, to suggest some work which they had to do.[60]

For Heisenberg, at least, the contrast to the two other most important centers for the training of young theoretical physicists at the time—the institutes of Niels Bohr and Max Born—was striking, if only because of the sheer number of hours that Sommerfeld spent with students. Bohr, always disorganized and slightly harried, would drop in on his pupils only occasionally, when he had a specific question or an issue that required discussion.[61] Born was more organized, but as a result meetings had to be scheduled in advance, with special appointments to discuss a particular problem, for example, made for the next week.[62] Insofar as Sommerfeld would spend lunchtimes most days at the famous Cafe Lutz, discussing problems with students and colleagues and—to the chagrin of those serving them—scribbling formulas in pencil on the marble tabletops, it becomes clear that more than half of every day involved the company of his pupils. This alone gives a simple answer for part of Sommerfeld's pedagogical success.

The "seminary" was more than a waiting room for meetings with Sommerfeld. Traffic went both ways, with Sommerfeld also dropping by with news of recent developments for collective discussion. Epstein, in fact, described talking as Sommerfeld's method of working: "Every problem he helped, he helped by chewing it over with his surroundings, students or assistants."[63] In his absence, the room functioned "as a kind of market place in which to exchange views about the most modern developments. One got a feeling of a very exciting and interesting development taking place about which you had to hear the latest news every morning." If the seminar was the first step toward training and practice in research, then the seminary was the next. As news arrived, students would gather around the blackboard, interrupting one another as they sought to analyze the problem together, seeking a "common opinion on the recent developments."[64] That this opinion was indeed common can be discerned from another of the peculiar practices of the group. Authorship, in a situation where so much communal conversation was involved, became an issue as entangled as the levels that Sommerfeld and his students sought to tease apart. As a result, Sommerfeld decided the issue himself, parceling out the topics among his students, either listing them as co-authors (even when the writing of the manuscript was mostly his task) or (as in the case of the various editions of *Atombau*) listing their contributions in his acknowledgments.

Perhaps the Wednesday Colloquium and the "seminary" are better described as "semi-formal" learning institutions, for a genuinely informal pedagogical economy also existed at Munich. The institute's relaxed atmosphere struck participants and observers alike. "I now understand," wrote Einstein in 1909, "why your students are so fond of you. Such a pleasant [*schönes*] relationship between professor and students probably exists uniquely there. I want to take you entirely as an example."[65] Although he stood ramrod straight and sported a dueling scar that led students to joke behind his back that he looked like a leader of the Hussars, Sommerfeld's gentleness with students belied his Prussian bearing. His teaching style was far from militaristic, and he seems to have eschewed the exercise of patriarchal power in favor of a more sympathetic authority. Old-fashioned in both politics and morality, Sommerfeld had rather firm ideas about how students should behave, counseling Heisenberg, for example, to give up playing chess: "If you do have that kind of effort, then you'd better do physics; if you want to have some recreation, you can go skiing."[66] This paternalism left its mark. Even the famously blunt, often impolite Pauli persisted in referring to Sommerfeld as "Sie" rather than "du" and for years would rise from his seat when his former professor entered the room.[67] Sommerfeld encouraged such formality, his relationship with his students—certainly in the later years of the school—perhaps best captured by Bethe:

> Sommerfeld was approachable, but at the same time he was very much the *Geheimrat*—very much the distinguished professor. His dignity was inborn, and was accompanied by a quiet sense of humor. A famous story has it that one of the foreign visitors in speaking to him, addressed him many times as "*Herr Professor.*" After a few weeks, the visitor was told that Sommerfeld was really *Geheimrat*, whereupon he addressed him, correctly, as "*Herr Geheimrat.*" Sommerfeld is reported to have acknowledged the change by saying, "You have really learned a lot of German."[68]

If Sommerfeld was a stickler for social niceties, much of his formality disappeared when he was discussing work, at which time he gave the impression, as Otto Scherzer remembered it, "not of a professor talking with a student, but of two equals discussing a most vital and interesting problem." Most of his students would remember conversations lasting long into the night in Sommerfeld's home, Otto Laporte recalling that official work would end around 10 p.m., when professor and pupil would stop and drink a glass of wine together.[69] As Debye described the Aachen years, he and Walter Rogowski were used as sounding boards for new ideas:

> [Sommerfeld] invited us to come to his house. We came to his house in the evening at 8 o'clock, had the evening meal, supper. And then you sit in his room. And in his room he began to talk. He asked you about it, although you did not know anything about it. He tried it out, so to say. And in this way, I learned a lot. But you sat there until 11 or 12 o'clock in the evening and you talked and he talked. The sessions were perhaps twice a week or so. It was not regular. You might meet him and he would say: "Oh come up, I have to talk about something."[70]

Sommerfeld himself emphasized the importance of pedagogical relationships that lasted well beyond the lecture hall. "Personal instruction in the highest sense of the word," he wrote in a 1949 article that looked back on his teaching career, "is best based on intimate personal acquaintanceship. Ski trips with my students offered the best opportunity for that."[71] These ski trips—uncommon in their mingling of an Ordinarius professor with a collection of his pre-dissertation students—became an integral part of life at Munich, with Sommerfeld turning up to lectures in his skiing clothes, and with visitors to the institute dragged along to the slopes to join scholars from nearby universities.[72] Ewald described the sight around Easter, when members from Sommerfeld and Röntgen's institutes were joined by colleagues from the technische Hochschule in Munich as well as Wilhelm Wien, his family and his assistants from Würzburg, Gustav Mie and Christian Füchtbauer from Greifswald, and—once—Laue from Berlin:

In the morning a long row of about twenty skiers stretched out across the mountain, often in pairs, conversing heatedly, in so far as breath to do so was available, or even staying standing; ragged snatches of conversation floated over the mountain ridge—six-vectors, photoelectric effect, Einstein, displacement law, $h\nu$—and let those eavesdropping guess what had stimulated their minds. In the evenings people gathered for dinner under the red glow of the petroleum lamp and afterward came the time when one could speak with Sommerfeld about more serious problems which required paper and pencil.... Many a one of us *Doktoranden* was able, in this way, to discuss in total relaxation his problems with his professor—often with the enrolment of others of those present from whom one could expect advice.

"This free exchange of ideas," Ewald concluded, "without any formality or restriction was the ideal of an academy."[73]

In a speech given in 1948, on the occasion of the first general assembly of the international union of crystallography, Ewald described the process by which he selected the dissertation problem that led him toward his part in the discovery of the X-ray diffraction of crystals. Under Sommerfeld, choosing a topic was not a matter of selecting a single problem within a range defined by a particular research program. Rather, students were offered any one of a myriad of problems suggested by Sommerfeld's wide-ranging lectures:

A large central table with green cloth surface was the usual place for conversations of the Professor with his students. But on this occasion we sat at the cherry-wood desk nearer the window. Sommerfeld took a foolscap sheet of paper out of the drawer and I saw a list of some ten or twelve research problems written out in his large clear handwriting. He discussed and explained them to me one by one. Calculation of self-induction by solenoids for alternating currents; propagation of radio waves over a surface of finite conductivity; an unsolved problem of gyroscopic theory; a new attempt at explaining the instability of Poiseuille flow, and further subjects. Each subject had its own merit and its own type of mathematical technique, and Sommerfeld pointed them out.[74]

In chapter 1, I argued that, in spite of the wide variety of topics that Sommerfeld studied in Munich, common problems nonetheless provided a form of intellectual cohesion. Sommerfeld might have drawn problems from mathematics, physics, or engineering, but he often examined physical phenomena deemed to be analogous to one another (the shock waves accompanying a decelerating object, for example) in a multitude of different contexts, offering solutions to questions in fields ranging from hydrodynamics to quantum theory. It was no doubt in such overlapping contexts that "tricks" for solving certain kinds of problems became particularly useful. The discussion above, however, would suggest that the pedagogical economy of the Sommerfeld School can offer a yet broader reason for considering Munich the site of a coherent school, one capable of producing a Sommerfeldian "spirit." In both formal and informal contexts, social and institutional cohesion produced a "research school" even in the absence of a unified research program.[75] Students were bound to one another and to their professor during their ski trips, which culminated in long evenings spent discussing physics problems; in the weekly colloquia and the skittles matches that followed; and in daily meetings outdoors at cafés and indoors in the seminary or Sommerfeld's office, where they analyzed contemporary problems to arrive at a "common opinion." But they were also bound by practices—by problems and their shared solutions—and by a method of presentation according to which each problem had to be solved in full, so that each member of the seminar could follow every step. The same logic extended to dissertations and publications, so that the range of problems solved in the school did not hinder a common understanding. There was indeed no single unifying program, no set of fast, simple, easily communicated laboratory techniques, and no group of tacit knowledges forged at the bench and over the burner, but there were nonetheless extensive common bodies of knowledge, shared through daily interactions that forged both intellectual and social bonds. These bonds created a multi-headed entity that was—and was known as—the Sommerfeld School.

Foucault and the Physicists: Cultivating Creativity and the Limits of Disciplining

In a recent essay, the historians of physics Andrew Warwick and David Kaiser offered a new framework within which one can understand the practices and processes of pedagogy in the modern sciences. Combining the analyses of two well-known philosopher-historians, Warwick and Kaiser provide what they semi-facetiously call a "Foukuhnian" perspective—one that uses arguments of Thomas Kuhn to fill lacunae in the thinking of Michel Foucault and vice versa. Kuhn, they posit, alerted historians of science to the importance of the relationship between scientific training and professional scientific research. By identifying the practice of research as puzzle solving, Kuhn illuminated the significance of the solution of exemplary problems by the science student. Students trained in the solution of progressively harder exemplary problems require and gradually acquire the ability to recognize novel problems as variants of problems already solved. It is precisely this skill that the practitioner of normal science requires. In this new understanding, problem solving, rather than the retention and reproduction of theorems, data, or laws, is the central point of instruction.

But Kuhn leaves unanswered the question of how one learns to solve problems at all.[76] For this, and in order to historicize particular forms of training, we may turn to Foucault:

Where Kuhn's account of pedagogy was confined to occasional and vague references to the contents of canonical treatises and textbooks, Foucault's evoked a much richer and interactive nexus of institutionalized gazes, bodies, gestures, architectures, routines, incitements, examinations and punishments. The latter approach constitutes a powerful resource in writing a pedagogical history of modern science and technology since it posits training as a general mechanism for the active production of knowing individuals that recognizes no natural distinction between the mind and the body, nor, by implication, between theory and practice. In this sense Foucault points to a level of analysis at which it should be possible to historicize the processes by which specialized technical competencies became the common preserve of widely extended communities of practitioners—the phenomenon Kuhn referred to as normal science.[77]

As Warwick and Kaiser note, Foucault complements Kuhn as process to product, Kuhn focusing on the end result of training and research, Foucault on the means by which students and researchers are made in the first place. And yet, compelling as this reading is, it is not to be assumed that Foucault is as vague about the end results of training and disciplining as Kuhn is about the mechanisms that lead to such results. "Discipline," in Foucault's words, "produces subjected and practiced bodies, 'docile' bodies. Discipline increases the forces of the body (in economic terms of utility) and diminishes these same forces (in political terms of obedience)."[78] In other words, the perfectly disciplined body—the end point imagined by Foucault—is simultaneously the perfectly obedient body. It is in this context that Foucault evokes the automata that gained such fame in the age of enlightenment, which "were not only a way of illustrating an organism, they were also political puppets, small-scale models of power: Friedrich II, the meticulous king of small machines, well-trained regiments and long exercises loved them."[79] One should thus not group Foucault with Kuhn in "provid[ing] hints of how to avoid a sterile educational determinism, which would treat the products of pedagogy as mere automata, destined only to mechanically repeat what they had been taught."[80] It is true, certainly, that Foucault depicts disciplining as part of a "positive economy," one capable of "extracting from time, ever more available moments and, from each moment, ever more useful forces."[81] Yet the tradeoff is that the mechanism is made "more obedient as it becomes more useful, and conversely."[82]

The point may be put this way: disciplining is capable (perhaps) of producing the perfect student, but the perfect student—in the sense that they are perfectly disciplined—is not necessarily the perfect researcher. As Warwick notes of the students trained at Cambridge, "many young wranglers, left more or less to their own devices after graduation, failed to appreciate the following fundamental difference between problem solving and significant research: when solving examination questions, students were always, in [George] Darwin's words, 'guided by a pair of rails carefully prepared by the examiner,' whereas in undertaking a truly original investigation no such guiding rails existed."[83] The need to step beyond the guard rails is an integral part of the process by which one becomes part of the professional research community. Sharon Traweek, in her anthropological study of high-energy particle physicists, argues that independence is seen as an essential attribute of post-doctoral students in this field. "The situation demands a "careful form of insubordination" on the part of the post-doc: they respect the tacit instructions of their elders by not following their explicit instructions, and this must be done with considerable delicacy."[84] The undergraduate and even the graduate student trained to obey is thus required at some point to disobey in order to "make it" as a researcher. There is no scope, however, within the logic of Foucauldian disciplining, to see such independence as anything other than a *failure* of discipline.[85] If we wish to understand how students learn to go beyond mimicry (or at least very careful and scripted guidance), therefore, and how they become researchers capable of contributing novel results to their community, we require more than an understanding of the means by which they have been disciplined.

Where Warwick and Kaiser are undoubtedly correct, however, is in understanding training to underlie even the apparently most improvisatory exercises. Jazz music, in their example, provides an ideal case for proving the importance of disciplined training. While jazz performance requires the ability to improvise, such creative moves are best understood as deviations from a base laid down by hours of practicing "certain rudimentary elements upon which they draw when performing their improvisations... often in highly formalized pedagogical settings." Within a discipline, in other words, one requires and acquires a disciplined creativity.

And yet the choice of a single example tends to obscure what may be the most important point. Jazz is not the only form of musical performance, after all. Orchestral performers may be asked to play with flair and feeling, but they are not required (for the most part) to improvise. Both jazz and orchestral musicians train, and both are very disciplined, no doubt, but jazz musicians train with the explicit aim of improvising on the basis of such training. One must assume, therefore, that their training regime differs from that of orchestral performers, and that some modes of training and practice work better (or are intended to work better) at supporting improvisation or creativity than others. In other words, there are different cultures of creativity, even where that creativity is based on disciplined training.

Whereas the Cambridge undergraduate system was a near-perfect instantiation of Foucauldian discipline in action, Sommerfeld's remained a mixture of the newer forms of training, particularly in its focus on problem solving in both pedagogy and research, while nonetheless retaining elements of older forms of apprenticeship. This is not to argue that Sommerfeld's students were in any way "more creative" than those of Routh or any other Cambridge tutor. Innovative research was produced in each site, but it was only in Munich that training for that research was formalized and institutionalized. Students who had graduated from the Tripos essentially had to seek guidance on their own, James Clerk Maxwell, for example, making much of his epistolary relationship with William Thomson. Within Cambridge itself, post-Tripos training in theory could be difficult to find, a point perhaps made most vividly by the fact that even Maxwell devoted little time to the matter. In the 1870s, having taken up a professorship at Cambridge, he concentrated his pedagogical efforts on establishing the

Cavendish Laboratory as a center for research in *experimental* physics, contributing little to the training of the next generation of researchers into the field he had opened up with his 1873 *Treatise on Electricity and Magnetism*.[86] A focus on the formalized elements of Cambridge training, in other words, does not suffice to explain the success of Cambridge scholars as creative researchers, rather than merely talented students. Discipline is necessary to understand the successes of Cambridge and Munich, but it is sufficient for neither.

To understand the differences between economies in Munich and Cambridge, it is worthwhile, therefore, to re-introduce Foucault's distinction between "initiatory" time and "disciplinary" time. Each is associated with a particular kind of training: the first with that of the apprenticeship, "an overall time, supervised by the master alone, authorized by a single examination"; the second with the new pedagogical practice, "specializing the time of training and detaching it from the adult time, from the time of mastery; arranging different stages, separated from one another by graded examinations; drawing up programs, each of which must take place during a particular stage and which involves exercises of increasing difficulty; qualifying individuals according to the way they progress through these series."[87] Routh's system of training was profoundly disciplinary. The Tripos examination acted as a boundary that marked the end of the time of training. To aid his students in passing, Routh divided them into groups, ranked by stage, with three groups by undergraduate year, a fourth for those about to take the exam. Each group was further divided, by ability, to produce between twelve and sixteen classes. Every two weeks, students were given a timed examination, after which they were publicly ranked. As the students were scaled and ordered, so too was the knowledge imparted to them, with concepts, techniques, and questions becoming increasingly more difficult, until the Tripos itself, in which questionists could expect to be asked to solve unpublished research problems.

Sommerfeld's system, however, was a hybrid. It was the dissertation, the "masterwork," that marked the completion of this stage of learning, considerably more initiatory by this point than disciplined. Hybridity entered at the earlier stages of training. Sommerfeld's students learned skills initially through the solution of graduated problems, a mode of training that was absent—at least according to observers like Pearson—in Germany only two decades before. But progression was not marked by examinations. As in Routh's system, students solved problems every week, and, if those listed in the back of Sommerfeld's texts are any guide, they gained slowly in complexity and in the knowledge required to complete them. But there was no timed examination, the process moving ahead instead as a "conversation" between assistant, professor, and student. If there was competition, it was not institutionalized: no rankings appeared afterward. Without the Tripos, Sommerfeld's students, although well trained in the art of problem solving, had no reason to put the effort into acquiring the skills of a Cambridge Wrangler.

It was in Sommerfeld's seminar that skills imposed by discipline and training began to be put to an end other than discipline itself. In the seminar, students learned to read research papers, studying and critiquing arguments and producing answers to minor research problems themselves. The seminar, in other words, acted as a transitional, intermediate training ground on the way to original research. The sign of this change was the shift in Sommerfeld's teaching techniques. No longer lectures supplemented by problems sessions, this pedagogical environment was one in which students themselves presented work or analyses for group discussion. Together with their interaction with one another and with Sommerfeld in the "seminary," one might mark this as both the end of disciplinary time and the beginning of initiatory time. In their dissertations, students worked on problems branching off from Sommerfeld's own work, choosing from a range of problems on offer or selecting their own, as Ewald did. Meeting often with Sommerfeld, sometimes working closely enough to co-author papers with him, this process was one where students learned, from Sommerfeld himself and from other doctoral candidates, through emulation, critique, and comment, what it was to be a researcher. In the passage through the seminar, students moved from the training ground to the guild; their teacher no longer *Herr Professor*, but rather *Doktor-Vater*; the aim, no longer that of correction, of disciplining deviance, but of creation, of inspiring independence.

The present-day American academic will recognize in Sommerfeld's system something strangely familiar. Without the institutionalization of the distinctions, the trajectory of a student on the way to the dissertation looks much like that of the United States today. Students complete undergraduate training, mostly through the progressive solution of more and more difficult problems, they then move to graduate school, where several years are spent in graduate seminars, learning more advanced topics, but also learning the basics of research, sometimes (but not often) being asked to present their own analysis of published papers. Finally, if they make it, students are allowed to undertake original research under the gaze of a watchful supervisor.

The novelty of such a system around 1900 should not be underestimated. Few countries, if any, had a form of educational training that could rival that given those studying for the Tripos. In the eyes of many, problem-based learning was a peculiarly British enterprise. Conversely, few if any countries could match Germany's research imperative, the model for many nations of the modern era. In Sommerfeld, as Klein's student, we see how both traditions could have been brought together to produce students both skilled in problem solution and dedicated to problem solving as original research. If the sheer amount of time that Sommerfeld spent with his students is one reason for his school's success, and the collegial and communal atmosphere in Munich is another, then the pedagogical economy of which Sommerfeld was a part, one devoted to the production of disciplined creativity, is surely a third.

The preceding critique of Foucauldian notions of discipline in the historical reconstruction of pedagogical practice has applications well beyond the training of physicists. The crucial question it poses may be phrased thus: How much of the training of future members of a scientific discipline can be explained by evoking notions of disciplining? The answer I have given is: Only the first, disciplinary phase. Modern scientific disciplines still have within them a strong remnant of older guild practices, and disciplining does not work to explain the process that, Foucault argued, it came to replace. Disciplining, most specifically, cannot produce people who themselves produce new knowledge and it is the production of novel knowledge that distinguishes the researcher from the student.

The point is worth emphasizing. It is not true that the process of disciplining cannot produce new knowledge.[88] As Foucault notes, an important element associated with the examination is the construction of a case history. These case histories, collected, tabulated, arranged in series, molded into statistics, become the very base of the modern sciences both of the individual and of the population, the twin sides of the forms of knowledge that inscribe "bio-power." But if this is knowledge *about* soldiers, pupils, the insane, the ill, the criminal, it is not knowledge that they produce. It is, rather, drawn out of them. The criminal does not, through his disciplining, by means of the panoptic gaze, become a penologist, but the penologist, in the disciplining of criminals, may thereby better understand the nature of criminality.

Kathryn Olesko's study of training in Franz Neumann's "Königsberg seminar" in nineteenth-century Germany provides at once one of the best examples of disciplining in the training of new physicists and one of the best examples of the limits such disciplining must impose. Olesko follows Jan Goldstein in suggesting that Foucault's arguments may be fruitfully applied to training in the modern professions. Discipline then connotes the "rigorous 'disciplined' training to which the professional himself has submitted...and through which he has gained mastery over a body of knowledge as entailing a serious commitment or higher calling."[89] In Olesko's definition, discipline then means "training the mind to follow certain rules of investigative protocol and rigorous techniques of investigation. In the physical seminar at Königsberg, Neumann disciplined his students in the mathematical and measuring methods of physics."[90] Through detailed, repetitive, time-consuming tasks, Neumann's students came to embody an "ethos of exactitude." The outcome of this disciplining, however, was precisely a growing loss of innovation. Discipline itself, rather than the exploration of new fields of knowledge on the basis of such disciplined training, became the end point of their investigations: "So much did the perfection of technique dominate seminar exercises that students began to consider the rational execution of technique an important subject of investigation in itself. Hence their publications reflected less the expansion of theory or the discovery of something new than the problems they encountered in practicing physics, especially in processing data."[91]

This limitation of the notion of disciplining in describing and explaining the practice of pedagogy in schools known for their ability to train researchers suggests a new tactic for historians of education. With some examples now of training as part of "disciplinary time," we need to return to older studies of "initiatory time" with this background in mind. How did schools in the twentieth century—like Sommerfeld's—manage the transition from a necessary drilling in basic skills to an equally necessary sense of independence and improvisation? What techniques have aided in the production of disciplined creativity? How, in other words, were some schools able to produce the jazzmen of science?

3 The Kaiser's Physicists: The Sommerfeld School Goes to War

We had left lecture room, classroom, and bench behind us. We had been welded by a few weeks training into one corporate mass inspired by the enthusiasm of one thought...to carry forward the German ideals of '70. We had grown up in a material age, and in each one of us there was the yearning for great experiences, such as we had never known. The war had entered into us like wine. We had set out in a rain of flowers to seek the death of heroes. The war was our dream of greatness, power, and glory. It was a man's work, a duel on fields whose flowers would be stained with blood. There is no lovelier death in the world...anything rather than stay at home, anything to make one with the rest.

—Ernst Jünger[1]

In September 1916, one of Arnold Sommerfeld's former students, Ludwig Hopf, wrote him from Aachen. Hopf, who had served both as a soldier in the infantry and as a driver, had just heard word that he was to be transferred to a military-scientific posting with the Air Force [*Fliegertruppe*]. It was this upcoming relocation that had pulled him from the front to nearby Cologne just in time to witness the birth of his third son, named Karl Arnold, after Sommerfeld. Momentarily freed from the necessity of transporting men and munitions in a heavy truck—a task rendered more treacherous by night without lights—Hopf reported that he had gained his first opportunity since the outbreak of hostilities to learn of what had been happening in the world he had previously occupied. "Of science," he wrote, "I've known only things by hearsay for the last year and a half. Even in the service time spent here I couldn't find peace and quiet in order to learn or to do something comfortably. You can imagine what it would mean for me now; I have a burning hunger for physics." The latest mail had brought a volume of the *Annalen*, which contained a summary of recent developments:

For me, everything is new; since before the war I knew of Bohr's model and during the war I had no direct contact with what went on in science. What a new world has been developed! Truly there is now laid a breach in the thickest wall which had until now narrowed our physical horizons. In general, for physics a great time has really dawned.... I firmly believe that all this will continue to work longer in the development of human generations that the whole senseless war, the end of which can't be foreseen and which changes absolutely nothing in the world.²

At one level there was a profound disjunction between Hopf's life as a student in Munich, with its sunshine and beer gardens, the meetings in local cafes with students and faculty in the summer, and ski trips in the winter, and the grimness of his wartime experiences. As an asthmatic he suffered particularly badly from conditions in the infantry. A rather rotund young man, he had earlier recounted with humor the rough military discipline he endured, and the understandable rapidity with which he grew tired of being referred to as "that fat one there!" by the commanding officer of his training unit.³ While initial days as a driver in Flanders afforded him and his fellow soldiers little to do other than to enjoy the countryside and purchase cheap eggs and butter, later pushes resulted in overwork with little food.⁴ Yet, at another level, especially after his posting to the flying troops, the wartime experiences of Hopf and many other Sommerfeld students were more like a continuation than an interruption of their civilian lives as theoretical physicists. Many members of the Sommerfeld School were involved in military activities that had a direct bearing on their dissertation research in Munich. As the Kaiser's physicists, that is, these young theoreticians had to change comparatively little in order to take up the practical, problem-driven military work they were assigned.

The first section of this chapter responds to the relative dearth of literature on "Great War physics" by sketching out the tasks that physicists took on in World War I, in order to understand how their experiences differed from other soldiers and civilians, and to describe the organizational forms that their participation took. This study of physicists at war is then turned on its head. Rather than using the experiences of theoretical physicists as a means to understand aspects of the war, war experiences are used in the second section as a means of understanding aspects of theoretical physics in the years before 1920, as the "physics of problems" style of the Sommerfeld School was turned to the solution of the problems of war.

Two elements bind this chapter to the preceding one, bridging the ostensible chasm between the prewar and intrawar periods. It follows a common cast of characters, of course, the members of a school. It also follows precisely those elements that made a disparate group of young men into a school in the first place by tracking the problems that they shared in peace as those problems were utilized in war. The pupils of the Sommerfeld School carried with them—as they "left lecture room, classroom, and bench [not to mention café and ski lodge] behind them"—not only a set of already extant, well-defined questions, but techniques and practices suited to answering those questions. Above all, they bore with them the true brand of their training, the means of rendering new questions into a form recognizable within the scope of a "physics of problems."

The Kaiser's Physicists

Given the enormous amount of attention paid to their activities during World War II, it seems surprising that so little has been written on physicists during the Great War.[5] This is, one assumes, largely due to the difference in the perceived effect of physicists in each conflict. If World War II, because of the Manhattan Project alone, is known as the physicist's war, then World War I was the chemist's and the technologist's war. It was the chemist's because of the need to produce replacement raw materials after the British blockade of Germany, and the ability of Fritz Haber and others to aid in the supply of materials for munitions and agriculture (as well as the development of chemical weapons).[6] It was the technologist's because of the new mechanical inventions that would revolutionize the field of battle: the tank, the submarine, the machine gun, the airplane, and the multiple advances in artillery, communications, and transportation. To the extent that physicists are mentioned, it is principally in the context of what was called "the war of the spirits"[7]—that is, the nationalist and often chauvinist and jingoistic war of words and ink that erupted between scientists in England and those in Germany. The signatures of Planck, Wien, and others have been noted on the "Appeal of the 93 Intellectuals," the document that affirmed the Germans' belief that theirs was a defensive war and denied the existence of German atrocities in Belgium. Sommerfeld, Stark, and Wien were also among the sixteen who added their names to the far more vitriolic "Proclamation against the English," which called for the boycott of English scholars and their works.[8] Of the wartime *activities* of physicists, however, in opposition to wartime *discourse*, comparatively little has been written.

If the lack of palpable effect is the reason for the historian's lack of attention, it is hardly a justification, unless we are to judge the significance of historical events solely on the basis of their outcomes. Physicists, like other Germans—whether scientifically trained or not—served their nation in large numbers, in whatever capacity they could. As Karl Herzfeld noted in a letter to Sommerfeld in 1916, "the general business [of physics] is naturally less intensive than in peacetime, because a great proportion of physicists are either busy in the field or with works for the military."[9] Wien, returning to his institute in Würzburg only two weeks after German troops entered France, discovered that it was empty: "The foreigners had fled and had, fortunately, reached the borders before the declaration of war, the Germans had hurried to their colors."[10] Participation in the war effort was enormously important to Germany's physicists, even if this participation apparently paled, in its results, next to that of other disciplines.

A measure of this importance may be seen from the efforts of Max Born, editor of the *Physikalische Zeitschrift*, to put together a list of physicists in military service. In February of 1915, Born wrote to the directors of physical institutes in German

Hochschulen asking for information on who was in the field, what their rank was, whether they had been honored with the iron cross and, for those who had died the *Heldentod* (hero's death), he requested a photograph. Born claimed: "Through such a publication shall be given the thanks that we owe to the defenders of our homeland; but at the same time it shall be manifested before the outside world that, like the whole of German science, physics as well stands together as one with the fatherland in need and danger."[11]

Given that it details those who already held positions in the early few months of 1915, this list provides an important insight into the place of physics in the very earliest phase of World War I.[12] What is striking, both about the list and the intent behind it, is that it is meant to represent the activities, not of physicists *as* physicists, but of physicists as loyal Germans. This is not about how physics the discipline was serving the Reich in its hour of need, but about how physicists were joining their compatriots in the trenches. The tasks that physicists were undertaking bears out the impression that they served considerably more as soldiers than as scientists.[13] This element, which seems peculiar in comparison to the utilization of physicists in World War II, reflects several factors particular to the case of the Great War. First, in the euphoria that surrounded the declaration of war in Germany—what has been termed "the spirit of 1914"—many of those who signed up in droves, including physicists, anticipated a great adventure, a cataclysmic cleansing of German society and Kultur, an escape from the tedium of civilian life into a purer, almost primeval existence.[14] They, like Ernst Jünger, "set out in a rain of flowers to seek the death of heroes." The war was their "dream of greatness, power, and glory. It was a man's work, a duel on fields whose flowers would be stained with blood."[15] This euphoria, not to mention the machismo, misinformation, chauvinism, and romanticism of the day, would have led many to take up the pistol over the pencil. Yet even if they had desired the chance to serve with their intellectual skills, the military was initially largely unconcerned by the question of what science could do for the war effort. When Wien suggested to the head of communications on the home front that all physicists be drafted so that their expert knowledge could be put in service of the Kaiser, the plan was refused on the grounds that the war would be too short for such a body to have a significant effect.[16] Even the chemists Fritz Haber and Richard Willstaetter were originally told that the government could find no use for them during the war.[17] A similar faith in the briefness of the war, coupled with an overweening faith in German military prowess led to the early decimation of units that would have included physicists and engineers. Almost all the technical officers of the Artillerie-Prüfungs-Kommission (APK, Artillery Testing Commission), for example, were sent to the front at the outbreak of war. The discovery that French artillery surpassed their own (and the horrific realization that German soldiers might be facing such artillery across the western front for some time) sent the German army scrambling in search of replacements.[18]

As the cartwheeling Schlieffen plan disintegrated, as the war dragged on past Christmas, as men became bogged down in the trenches, as the British naval blockade took hold, and as casualties mounted, the situation for physicists changed. The spirit of 1914 did not long last out the year, and correspondence shows more physicists attempting to transfer out of the infantry than volunteering for service in it. At the same time, not only did the army require more men to replace the fallen; it also required (or, perhaps better, realized that it required) a growing system of support. More telegraphers and wireless men, more X-ray technicians for the huge numbers of wounded and, especially, more men in the testing commissions for artillery, transport, and munitions. On the munitions side alone, World War I was like nothing European nations had experienced before. During the Battle of the Marne, German armies used more munitions every day than they had during the Franco-Prussian war in its entirety.[19] Stockpiles dwindled to nothing almost immediately, so that the war had to be waged using only what could be manufactured at the time. The requirement that this hurriedly produced material be assessed quickly pulled the scientifically trained from their places as soldiers, creating a tension between the need for experts and that for infantry that was never fully resolved. Thus, for example, Sommerfeld wrote to Max Wien in 1916, asking him to requisition a former student, Wilhelm Hüter, who had been injured in the field. Wien was supportive, but noted that he was not in fact allowed to requisition those who were fit for active military service. In particular, he was allowed "in no case to pull an infantry officer out of the trenches."[20] For the duration of the war, the military high command continued to value the foot soldier over the physicist.

Nonetheless, by September 1915 Paul Ewald, could speak of "a small army of physicists" who had gathered at the APK.[21] Presumably he had in mind the group that involved Max Born and Rudolf Ladenburg, who worked (like many allied physicists) on a variety of problems including that of sound ranging, that is, locating the position of enemy artillery on the basis of the report from the guns.[22] Fritz Haber's gas group contained, as one might expect, a number of physicists, as did the radiotelegraphic section of the Verkehrs Prüfungs Kommission (VPK, Transportation Testing Commission), headed by Max Wien. Among other tasks, Wien's group, set up in June 1915, was attempting to introduce effective wireless telegraphic systems to airplanes. While successful in drawing physicists away from duties close to the front, the project did not impress pilots. Many in fact refused to use the extra equipment. "One does not wish for the danger of flying," one officer was overheard to complain, "without adding to it the possibility of electrocution."[23] Wilhelm Wien joined his cousin in the communications division in Berlin in 1916 along with "a large number of physicists, mechanics, and glassblowers" who were all trying to produce less expensive and more reliable amplifier tubes [*Liebenröhre*] for radiotelegraphic work. Again, however, the effect of their work for the war effort as a whole was not clear. "I did not gain the impression," Wien wrote in his autobiography in 1930, "that we, like German physicists on the whole, had really achieved something of great significance for the carrying out of the war." In his view, at least, allied physicists had been entrusted with tasks of considerably greater military importance.[24]

Such research groups, whether military or industrial,[25] created local concentrations of physicists, which grew as men were either individually "reclaimed" from other units, or as men specifically volunteered. Max Born was one of these latter, simply turning up, by his own account, at Wien's unit in Doeberitz to become a soldier in the German army for the third time in his life. "After having been a dragoon and a cuirassier," he wrote in his autobiography, "I was now to become an airman."[26] Later, after drilling in the rain had left him without enthusiasm for Max Wien's group, he was requisitioned by his friend Rudolf Ladenburg and transferred to the APK. From there he made his own efforts to increase the size of the group, regarding this as his chance "of saving gifted young men from being wasted at the front."[27]

It should nonetheless be emphasized that, in contrast to World War II, there was no central concern that allocated physicists to particular tasks and, further, that the possible tasks to which they—as possessors of particular skills—were assigned went far beyond what might strictly be designated as physics. As the presence of physicists in Haber's group demonstrates, the growing importance of chemistry to the war effort trickled down to some closely allied branches of physics. So, too, with technological developments, especially those in communication technologies. Roughly 15 percent of those listed by Born in the *Physikalische Zeitschrift* were assigned to telegraphy or wireless telegraphy units, quite apart from Wien's research group. Sections of the medical corps also saw a large number of physicists in charge of equipment. Writing in the *Physikalische Zeitschrift* in 1915, Sommerfeld noted the services that a discovery in physics had made to the progress of the war:

In the storms of the European war, Röntgen's seventieth birthday will pass by unnoticed by many fellow scholars [*Fachgenossen*]. And yet we have right now at this time, where thousands and yet more thousands of our fellow men owe the retention of their limbs and the possibility of healing to Röntgen-rays [i.e., X-rays], where every field hospital is fitted out with a ready-to-use Röntgen apparatus (or at least should be so fitted out), every reason to be thankful to the discoverer of Röntgen-rays....[28]

Despite the fact that X-ray equipment had rapidly dispersed from the physicist's laboratory and had moved into the surgeon's hospital, physicists were still among those best able to use the machines. Sommerfeld noted in a letter to Paul Ehrenfest in January 1915 that three of his former students were occupied with the use of medical X-ray equipment in field hospitals.[29] All of them had been involved in the discovery of X-ray diffraction in crystals in Munich in 1914. Their expertise seemed to travel well, even if, as Ewald noted, there was not in fact a great call for the equipment near the front, where the pressure of time worked against the regular use of X-ray images. "I'll not complain of overwork, that's for sure," he wrote to Sommerfeld. "There was a greater need for Röntgen apparatus in Munich. But perhaps the good sir doctors are also training themselves more on the apparatus. Nonetheless, it lies in the nature of the operation, to send on everyone that doesn't lie dying. With them at most the dressings are refreshed—consequently [these are] only amputations and also urgent operations, for which the X-ray pictures aren't needed."[30]

If medicine seems like an obvious part of war, another site where one could find physicists—meteorology—seems much less so. Yet the successful use of aircraft (both balloons and airplanes) and of poison gas in World War I required an intimate knowledge of local weather developments.[31] A mistake at the battle of Verdun, for example, had seen French balloons blown into enemy territory to be hunted down by German soldiers. Fritz Haber's early solution to the problem of clearing out enemy trenches in front of advancing German shock troops was simply to release poisonous chlorine gas from canisters in the German lines. The problem was that the prevailing winds blew *toward* the Germans, a fact that led Emil Fischer to desire Haber's "failure from the bottom of my patriotic heart; for if he succeeds, the French will soon figure it out and then turn the tables, which will be very easy for them to do."[32]

The connection of meteorologists to airship companies explains the caption—"The Heroic Death of a Photogrammeter"—of the postcard depicted in figure 3.1, which was sent to Sommerfeld by the astrophysicist and mathematician Karl Schwarzschild. Meteorology formed part of training in astronomy at most higher educational institutions, so that Schwarzschild's position in charge of a weather station in German-occupied Belgium was one that made good use of his skills. As the commanding officer of such a station he would have come into regular contact with photogrammeters, men whose task it was to photograph friendly and hostile territory from balloons or planes, producing ready-to-use maps. Schwarzschild's personal interests in both ballooning and photography offer further background for the image.

Schwarzschild was not the only physicist to take up weather observation during the war. Rudolf Seeliger (son of the famous astronomer Hugo von Seeliger) wrote to Sommerfeld in 1915, telling his former teacher that, since he would soon have been inducted into the infantry, he had decided to volunteer for a position in an airship company. It was perhaps with Schwarzschild—the colleague and friend of both his father and supervisor—specifically in mind that Seeliger outlined his larger plan, to eventually find a place as a meteorologist.[33]

Clearly, then, physicists found a multitude of parts to play in the Great War. Yet, while it provides an overview of the range of roles for physicists in the service of the Kaiser, this general depiction of *Kriegsphysik* elides a useful division among three kinds of war physics. The first kind can be depicted as that which made use of skills that were common to many physicists. Thus, for example, Schwarzschild, before his death

Figure 3.1
Postcard, Karl Schwarzschild to Arnold Sommerfeld (undated). Note that the photogrammeter is skewered on one of his own maps, the coordinates marking the location of the death weapon. Running blood looks like it might also represent the paths of rivers. There is possibly a self-conscious irony. Despite the statement, the photogrammeter's death does not seem that heroic, perhaps an intended and self-referential jibe at those who were not sharing the fate of comrades in the trenches. DM NL 089 (059). Courtesy Deutsches Museum, Munich.

from a rare skin disease at the front in 1916, moved from meteorology to ballistics, calculating the trajectories of shells. He was a famously virtuosic mathematician, so one can assume he performed his tasks with exceptional facility, but they were nonetheless presumably calculations that many mathematically trained physicists could have made. So, too, with Born's sound-ranging work, which was in fact independently duplicated by allied scientists. As a contrast, the second category contains tasks that made use of the *particular* skills of certain physicists, in areas outside those of their normal field of research. For example, the X-ray crystallographer Ewald's service as a *Feldröntgenmechaniker* (field X-ray mechanic), or Schwarzschild's meteorological service. The third kind of war work was directly related to an area of expertise. In this category one can put Max Wien's prewar work on wireless telegraphy or Fritz Haber's chemical synthesis of ammonia.

What is striking about the Sommerfeld School is how many of its members, including Sommerfeld himself, fall into the second and especially the third categories, pursuing war problems that were intimately related to their prewar research. The second section of this chapter therefore narrows its focus from the activities of physicists in general to those of the Sommerfeld School in particular. Having determined in the preceding chapter that there was enough cohesion supplied by Sommerfeld's particular pedagogical practices and enough local social interaction to bond the group into a coherent collective in Munich, the next step is to delineate a set of commonalities carried beyond the bounds of the local—a set that was made up of the problems of physics on the battlefields of Europe.

The Sommerfeld School Goes to War

A few months before the end of the Great War, in the midst of his ongoing studies of quantum theory and spectral lines, Sommerfeld began writing a document detailing the extent of his work for the Kaiser Wilhelm Foundation for War Technology and Science [Kaiser-Wilhelm Stiftung für kriegstechnische Wissenschaft, KWKW]. The foundation had been established less than two years earlier as a means of combining the efforts of mainly civilian scientists for military work and Sommerfeld had been contacted in February 1917 by the head of the physics division, Walther Nernst, who asked him to undertake "theoretical investigations in the area of radio telegraphy."[34] Sommerfeld agreed, taking as his specific brief not only questions regarding the theoretical treatment of various shapes of radio antennas but also certain additional problems in ballistics. A year later, in March of 1918, he could report:

> My war scientific works were ordered [veranlasst] partly by the VPK [Verkehrs Prüfungs Kommission] now Tafunk [Technische Abteilung Funk] (Rittmeister Prof. Max Wien), partly by the Torpedo Inspection at Kiel (Captain v. Voigt and Prof. Barkhausen), partly by the APK (Oberstleutnant Koch).... My works are all of a theoretical (calculational) nature. They have taken me several times to Berlin, Kiel and once to Göttingen for a conference in Prandtl's institute. They fall into three groups, corresponding to the above places that have ordered them.[35]

All three sets of problems—two involving wireless telegraphy (on land, and at sea), the last involving the gyroscopic motion of spinning shells—were characteristic of the work of the Sommerfeld School. The importance of wireless telegraphy to the Sommerfeld School has been noted several times, yet it is worth emphasizing that it was also seen as critical to Sommerfeld's contemporaries. Upon recommending Sommerfeld for membership to the Berlin Academy of Sciences in 1920, Einstein listed three principal and significant areas of research: the analysis of diffraction phenomena through Maxwell's theory of electricity and light, the quantum theory, and the theory of wireless telegraphy, which "was powerfully advanced through his work and that of his students."[36] Coming so soon after the end of the military conflict, Einstein probably had both Sommerfeld's prewar and intrawar work in mind.

Wireless telegraphy had also been a research topic for Sommerfeld during his Aachen period, yet it received an added emphasis in Munich after 1906, as Sommerfeld and his students set themselves the task of explaining the operation of Marconi's recently patented horizontal directional antenna. A simple vertical sender radiates its signal equally in all directions, a fact that makes it undesirable if the receiving antenna lies in a particular direction, since much of the signal is wasted (figure 3.2). In addition, if the signal is not meant to be received by all (such as those sent from land to a submarine), a means of restricting the spread of the signal is required.

Early attempts to direct signals involved the use of parabolic mirrors, which focused the electromagnetic radiation into a given channel. The large size of radio waves, however, required the use of enormous mirrors (at least 300 feet in diameter), rendering this method largely impractical. Marconi's solution was to use a sender bent into an inverse L shape, with its vertical part considerably shorter than the horizontal. The signal from such a sender displays a characteristic intensity map, in the shape of an asymmetric "figure 8" (figure 3.3). The signal is more intense in the direction that points opposite to the horizontal arm of the sender. An antenna in the same shape, but facing the opposite direction, receives more strongly when it is lined up with the sending antenna, so that one can use the antenna both to enhance the signal and to locate a distant sender.

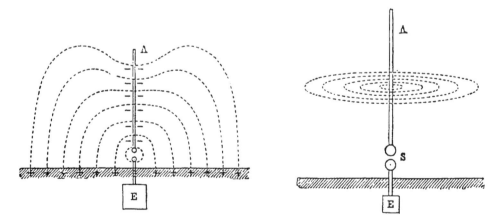

Figure 3.2
Left: Electric field around dipole antenna. Right: Magnetic field around dipole antenna. Note that both fields possess rotational symmetry. Source: J. A. Fleming, *An Elementary Manual of Radiotelegraphy and Radiotelephony for Students and Operators* (Longmans, Green, 1916), 150.

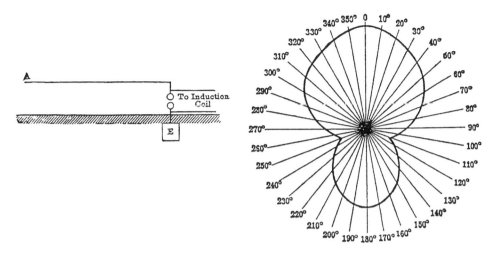

Figure 3.3
Left: Bent sender. Right: "Figure 8." Source: J. A. Fleming, *An Elementary Manual of Radiotelegraphy and Radiotelephony for Students and Operators* (1916).

This means of locating invisible distant senders has considerable military significance, since it can be used both by the intended recipient to tune in on a signal, as well as by hostile radiographers as a means of pinpointing the site of enemy transmissions.[37] An Italian pair, E. Bellini and A. Tosi, had begun the manufacture and sale of such directional antennas (using a different technique than Marconi's) by 1911. It was the critique of Bellini and Tosi's "sender theory" that formed the conclusion of Sommerfeld's report to the KWKW, which otherwise investigated the action of a particular horizontal antenna then in use throughout the German war zones in "direction-receiving stations" [Richt-Empfangs-Stationen, sometimes abbreviated RE Stationen].[38]

Sommerfeld's prewar engagement with the problem of directional radio telegraphy was thus exceptionally topical, especially given the fact that Marconi's bent directional antennas were still poorly understood. Indeed, Sommerfeld noted in 1911 that there were still those (including the doyen of German wireless telegraphy, Count Arco) who resisted the claim that the Marconi sender produced a directional signal at all. Arco suggested that the real reason for having a horizontal component to the antennas was merely that it offered a means of increasing the size of the antenna without the trouble of having to extend it vertically. Even among those who accepted the existence of the phenomena and sought a theoretical means of explaining it, there was a similar lack of consensus. In 1908, the English professor of electrical engineering at the University of London, John Ambrose Fleming, put forward his own theory, but noted that his "almost self-evident explanation of the action of a bent antenna has, however, not been accepted by German writers."[39] His explanation, and the objections to it, were reprinted verbatim in the (partially revised) third edition of the work in 1916.[40]

Fleming's explanation—essentially a geometrical argument that summed the effects of a closed and open circuit in close proximity—was indeed "almost self-evident." Yet in being so, it ignored what Sommerfeld and Jonathan Zenneck, professor at the Munich technische Hochschule, saw as the paradox of the horizontal antenna.[41] A signal is received when oscillating magnetic field lines cut the plane of the antenna. For a vertical sender, magnetic field lines run parallel to the earth's surface, and thus perpendicular to a receiving vertical antenna. For a horizontal antenna, the magnetic field lines are parallel to the antenna, and thus do not cut its plane as they expand and contract. A horizontal antenna should be incapable, by this reasoning, of receiving a signal from a vertical sender. The solution to the paradox came, in Zenneck's reasoning (which Fleming cited), by taking into account the fact that the earth's surface is a not a perfect conductor. The effect of this is to tilt the incoming waves, thus creating a horizontal component, which is larger according to the poorness of the conductivity of the surface over which it travels. Over land, the horizontal component is large, over sea water (a good conductor), the inclination to the horizontal is small.

Sommerfeld had made use of Zenneck's argument as early as 1909, in a paper "On the Spreading Out of Waves in Wireless Telegraphy."[42] Standard analysis had tended to ignore the actual passage of the electromagnetic waves over given surfaces, concentrating on sender and receiver theories. The dominant view, according to Sommerfeld, assumed that the Earth was a perfect conductor, a surface of zero potential, and that one could therefore essentially speak of the passage of radio waves as one would of any other "space waves" [*Raumwellen*], like those of acoustics or Hertzian electromagnetism. As Sommerfeld demonstrated, however, the effect of the imperfect conductivity of the earth was not negligible, and resulted in the production of "surface waves [*Oberflächenwellen*]," like those seen in water or in seismic activity. It is these surface waves that are inclined to the horizontal, minimally over sea water, by an angle of about 30 degrees over dry land.[43]

Sommerfeld's 1909 paper provided the background for a problem that Sommerfeld assigned to one of his students, Hermann von Hoerschelmann, which concerned "The Mode of Action of the Marconi Sender in Wireless Telegraphy." Hoerschelmann demonstrated that the directional effect of the bent antenna could be modeled via the action of two vertical antennas, affixed along the line of the horizontal. These were assumed to not be in phase and to interfere with each other, producing the characteristic "figure 8" pattern of the Marconi sender. Demonstrating the links between the peacetime and wartime work of the Sommerfeld School, Hoerschelmann's dissertation was then cited by Sommerfeld in his report to the KWKW on "The Directional Action of a Horizontal Antenna via a Visualizable Representation." There Sommerfeld noted an added effect, which Hoerschelmann's dissertation raised as a possibility. The two-vertical-antennas model, to the extent that it represented the action of the Marconi bent antenna completely, suggested the existence of *three* components to the received signal: the two that were assumed in standard analysis (a vertical and a radial component) and a "side component" perpendicular to them both. If such a "side component" existed, it would cause deviations from the assumed directionality of the incoming signal, and the deviations would be larger over dry, poorly conducting surfaces. At the close of his report, Sommerfeld pointed to new measurements from the technical section of the radio research unit [*Tafunk*] in Strassburg that seemed to affirm his prediction.

While the prewar work of students was being utilized for the war effort, students at war were continuing their research. Four young men, including von Hoerschelmann, completed their dissertations under Sommerfeld in 1911. All worked on topics related to wireless telegraphy. An American, Hermann March, studied the "Spreading out of Wireless-Telegraphic Waves on the Surface of the Earthly Sphere [*Erdkugel*]." Wilhelm Lenz completed his dissertation on "The Electromagnetic Alternating Field and its AC resistance, Self-Induction, and Capacitance" in March. Though Lenz's title made no mention of wireless telegraphy, the connection is made clear by his publication, the same year of his dissertation, of an article with similar analysis, titled "A Supplement to the Report by J. W. Nicholson on the Effective Resistance of a Coil." Nicholson was an author often cited by Sommerfeld for his work on wireless telegraphy, and Lenz's article was published in the *Yearbook for Wireless Telegraphy and Telephony*, edited by Max Wien. During the war, Lenz became a *Funker* (radio telegrapher). He wrote Sommerfeld regular reports concerning his research, both officially requested and personal, on radio antennas. One of these in particular made it clear that both he and his former teacher, in northern France and Munich respectively, were continuing to converge on similar topics:

You are right in your supposition that we are also occupied here with the RE Deviations [i.e. Richtung-Empfangs, or Directional Receiver Deviations] from directed senders. Our observations are not very marked, yet from another side deviations of up to 12 degrees have been observed. How far these can be trusted remains to be seen.
We have made preparations in order to be able among other things to test this question. The effect only seems to be expected at a great numerical distance and this in general is probably unavailable due to the great wetness of the earth here in the North. Experiment will presumably bring clarity to the issue.[44]

The fourth of the 1911 cadre, Wilhelm Hüter, completed his dissertation on "Measurements of Capacitances of Solenoids" in July. This, as Sommerfeld pointed out, was intended to be the experimental counterpart to Lenz's project. Hüter and Lenz clearly bonded over their shared task, for it was at Lenz's repeated urging that Sommerfeld wrote to Max Wien, asking that Wien requisition Hüter and pull him out of the trenches. This was achieved, in spite of military regulations that forbade the "reclaiming" of those fit for active infantry duty, and Hüter ended up in Wien's wireless telegraphy research unit. Dissertations with Sommerfeld prepared one well for a military posting with the Kaiser's physicists.

As with wireless telegraphy, Sommerfeld's other area of research for the KWKW, gyroscopic phenomena in ballistics, also had a history that preceded his arrival in Munich. Three of the four volumes of the *Kreisels* textbook that Sommerfeld co-authored with his former teacher, Felix Klein, were completed by 1904. The fourth volume, however, which contained the practical applications of gyroscopic theory—to ballistics, to analysis of the motion of torpedoes, to the action of gyroscopic compasses—was published in 1910. Sommerfeld had, by then, passed the editing tasks on to one of his students, Fritz Noether, who completed his dissertation on "Rolling Motion" in 1909. Between that time and the end of the war, Noether divided his research interests quite evenly between two of the central mathematics problems of the Sommerfeld School: turbulence and gyroscopes. In 1919 he published an article in the *Artilleristische Monatsheft* (*Artilleristic Monthly*) that dealt with the "Analytical Calculation of Shell Vibration," which drew, as he pointed out in his first footnote, on an unpublished manuscript of Sommerfeld's.[45] It was presumably this manuscript that Sommerfeld had produced for the KWKW.

Spinning shells travel in a more easily predictable path than ones that do not spin. Think, for example, of the smooth motion of a well-thrown American football. It is the spinning of the shell that connects ballistics with the study of gyroscopes. Yet, in spite of the smoothness added to the shell's flight, the rotating motion induces other resistive effects. The air through which the shell moves causes the shell to vibrate and alters its flight path. This phenomenon, Noether noted in his article, had only been treated before in a limited fashion, either qualitatively in Sommerfeld and Klein's *Über die Theorie des Kreisels*, or quantitatively in the textbooks and articles of the "Pope" of German ballistics, Carl Cranz, but restricted to very flat flight paths, or to graphical solutions. However, the experiences of the war, "the praxis of ballistics," had raised questions that now required a full analytic solution of the problem. It was such a solution that first Sommerfeld and then Noether sought to provide, albeit with only limited practical success. While their mathematical analysis was the most extensive, and offered the reader "a high spiritual enjoyment," the pair's work was, at least in Cranz's view, still incomplete in its consideration of the full physical detail of the problem.[46]

To these immediate extensions of the programs of the Sommerfeld School, these "war problems" which were variations on peacetime themes, one can add a number of less direct military uses of the skills of Sommerfeld's students. Like many other physicists, those who had been part of the Munich school for theoretical physics adapted their scientific qualifications as they adopted new roles in the military. Ewald and Friedrich, within a few years of their discovery of the X-ray diffraction of crystals, were studying X-ray images of the shrapnel wounds of countless soldiers. Taking time away from this in 1916, Ewald wrote to his Doktor-Vater reporting that he and another Sommerfeld student, Otto Blumenthal, visited the nearby weather station often and at present both were doing work on the conduction of heat in various media.[47] Noether, while he contemplated ballistics problems, and continued publishing on turbulence, also wrote to Sommerfeld reporting that he had turned his mathematical talents to the analysis of photogrammetric images. It was, in his words, "more geometry than theoretical physics," but he clearly grasped at whatever was available.[48] The same motive was presumably behind Lenz's efforts, when he wrote to Sommerfeld in late January 1916, offering a dimensional analysis of the pitch of incoming shells.[49] A year earlier, the English had succeeded in blowing up a large munitions depot in the German-occupied town of Lille. The explosion was enormous, with whole chunks of wood, rather than mere windowpanes being blown out, and the rage of the local population against their English "allies" threatened, as Lenz commented, to outdo their hatred of their German enemies.[50] Lenz was rather calm about the matter, offering the somewhat laconic comment that the constant shooting wouldn't be so bad if one didn't fear that more munitions depots would thereby be hurled into the air. Lenz had slept late and had not seen that explosion, but he observed the attempted shelling of his camp closely. He first saw the flash of the guns in Armentieres, 15 kilometers away. About 50 seconds later, he heard two cracks, spaced a second and a half apart, one caused by the "head" and the other by the "main" sound wave. Five seconds later, the whistling of the shells could suddenly be heard, their loudness rising rapidly to an apparently constant level.[51] Again, one can point to the impact of training at the Sommerfeld School, since Lenz was consciously examining the sonic wave phenomena that other Sommerfeld students—Hopf in particular—had previously analyzed. Given the element of danger involved in doing so, however, and the lack of practical utility of his actual calculation, one can only assume that the boredom was stultifying.

In between these two extremes, between the smooth continuance of prewar problems and the adaptation of prewar skills to new militarily related ends, lay a third path. In the case of Ludwig Hopf's work on aircraft design and testing, it was not the actual problem upon which he had worked in Munich that he brought to the *Flugzeugmeisterei*, nor was there a complete disconnect between his dissertation and his war work. In examining Hopf's aerodynamic studies closely, it becomes clear that he took from Sommerfeld not a problem and a set of solutions, but a way of choosing problems, a way of seeking out solutions. As with his work on the hydrodynamical theory of turbulence, and in accordance with the practical edge of the "physics of problems," Hopf sought to fuse praxis with theory, choosing as we shall see, not the mathematician's approach to the theory of flight, but the engineer's.

As with so many research groups, the aircraft command unit at Charlottenburg had come into being as a result of the failure of the Schlieffen plan to end the war decisively and rapidly in favor of Germany's armies.[52] As the first volume of the classified technical reports of the group noted,

Scientific research work into the nature of flight [*Flugwesen*] had been interrupted, at the beginning of the war, as a result of the calling up for army service of the numerous men trained in this special area. The long duration of the war and the rapidly rising requirements [brought about by] the achievements and qualities of aircraft led not only to the recalling of postponed, and the tackling of new, scientific investigations, but also the calling up of those powers, the advice and knowledge of which could serve for the construction of airplanes, in concert [now] to a further extent than earlier.[53]

Among those called to take up the problems of aircraft design and analysis was Hopf. His path to Charlottenburg was indirect, but made a certain sense. After completing his dissertation work with Sommerfeld on hydrodynamic problems and a short assistantship with Einstein, whom he accompanied when Einstein was called from Zurich to Prague, Hopf moved to the technische Hochschule at Aachen. There he served as an assistant first to H. Reissner and then his successor, Theodor von Karman, both of whom had published widely in aerodynamics. Reissner, in particular, was among those requested in 1916 to look at the problems of flight by the KWKW. At the outbreak of hostilities, Hopf had just been made Privatdozent, so that in a period when theoretical work on airplane flight was in its infancy, he would have been one of a handful of physicists and engineers with the training and background sufficient to deal theoretically with the problems that war had raised for science.

In the letter that began this chapter, Hopf had just heard that he was to be transferred to the *Fliegertruppe*, a move that had caught him by surprise, since it had occurred, as he put it, "without the least effort on his part." At the time he had not been sure about his good fortune. As he noted, "the driving corps [did] not gladly give up a man ready for active military service."[54] Less than a month later, however, he was happily ensconced with others working on "*Kriegsphysik*," and he recounted his activities to Sommerfeld with some enthusiasm:

I am at the testing institute and hangar of the air troop in Adlershof and have been working together with Dr. Everling on scientific questions of flight technology, experiments and theories. Happily the value of theories for these technical problems is estimated very highly there; there are many and interesting tasks and I am learning a great deal. My position is a very independent one and I am not at all bothered with military matters. At first it required a little trouble in order to get the rusted wheelworks of the brain back in business, now, but still slowly, the motor is thawing and I hope I'll not merely learn here, but will also do something for our flying work [*Fliegerei*]. I'll certainly be able to create more than as a driver, in any case.[55]

His brain clearly thawed quickly, for within months Hopf was—true to the collaborative intentions of the air command—co-authoring articles on aerodynamic problems, especially those of aircraft stability. It was a mastery of the material achieved at a remarkably fast pace, for while one can assume that Hopf had a certain familiarity with aerodynamic problems and their methods of solution, and was no doubt better prepared than most, he was far from being an expert in the subject. His own prewar research was significantly different to his military studies. With Einstein, he had published papers on radiation theory and on the calculation of probabilities, and in 1913 he published an article summarizing present understandings of relativity theory in the journal *Aus der Natur*. Apart from these, Hopf seems to have continued the studies of the problem of turbulence that he began in Munich. In 1914 he wrote to Sommerfeld reporting that he had only just broken off his "peacetime battle with the English in the area of Hydrodynamics."[56] In May of 1915 he noted that he was pursuing the turbulence problem once more. He wanted, he wrote with grim humor, to return to it in peace, hopefully without having suffered the Heldentod first.[57]

At a superficial level there appears to be a connection between these papers and those of Hopf's time at the Sommerfeld School. A formal analogy exists between hydrodynamics and aerodynamics: to the extent that air can be considered a simple fluid, the latter reduces to the former. And yet a closer study of the content of Hopf's wartime aerodynamics makes it clear that he draws little on classical hydrodynamic theories. His analyses of stability are largely considered in mechanical terms, that is through a study of the balance of forces on the center of gravity of particular aircraft designs. "Lift" and some other hydrodynamical phenomena are thus reduced to yet another mechanical force and their fluid-dynamical cause is not considered.

The distinction between the two approaches is made explicitly clear in the textbook *Aerodynamik*, published in 1922, which Hopf would co-author with R. Fuchs.[58] It was, at the time, the definitive textbook on the subject, one that sought, as they put it "to make use, not only of the particular experiences which we could gather as leaders of the aerodynamic section of the of the aircraft command, but also to make all the material of aerodynamic science into a single textbook."[59] The text was divided into two roughly equal sections, the first dealing with "Forces of Air" [*Luftkräfte*], the second with "the Movement of Aircraft" [*Die Bewegung des Flugzeugs*]. As the authors note, "the task of giving an explanation of the forces of air is a hydrodynamic one, the problem of the movement of a fixed body in an incompressible fluid."[60] While easily stated, the problem was nonetheless "one of the most difficult and unclear in all of mechanics," and its present partial solution owed much to the fact that it was essential to an understanding of flight technology. Even now, the authors make clear, solving equations required a nimble balancing act between finding expressions simple enough to evaluate on the one hand, and having these be representative of reality on the other. The smallness of frictional forces suggests that one can replace the "insuperably difficult differential equations of frictional fluids with the simpler ones of frictionless [fluids]." Yet the equations one arrives at are not single-valued [*eindeutig*] without the further assumption that the fluids are eddyless. The assumption of eddyless fluids, however, runs counter to experience and eddies do not arise in frictionless fluids.[61] To avoid the difficulty, one is required first to analyze solutions that do not contain frictional components, but which have "such eddies [already] within them so that the forces of air can be calculated in agreement with experience." Then one makes use of equations with frictional components to demonstrate that it is precisely these "characteristic eddy phenomena" that arise at those places where the frictional effects do not remain small, that is, at the surface of the fixed body.[62]

In contrast to this mathematical ducking and weaving, the analysis of the movement of aircraft proceeds at a far less abstract level. Hopf put it as follows:

More primitive are the aids of theory which deal with the movement of aircraft and which pursue purely practical aims. Here the forces of air are inserted as experiment supplies them, either in purely empirical dependence on each other and on geometrical relations, or approximated through simple relations. Hydrodynamic questions are thereby completely expelled and it remains only to consider the aircraft as a rigid body under the influence of certain given forces.[63]

These two sections of the text were intended to appeal to two different sets of readership. Thus, the authors note that the chapters on the "physical theory of air forces" are of interest in particular for the scientist [*Wissenschaftler*]. Middle chapters, spanning both sections provide "calculational foundations for the balance of aircraft," while the latter part of the text, entirely in the section of the movement of aircraft, supplied information of "the dynamics of the aircraft, to which the task falls of giving guidelines for the further development of aircraft—perhaps in the future still considerably stronger than at present."[64]

Three further aspects of the text need to be emphasized. First, while the two authors claimed to have worked together in nearly every part of the text, the writing was divided between them. As the authors inform the reader, Hopf wrote the introduction and the section on the movement of aircraft (section 2), Fuchs wrote section 1, on the forces of air. All quotations in the text above have been taken from either from the authors' joint "Vorwort" or from those sections written by Hopf.[65] Second, as a means of classifying Hopf's work for the *Flugzeugmeisterei* in terms of the hydrodynamics/praxis split that he and Fuchs offer, it is worth emphasizing that all citations of his papers occur in the second, non-hydrodynamic section. Hopf's papers, including "On the Static Stability of Double-Deckers," "The General Longitudinal Motion [*Laengsbewegung*] of Aircraft," "Pressure-Point Migration and its Influence on Stability," and "The Form of a Wing and its Influence on Wing Forces," all treat the aircraft, or its components as "a rigid body under the influence of certain given forces." The hydrodynamic origins of these forces is rarely, if ever, pursued. Third, in spite of the emphasis on praxis in Hopf's part of the text, the approach is nonetheless theoretical in its nature. The "aids of theory" may be "primitive" as Hopf put it, but the aim was to improve their utility, rather than to avoid them. The latter point, crucial to an understanding of the motives behind Hopf's choice of problems, was made explicit in an article co-authored by Hopf and Fuchs and published in the third volume of the technical reports of the *Flugzeugmeisterei* in 1918. There the authors introduced a discussion of the longitudinal motion of aircraft with the following description of their field:

All aerodynamical calculations, which lie at the base of the design and assessment of aircraft, relate to uniform stable [*stationär*] motion. There are, however, a large number of practically more important questions, that cannot be dealt with through these kinds of calculations.... If *Aerodynamik* is to become a wider-reaching foundation for *Flugtechnik*, it must turn to the questions of accelerated, disturbed motion and explain the processes [*Vorgänge*] of a chance or intended disturbance, of a steering deflection or some other alteration to the state of flight. Our experimental knowledge suffices, to some extent, in providing the mathematical analysis of these phenomena with the necessary foundations; where that is not yet the case, close theoretical study can at the moment provide guidelines for still-necessary investigations via models and in flight.[66]

One can now begin to see the connection between Hopf's wartime pursuit of aerodynamic problems and the dissertation research he completed under Sommerfeld's supervision. In both cases, he sought to bridge a gap between analytically simpler, but idealized mathematical treatments on the one hand, and more practical, empirically based efforts on the other. Questions involving the movement of aircraft motivated a kind of theory that was both inflected and informed by praxis, a theory located in a middle ground connecting the mathematics of hydrodynamics and the practical, model or operator-based knowledge of *Flugtechnik*. Hopf's dissertation work on turbulence bridged a similar schism between theoretical hydrodynamics and technical hydraulics. During the war, the gap between the theory of flight in hydrodynamical terms and its practice in the hands of pilots, inventors and engineers was even greater. Again, however, Hopf's approach was to offer, as Sommerfeld would repeatedly in his own work, a compromise, a theorized praxis and a practical theory. By largely abandoning hydrodynamics during the war, Hopf had actually turned away from the letter of part of his dissertation research, but his ongoing pursuit of a means of mediating between theory and *Technik* demonstrates that he, like other members of the Munich school, sought to solve the problems of war in a Sommerfeldian spirit.

Conclusion

The four years of the Great War constitute one of the most fruitful periods in the intellectual development of theoretical physics. Niels Bohr put forward his model of the atom in 1913, and a string of analyses followed rapidly upon it. Sommerfeld's significant contributions to the development of the quantum theory of spectral lines—the topic of chapters 5 and 7—were all published during the war. His textbook *Atombau und Spektrallinien* appeared a year after the war's end. Over the same period, Einstein's major formulations of his general theory of relativity appeared, provoking reactions that ranged from confusion to skepticism to wonder.

Those in the field were aware of the almost daily developments in the quantum and relativity theories. Sommerfeld's students often received offprints of his papers, and Hopf's example shows that, however late, news did trickle to the front. In the letter that began this chapter, written in 1916, Hopf enthused both about the quantum theory after the introduction of Bohr's model ("What a new world has been developed!"), and about Einstein's "gravitation theory," which he claimed "has now certainly experienced a crowning, which far exceeds all that was still, only a short time ago, believed possible."[67] Letters to Sommerfeld after 1915 were full of compliments for his work, and those of his students still in Munich. "I have been sunk in astonishment today," wrote Lenz in May of 1916, "about progress with the quantum through Epstein's preliminary [*vorläufige*] communication to the *Phys[ikalische] Z[eitschrift]*. In the face of these unheard-of successes, the consequences of your fundamental assumptions, I have been tempted to doubt in the reliability of mathematics."[68] In another letter he noted that "molecules certainly appear to lift their last veils extremely quickly in Munich on the Isar."[69]

Nonetheless, those involved in *Kriegsphysik*, in spite of their interest in developments on the home front, were rarely able to work on these matters themselves. Testing commissions did not require quantum theorists, the flying corps did not seek out specialists in relativity. Moreover, the majority of those contributing most actively to the growth of the quantum and relativity theories—Einstein, Planck, Lorentz, Bohr, Ehrenfest—were, either because of age, nationality, or political objection, uninvolved in military service.[70] Few with wartime postings were able to emulate the achievements of Karl Schwarzschild, who famously contributed two seminal papers to the fields of the quantum theory of spectral lines and of general relativity theory from the eastern front. It is thus perhaps no wonder that the (mainly intellectual) histories of physics for this period tend to trace the development of ideas with the war as background, usually as a hindrance to the "real" task of theoretical physics.

To the extent that the war has been discussed in the context of the activities of theoreticians, this has largely occurred in relation to wartime *discourse*: who opposed the war, who supported it. As with the content of the kind of theoretical physics, the emphasis has thus been on "principles." Einstein's principled (and lonely) stand against the war is a well-known fact. Planck's initial support and later (partial) repudiation is almost as famous. Yet it should by now be clear that theoreticians did considerably more than simply talk about the war. They were active participants, both as soldiers and as scholars. If one is willing to countenance the existence of parallel forms of theoretical physics, like Sommerfeld's "physics of problems," it becomes evident that the war was not an interruption to the development of theory, but a continuation of it. One needs not merely to understand physics in the context of a war, but also war as the context for physics.

Sommerfeld's students contributed to the field of physics from the field of battle. Indeed, an understanding of the problems of war brings into sharp relief the nature of training in the Sommerfeld School. So many students moved easily into areas of war research and practice that made explicit use of their particular skills and training that it becomes clear that the practical, problem-driven approach characteristic of the war was developed and honed during peacetime study in Munich. Bound by shared problems and social ties before the war, members of the Sommerfeld School carried his "spirit" with them as they left the Bavarian capital, placing the skills of "physicists of problems" in the service of the Kaiser.

II

4   The Practice of Principles: Planck, Experiment, and the "Thermodynamic Method"

In 1894, five years after arriving in Berlin to take up the chair in theoretical physics that Gustav Kirchhoff's death had left empty, 36-year-old Max Planck was elected to membership of that city's eminently prestigious academy of sciences. Put forward by Hermann von Helmholtz, August Kundt, and Wilhelm von Bezold, Planck was a worthy if not exceptional candidate.[1] Certainly no one at the time could have imagined that he would become the "dean and definition of theoretical physics in Germany."[2] However, 1894 marked a turning point in Planck's research. For the previous several years he had made substantial, if not always entirely novel, contributions to a new field: thermochemistry. For more than two decades afterwards he would work on topics connected to the problem of radiation in a black body, research that would win him a Nobel Prize in 1918.

In front of his new colleagues in the Berlin academy, Planck sought briefly to outline his view of the history and future development of physics. Compared to the physicist of an earlier period, he claimed somewhat gloomily (and with arguable accuracy), the theoretical physicist today had disproportionately difficult tasks ahead of him. Around the middle of the century, a single, firm and certain goal had existed for all those that looked to the natural sciences for simple, great thoughts, for an all-encompassing worldview. This goal—the founding of all natural phenomena on mechanics—had seemed reachable back then in the early years after the discovery of the energy principle. Now, however, "an impasse, a certain disillusionment, has entered into this striving focused directly on this highest goal."[3]

The problem at the heart of this "impasse," according to Planck, was one that would have been familiar to all those listening. The mechanical method in physics produced too many solutions, so that finding the one true representation of nature became an impossible task. Yet, while the realization of this problem may have made the task of the contemporary physicist more difficult, one mode of solution already existed. One could bypass mechanics altogether, much as it had recently been bypassed in several areas of physics:

The entire new development of thermodynamics has occurred independent of the mechanical theory, solely based on the two main principles of heat theory. Moreover, the fundamental relations between electrodynamics and optics, between galvanism, chemical affinity and thermodynamics have been founded entirely without reference to the mechanical nature of the relevant processes. In exactly the same way one can hope that we can also find out a deeper explanation about those electrodynamic processes which are directly dependent on the temperature, as they are manifested, in particular, in heat theory, without first having to take the laborious detour via the mechanical interpretation of electricity. There then remain, however, as a solid starting point, only a few propositions, above all the universal principle of energy.[4]

In other words, rather than looking for detailed mechanical models of electrical phenomena, one could instead begin with far more general principles and derive results from these. Such had been Planck's approach to physics, beginning with his dissertation, in 1879, "On the Second Law of the Mechanical Theory of Heat." His nominators for membership in the society noted the sheer number of different themes, investigated over a 15-year period, which flowed from Planck's deployment and development of that principle. The conditions under which different states of the same substance could co-exist, the equilibrium states of mixed gases, the course of chemical reactions, dissociation, thermoelectric forces, and the electromotive forces of electrolytes "were all developed from a single general principle, the validity of which can scarcely still be doubted, and which as a result confers a higher degree of credibility upon the results obtained."[5]

Planck's *Antrittsrede* was both brief and well constructed. While it outlined his own approach to physics and emphasized its relative novelty and power, it also implicitly invoked connections to the recent work of older, more prominent colleagues in the audience. It was Helmholtz, after all, that Planck would later credit for his "path-breaking" work "in the development of pure thermodynamics, i.e. that theory of heat that eschews any special kinetic hypotheses and restricts itself to the application of the two fundamental laws [of thermodynamics]."[6] And, however path-breaking, that work was not to be seen as iconoclastic. In common with Helmholtz, Planck did not advocate abandoning mechanics altogether. Having effectively discredited any universalist methodology based on mechanical modeling, he nevertheless concluded his speech by arguing that the "fundamental interconnectedness" of all nature's forces was based in unity, indeed in identity, and the best means of uncovering that identity lay in mechanics. Writing to the Munich theoretical physicist Leo Graetz, in 1896, he made explicit his disagreement with a position like that of his student Ernst Zermelo, who "believes that the second law, considered as a law of nature, is incompatible with any mechanical view of nature." In Planck's view, the problem was not mechanics in general, but atomistic mechanics more specifically. Whereas the latter seemed to violate the absolute character of the second law, a continuum mechanics might not have to: "The problem becomes essentially different, however, if one considers continuous matter instead of discrete mass points like the molecules of gas theory. I believe and hope that a strict mechanical significance can be found for the second law along this path, but the problem is obviously extremely difficult and requires time."[7]

In terms of ontology, then, Planck was a proponent of the mechanical worldview, at least until the early twentieth century. Methodologically speaking, however, it was not mechanics but thermodynamics that provided the model for his scientific practice.[8] Where Sommerfeld built his theoretical physics by drawing in disparate elements from multiple disciplines, Planck sought to expand one part of physics to encompass the whole. Thermodynamics came to be the synecdoche for Planck's theoretical physics. Delineating what this meant for his theoretical practice is the concern of the first two sections of this chapter, which offer a detailed analysis of Planck's "thermodynamic method" across the course of the first two decades of his professional career. This study of the "practice of principles," with its emphasis on abstract and general solutions, thus provides a contrast to the analysis of Sommerfeld's practices of detailed problem solving in preceding chapters. At the same time, it offers the background necessary for the explicit comparison between each scholar's approach to the quantum theory which is the subject of chapter 5.

Planck was hardly alone in emphasizing the power of principled thinking in physics at the end of the nineteenth century.[9] The unity that the grand principles of thermodynamics and mechanics seemed to offer amidst a wealth of new results appealed to many.[10] The desire to avoid detailed micro-mechanical modeling was also a common trope in writings of the time. This commonality on specific issues could occasionally produce strange bedfellows, as when Planck declared his enthusiasm for the philosophy of the phenomenologist Ernst Mach and (much more briefly) the physics of the group known as the "energeticists."[11] However, Planck differed from previous proponents of analyses based on principles in at least one fundamental way. Planck was the first of a new breed; the first theoretical physicist to perform virtually no experiments himself, and the absence of experimental practice from his work shaped his theoretical analysis in fundamental ways. This would be true of Sommerfeld as well, perhaps only the second German theoretical physicist in the modern sense of the term. But where Sommerfeld embedded himself in experimental detail and data as part of his problem-solving theoretical practice—at times to the chagrin of his colleagues—Planck largely avoided the theoretical analysis of experiment or the extensive use of numerical data. It was not that experiment was unimportant to Planck, nor that he wished for a genuine separation between experimental and theoretical physics in disciplinary terms, but rather that the thermodynamic method offered a means of downplaying the significance of experimental detail within the context of theoretical practice.[12] The mechanism through which processes occurred could be ignored in one's theory and hence the possibility of their experimental realization was a question that, for Planck at least, could be sidestepped.[13] Even those customarily more interested in experimental detail noted the power of theoretical thermodynamic reasoning in this respect. Jacobus van 't Hoff, in a discussion of the application of Avogadro's law to dilute solutions, put it as follows:

> ... it may be remarked that any notion one may form as to the mechanism producing osmotic pressure, or the action of semipermeable membranes, is without influence on the reasoning. Thus, the question of whether the pressure is produced by the solvent or by the dissolved body can be left out of consideration; so too, whether it is dependent on collisions or by attractive forces. The action of the membrane, too, whether it is as a sieve, or by means of absorption, is indifferent. All this is the case because the proof to be given is based on thermodynamics, and is consequently free from assumptions on the mechanism.[14]

That it was thermodynamics that would provide Planck's ideal for a physical methodology is, thus, essential. Many might agree that principles must be both absolute and all-encompassing; some might emphasize general principles over microscopic modeling, but few would take this point as far as Planck, who used the generality of thermodynamic thinking to produce what he regarded as a novel space for theoretical analysis, a distinct way of "experimenting theory."[15] The final section of this chapter explores Planck's most coherent articulation of his position on this issue—one that allowed the dissolution of methodological difficulties that had plagued his early work in thermochemistry, and that rewrote the past and present of his discipline to produce a vision of its ideal future, one freed from the "anthropomorphism" of experimental practice.

Ideal Processes in Thermochemistry

Looking back toward the end of his life on the events of the late 1880s and the early 1890s, Planck still retained a sense of the discomfort his work on dissociation in dilute solutions had brought him. His studies had led to a genuinely new result, but in pressing his claim he became embroiled in a sharp priority dispute with the chemist Svante Arrhenius, who sought to force Planck to retreat from statements that implied that he had provided a more general and satisfying solution than the Swede. Both men had concluded that dilute solutions contained not only the solvent and molecules of the solute, but also dissociated components of that solute. Yet, while Arrhenius insisted that those components were *ions*—that the process involved was ionic dissociation—Planck's treatment offered no comparable insight into the nature of the dissociated products.

Arrhenius, himself looking back in 1912, clearly took a certain amount of satisfaction in reporting Planck's capitulation. Although he noted that the support of men like Planck and Boltzmann had "helped in a high degree to protect the new theory from the attacks of physicists," and that "by their great authority they also gave a strong support to the new ideas in the eyes of chemists and of scientists in general," he also quoted what he portrayed as Planck's "concession" to his own position. "Planck also conceded (1892) that by means of thermodynamics 'nothing could be demonstrated regarding the qualities of the dissolved molecules, either in respect to their chemical or electrical properties, and that to this method could be ascribed no convincing conclusion, but only a heuristic signification.'"[16] Planck's own memory of the "failure" (or not) of thermodynamics in this case, and of the extent of his concession, was somewhat different. "Arrhenius," he recalled with some ire, "challenged the admissibility of my demonstration in a rather unfriendly fashion, while he emphasized that his hypothesis applied to ions, that is to electrically charged particles, to which I could only reply that the laws of thermodynamics are independent of the question of whether the particles are charged or not."[17]

If personally painful, the exchange nonetheless proves deeply illuminating for an attempt to understand the details of Planck's thermodynamic method. Pushed to justify his position, Planck made explicit what he saw as the true novelty of the approach to physics that he (and others) were advocating. And, as the quotation above might suggest, he did not, in fact, abandon this approach in 1892. Pushed again, in the first years of the twentieth century, to defend the assumptions he had made in the work on thermochemistry, Planck reiterated his defense of what he had termed "ideal processes"; assumptions that required no realizable experimental verification in order to be used by the theoretician. If the debate between Arrhenius and Planck was, in part (as Arrhenius would suggest) about the place of experiment in theoretical practice, then Planck made few concessions, arguing instead in favor of theoretical methods that could operate independent of their actualization—even approximately or momentarily—in the physical world of the laboratory.

Ideal Gases and Dilute Solutions
In making his foray into the territory of chemists in December 1886, Planck was aware that he needed to make some adjustments to his earlier analyses. The level of mathematical sophistication that was characteristic of his previous papers would, no doubt, have intimidated many potential readers, and his language needed to be shaped to fit into a new environment. It was not easy, he noted, for a theoretical physicist to adapt to the modes of thinking and expression of chemists, but he had been heartened in the endeavor by the "clear, enormous fruitfulness" of the methods he had to offer.[18]

Among the most difficult adjustments would, one suspects, have been the need to adopt the language of chemical molecularity. Planck had studiously avoided all such assumptions in the work that had followed from his dissertation. The reasons for this were twofold. On the one hand, the eschewal of all detailed analysis of the molecular properties and structure of matter followed from Planck's emphasis on the generality of thermodynamic analysis. "What interested me most in physics," he would write in 1943, "were the great general laws that possess meaning for all natural processes, independent of the characteristics of the bodies that took part in the processes.... Hence the two laws of thermodynamics captivated me to a particular degree."[19] In addition, molecularity was clearly tied to atomicity in Planck's mind and he was aware of Boltzmann's proof that an atomistic statistical mechanics produced a non-absolute second law of thermodynamics.

Planck's 1880 Habilitationsschrift, on states of equilibrium of isotropic bodies at various temperatures, took as its aim the representation of the influence of temperature on the elastic forces within a body, hence relating the theory of elasticity to the laws of thermodynamics. He emphasized, however, that his theory provided a way of achieving this aim "without one being required to make special assumptions about the molecular characteristics of the bodies: it satisfies much more the assumption that these are fulfilled *continuously* by matter."[20]

These issues of molecularity, atomicity, and continuity recurred in "Evaporation, Melting, and Sublimation," a paper written in December of 1881. There, too, Planck's aim was to demonstrate the results that proceeded from the two laws of the mechanical heat theory "utterly independently" of any assumptions about the inner composition of the body. It was important in general, he argued, to proceed as far as possible in this analysis with these principles alone, turning to molecular hypotheses only when other suggestions were completely exhausted. Physics was far from currently being in such a position.[21] The paper's final lines, over-dramatic and oft-quoted, laid out Planck's position on the question of the atomic theory explicitly. "The second law of the mechanical heat theory, consistently carried through," he argued, "is incompatible with the assumption of finite atoms. It is hence to be expected that in the course of the further development of the theory it will come to a battle between these two hypotheses, which will cost one of them their lives." Predicting the outcome of this battle with certainty would be premature, but various indications suggested that the decision would go to the assumption of continuous matter.[22]

It is thus to be expected that, although Planck spoke of (chemical) molecules in the work he took up after 1884, he placed certain limitations on their applicability to all areas of physics. Indeed, although he promised to adopt the language and modes of thought of chemists, it is hard to read his papers in the field as more than the extension of previous analyses to a new area, one where Planck seemed more interested in teaching than learning. The effect was to draw chemistry under the umbrella of the laws of thermodynamics, rather than to truly place physics in the service of chemists. The first of three papers "on the Principle of Entropy Increase" began with the observation that the second law of the mechanical heat theory was, like the first, not restricted to thermal processes. Echoing his earlier statement, he noted how far one still could go in carrying through the extension of the principle to all physical and chemical phenomena:

Proceeding from the thought that it is of high interest for the rational development of every [form of] natural knowledge to become acquainted as completely as possible with the totality of the laws which, to this point in time, are preserved in one hypothesis.... I have the intention to extend a little further in several successive treatments the series of conclusions that can be drawn from the Carnot-Clausius principle *in and of itself*, i.e. without reference to certain conceptions of the essence of molecular motions, merely taking as a basis the principle of the conservation of energy.[23]

A series of results followed in the subsequent papers, but it was the final section of the last article that was to prove most significant. Having developed a number of results for gases Planck turned to the analysis of dilute solutions. The move was a natural one, for the investigation of the close connection between these two states had been a matter of considerable importance for the recent development of physical chemistry. Van 't Hoff had made a number of advances in the theory of solutions by noting the "analogy" between "ideal gases" and "ideal [very dilute] solutions" and using the results from the thermodynamic analysis of the former to understand the latter. Van 't Hoff's analogy, however, had been justified by a mixture of empirical and theoretical reasoning and Planck's aim in his own work was first to demonstrate its validity (indeed, its necessity) on purely thermodynamic grounds before turning to the consideration of more specific chemical questions.

Planck began by noting that, for a sufficiently dilute solution, where the number n of molecules of the solvent was much larger than the number of molecules n_1, n_2, n_3,... of the different solutes, the energy U and the volume V were linearly dependent on the number of molecules, or

$$U = nu + n_1 u_1 + n_2 u_2 + \cdots, \tag{1}$$
$$V = nv + n_1 v_1 + n_2 v_2 + \cdots.$$

The physical significance of the equations was that, under these conditions, further dilution at constant temperature and pressure produced neither further enthalpic change, nor volume contraction or dilation. The changes that occurred were only those that one might measure if one added more solvent in the absence of any dissolved substances. For Planck, these equations functioned as *definitions* of the meaning of the term "dilute solution" rather than theoretical approximations or idealizations of laboratory substances. Were one to notice deviations between these conditions and those observed in the laboratory, then it should be concluded that the solution was not yet dilute enough. The dilute solution had gone from an empirical to a theoretical substance.

From these expressions, at constant pressure, p, and temperature, T, the entropy, S, given by $dS = (dv + pdv)/T$, can be written as a linear sum of $n + 1$ terms. Since the n's are all independent of one another and the u's and the v's are independent of the n's, one can assume the existence of functions s, s_1, s_2, s_3,... that depend only upon the temperature and the pressure, defined by

$$ds = \frac{du + pdv}{T}, \quad ds_1 = \frac{du_1 + pdv_1}{T}, \cdots.$$

Integration then gives

$$S = n(s + N) + n_1(s_1 + N_1) + n_2(s_2 + N_2) + \ldots, \tag{2}$$

where N, N_1, N_2,... are constants of integration, dependent only on the number of molecules and where, conversely, s, s_1, s_2, s_3,... are independent of n. The question then becomes one of determining the constants of integration. The value of the N's will remain unchanged if the number of molecules, n, n_1, n_2,... is constant for varying temperature and pressure. Planck, knowing this, made a key assumption, relating the dilute solution to an ideal gas, not by analogy, but by a thermodynamic process. The results from his earlier studies of gases, he claimed, could then be transferred directly:

Let us now for constant n make T very large and p very small, then all the matter contained in the solution as well as the solvent itself will arrive at the state of an ideal gas, because through the appropriate increase of temperature and decrease of pressure the bond of the molecules among themselves can be arbitrarily relaxed; hence the whole solution takes on the properties of a mixed ideal gas and we can bring into application here the expressions for U and V that we had established in a previous section for such a mixture:

$$U = n(cT + h) + n_1(c_1T + h_1) + n_2(c_2T + h_2) + \cdots$$

$$V = \frac{T}{p}(n + n_1 + n_2 + \cdots)$$

In doing so, incidentally, it is completely inconsequential [gleichgültig] *if the given state can really be arrived at experimentally, and certainly whether it represents a stable state of equilibrium or not; because these expressions are completely independent of this* [question].[24]

The final point in the lines cited above, introduced so casually, was, in fact, anything but incidental. Planck had clearly seen an apparent objection, and an obvious one at that. If one considers the case described (a dilute solution, say, of NaCl in water), lowering the pressure and increasing the temperature does not automatically produce the result required. In most common cases, the water will vaporize, leaving a solid salt. For Planck's process, however, one requires both the salt and the liquid to vaporize *and* to maintain their molecular integrity as compounds. Whether it was at all possible to carry out such a procedure cannot have been clear to Planck, nor was it clear, as we will see, to later opponents. The argument, however, was a thermodynamic one and the details of the process, including the very possibility of its experimental realization, did not matter for Planck. It was thermodynamically possible and hence the results followed.[25]

With the relation between the two states established, Planck could now equate the terms in the expressions for U and V above with those for equation 1, and insert the values thus arrived at into equation 2[26]:

$$S = n[(c+1)\log T - \log p + k + N] + n_1[(c_1+1)\log T - \log p + k_1 + N_1] + \cdots.$$

A comparison of this with an earlier expression for an ideal gas produced an equation relating the entropy to the *concentrations* of the individual substances within the solution, $C, C_1, C_2, \ldots$, where $C_k = n_k/(n + n_1 + n_2 + \cdots)$:

$$S = n(s - \log C) + n_1(s_1 - \log C_1) + n_2(s_2 - \log C_2) + \cdots.$$

After further substitutions from equations earlier derived for an ideal gas, Planck arrived at an expression for the relationship between the concentrations of the various substances at equilibrium and their thermodynamic variables.

$$\sum v \log C + v_1 \log C_1 + v_2 \log C_2 + \cdots = \sum v\varphi + v_1\varphi_1 + v_2\varphi_2 + \cdots. \tag{3}$$

Here $\varphi = s - (u + pv)/T$ and $v, v_1, v_2, \ldots$ represent the change in the number of molecules for a given reaction. In the cases considered by Planck, namely the evaporation or freezing of a dilute salt solution in water, the chemical reactions may be written $H_2O + M \to M + (H_2O)'$, where the prime denotes a change of phase and the salt denoted by M is assumed to have molecular weight m_1, so that the total mass of the salt in solution is $n_1 m_1$.[27] In these cases, the values for the v's are, respectively, $v = -1$, $v_1 = 0$, and $v' = 1$. Since in either of these cases, further, the water is the only substance in the gas or solid phase, $C' = 1$ and equation 3 may be written simply as

$$-\log C = -\varphi + \varphi'.$$

For $n_1 \ll n$, this becomes

$$\frac{n_1}{n} = \varphi' - \varphi. \tag{4}$$

Planck then considered an expansion of the φ's in powers of $(p - p_0)$, where p_0 is taken to be the pressure for a saturated gas in equilibrium with pure water at a given temperature, T. Assuming a very high level of dilution, the series can be broken off at first order:

$$\varphi = \varphi_0 + \left(\frac{\partial \varphi}{\partial p}\right)_0 (p - p_0),$$

$$\varphi' = \varphi'_0 + \left(\frac{\partial \varphi'}{\partial p}\right)_0 (p - p_0).$$

At equilibrium, $\varphi = \varphi'_0$ and equation 4 may be written as

$$\frac{n_1}{n} = (p - p_0)\left[\left(\frac{\partial \varphi'}{\partial p}\right)_0 - \left(\frac{\partial \varphi}{\partial p}\right)_0\right].$$

If one introduces the terms v'_0 and v_0 to represent the "molecular volume" (volume per molecule) of water at pressure p_0 in the gaseous and liquid states respectively, then a little manipulation following from the definition of φ gives

$$\frac{n_1}{n} = (p_0 - p)\frac{v'_0 - v_0}{T}.$$

Finally, Planck notes that $v'_0 \gg v_0$ and uses Avogadro's law, $pv = T$ (in units where Avogadro's constant is set to 1), to give an expression for the relative depression of the point of evaporation of water in relation to the concentration of a salt dissolved within it:

$$\frac{n_1}{n} = \frac{p_0 - p}{p_0}. \tag{5}$$

At this point, as Root-Bernstein notes, Planck makes use of a piece of numerical chemical data for the first time in the paper.[28] For a 1 percent solution, by mass, of salt in water (molecular weight = 18), this may be written

$$5.6\, m_1 \left(\frac{p_0 - p}{p_0}\right) = 1.$$

Using analogous arguments and somewhat more chemical data, Planck also arrived at an expression for the depression of the freezing point of a dilute solution of salt in water:

$$\frac{m_1(T_0 - T)}{18.2} = 1.$$

Each of these expressions, Planck noted, was identical to those discovered earlier by van 't Hoff—using a completely different method based on the first, rather than the second law—except that the chemist had introduced an extra term to bring the theory in line with experimental results. Van 't Hoff had multiplied the right hand side of each expression by a term he labeled i, which was a constant, dependent on the particular salt used in the solution. It was toward an explanation of this term that Planck turned in the final section of the paper, titled "Apparent Deviations from the Theory." In offering such an explanation, however, Planck seemed to know his conclusion from the outset. A great many approximations had been made in his derivation, not the least of which flowed from his definition of a dilute solution, yet he rejected the suggestion that insufficient dilution might be the cause of the deviations of his results from those determined empirically. Without calculating the effect such a condition may have had on the expressions derived, he stated that the difference was too small to account for the size of the i term in each case.

That left his theory as the source of the discrepancy, but Planck declared that this was based on assumptions of such a fundamental nature that one could alter any one of them only with great difficulty. Of the three basic ideas upon which his analysis rested—the second law, Avogadro's law, and his thermodynamic assumption that all substances could be brought, through a sufficient temperature increase and pressure decrease, to the state of an ideal gas—Planck declared:

If one wished to give up [*aufgeben*] one of these assumptions, this would mean shaking precisely those laws that appear, among all of them, to us as the most solid foundation of all theoretical investigations, and those in which we see precisely the main advantage of the methods proposed by us. Before we decide to take such a step, we would have to offer up [*aufbieten*], in fact, everything, in order to proceed with other assumptions.[29]

The only recourse, then, was a reformulation of the meaning of the experimental data. The term denoted as m_1 in the equations above, previously understood as the molecular weight of the salt, needed instead to be understood as a composite term, representing the salt and the products of its dissociation, of masses $m_1, m_2, m_3, \ldots$ and in quantities $n_1, n_2, n_3, \ldots$. Equation 5 and the equivalent expression for the freezing point of water must then be written

$$\frac{p_0 - p}{p_0} = \frac{T_0 - T}{102} = \frac{n_1 + n_2 + n_3 + \cdots}{n}$$

The final paragraph of the paper was triumphant, as Planck noted that his positing of the existence of dissociation in dilute solutions accorded with similar claims made by Meyer, Ostwald, and Arrhenius, who had, however, done so on the basis of certain assumptions as to the molecular properties of the solutes. Planck's, however, was a victory for thermodynamics in all its generality. "In any case," he concluded, "this much is certain, that the assumption of the laws derived by us here offers the only possibility of bringing the general principles of thermodynamics, known and valued at this moment as the most solid foundations of research, in harmony with the results of experience."[30]

Ideal Processes

Arrhenius's paper, also postulating dissociation as an explanation for the *i* term, appeared in precisely the same volume of the *Zeitschrift für physikalische Chemie* as Planck's. At first an almost baffling case of "simultaneous discovery," the closeness in timing is, in fact, easily explained. Ostwald, the journal's editor, had been responsible for drawing Planck's attention to the most recent work in the field, especially that of Arrhenius and van 't Hoff, so Planck had essentially been led down the same path as the chemists, identifying the same problems as the key issues to be explained.[31] That said, Planck's and Arrhenius's explanations were dramatically different in detail. The former claimed decomposition without a discussion of the nature of the products in solution, seemingly assuming that the components of the salt were whole atoms or smaller molecules. Arrhenius, on the other hand, making use of his earlier work on ions in electrolytic solutions, explained the *i* term by assuming *ionic* dissociation and thus that the products in solution were charged.

For Planck, this extra information about the molecular properties of the dilute solutions seemed almost to detract from the generality of Arrhenius's explanation. Indeed, although his jockeying for priority in the months after the simultaneous publication of their papers was rather muted, his argument for the greater logical validity and scope of his approach was stated strongly. In a paper written in February 1888, Planck noted that he and Arrhenius had arrived at precisely the same conclusions and rather patronizingly put this forward as further proof for the correctness of his own approach. "If now also," he wrote, "the foundations of [Arrhenius's] comments—the deep [*durchgreifende*] analogy, that he sets forth between the behavior of the osmotic pressure in dilute solutions and the pressure of ideal gases—probably may not claim the rank of a completely valid proof, nonetheless the fact that completely independent of purely theoretical discussions, the same ideas have been stimulated and supported on the most variegated grounds from the chemical side, appears to me to be a sign for the fact that in this case as well the claims of the second law of the heat theory will again, in time, find complete acceptance."[32]

Arrhenius was not impressed. Writing to Ostwald at the beginning of February—and hence presumably before seeing Planck's most explicit comments on the matter—he put forward his own methodological critique, one that went to the heart of Planck's claims for the validity of the theoretical thermodynamic method. "I still cannot buy [the notion] that the basis upon which he determines the freezing point depression and the vapor pressure depression are so much stronger than those invented by van 't Hoff," he noted with some asperity. "He postulates *de facto* that the same laws that are valid for gases should be valid for solutions, but he has no experiments like the ones van 't Hoff has presented, and so it appears to me that his hypothesis is all the more shaky."[33] On the other hand, Planck had clearly reached another constituency. Precisely his emphasis on theoretical and mathematical rigor seemed to Helmholtz to be what was needed in the new work on the theory of solutions. In a letter written in 1891, Helmholtz noted that Walther Nernst had recently "thrown himself zealously" into the new work in physical chemistry, an area opened up by van 't Hoff and "advocated with great vigor" by Ostwald through the editorship of the *Zeitschrift*:

These theories have already proved to be of great practical utility, and have led to a multitude of demonstrably correct conclusions, although they imply some arbitrary assumptions which do not seem to me to be proven. The chemists, however, make use of this hypothesis (of the dissociation of a portion of the compound molecules of the dissolved salts) in order to form a clear conception of the processes, and they must be allowed to do this after their fashion, since the whole extraordinarily comprehensive system of organic chemistry has developed in the most irrational manner, always linked with sensory images, which could not possibly be legitimate in the form in which they are represented. There is a sound core in the whole movement, the application of thermodynamics to chemistry, which is much purer in Planck's work. But thermodynamic laws in their abstract form can only be grasped by rigidly trained mathematicians, and are accordingly scarcely accessible to the people who want to do experiments on solutions and their vapor tensions, freezing points, heats of solution, etc.[34]

Planck thus had positive and negative reasons to try both to articulate and justify his methods to a larger audience. An opportunity to do so came in the form of the 64th annual meeting of the Gesellschaft deutscher Naturforscher und Ärzte, held in Halle in 1891.[35] In reading his "General Remarks on the Recent Development of the Heat Theory," one notes rhetorical strategies very similar to those articulated three years later in his *Antrittsrede* to the Berlin Academy. On the one hand, Planck was determined to delineate his role in the production of a new theoretical approach to the physical world, one based on principles and to be opposed to one that emphasized detailed mechanical modeling. On the other hand, considerable effort went toward demonstrating that he was hardly alone in this new, progressive movement and that his methods needed to be judged in the context of similar attempts to deploy generalized thermodynamic arguments in the analysis of physical and chemical systems. Ultimately of limited success as a sustained defense of his position against determined critics like Arrhenius, the lecture nonetheless constitutes one of the clearest of all Planck's statements of his theoretical methodology.

Opening his comments to the combined physics and chemistry sections of the meeting, Planck described two distinct paths that the heat theory had taken since the middle of the century. The first proceeded solely on the basis of what Clausius had termed the "two laws" of thermodynamics, avoiding any representation of the nature of energy or heat. The second focused on the molecular processes of which heat was supposed to be a manifestation, seeking to determine the kinds of forces and motions that made up the energy of a body. Each path had its merits, Planck suggested, but it had to be noted that in spite of the physical insight and mathematical skill expended by two of the best recent proponents of the second method—Maxwell and Boltzmann—their exertions were not matched by the fruitfulness of their results.[36] Indeed, one of the most significant discoveries of recent years, the analogy between dilute solutions and ideal gases, had neither been discovered by means of the kinetic theory, nor could one expect much progress in the theory of solutions along those lines. In contrast, and almost paradoxically, it had been the first method—which avoided all "special molecular representations"—that had produced so many insights into the "world of the molecule."[37] Clearly, Planck informed his audience, the two laws alone could not have achieved such a striking result, and so the task of his lecture was to articulate what else had been involved in the achievements of thermochemistry. The answer, in essence, was the use of "ideal processes."

Planck proceeded to describe an array of examples of the use of the ideal in recent theoretical analysis. Interested in demonstrating how widespread such methods were, however—and hence in rhetorically enrolling the greatest number of allies—he tended to minimize what were, in fact, rather striking differences between the cases he described. Indeed, there were three quite distinct notions of idealization running through the lecture: what might be termed, in turn, abstraction, extrapolation from experiment, and extrapolation from theory. Of abstractions, an example was the notion of a reversible cycle, so commonly deployed in thermodynamic reasoning. Perfect reversibility was, in the natural world, a fiction, but such cycles, it was claimed, "provide theoretical research the strongest and simultaneously the most supple means of penetrating into unknown areas [and] they are therefore also to be regarded as the essential cause for its powerful advance in the modern age."[38] The deployment of processes that required an infinite amount of time (for example, the transfer of heat from one body to another at the same temperature) offered another example of such idealizations. Much like "frictionless" planes and "incompressible" fluids, such processes were part of the standard arsenal of the physicist.

More interesting were Planck's examples of objects that were extrapolations from experiment. The prime candidate in this category was the notion of a semi-permeable membrane.[39] In his "Osmotic Investigations," Wilhelm Pfeffer had made a membrane that allowed the water of a sugar solution to pass through it, but not the sugar molecules. Van 't Hoff, in pressing the analogy between the pressure of an ideal gas and the osmotic pressure of a dilute solution, had extended this idea to those cases where the practical realization of such an object was not necessarily available.[40] A similar situation existed for gases, where a glowing sheet of platinum could be used to separate hydrogen from oxygen and other gases. The American theoretician Josiah Willard Gibbs had extrapolated from this case and argued, on the basis of a theoretical semi-permeable membrane that could separate any gas from any other, that the total entropy of a mixed gas could be understood as the sum of the entropies of each of the individual gases, a result that Planck himself had used in his own analysis. "The enormous generalization that Gibbs has given to this tenet," Planck noted, "and which must, in and of itself, appear irresponsibly daring, rests clearly on the self-evident thought that the validity of so fundamental a tenet as that of the entropy of a mixed ideal gas, cannot depend on the arbitrary circumstance of whether we really have available in each individual case a suitable semi-permeable membrane." Either, he continued, the law is valid in general or it is not. If it is, "then clearly the proof in a single case suffices."[41]

It is in the third category, extrapolation from theory, that one might place Planck's own "ideal process," namely his argument that the connection between a dilute solution and an ideal gas consisted in the fact that one could—theoretically—transition from one to the other without the number of molecules involved in the process changing. "In reality," Planck noted, "such a process will admittedly often not be realizable, because in many cases, at high temperatures, as are necessary here, chemical transformations occur, and the molecules are thereby altered."[42] Indeed, the truth of Planck's claim for the existence of such processes in the natural world would seem to be honored more often in the breach than in the observance, and it should be noted that, unlike the case of the semi-permeable membrane, Planck gave no example of any situation in which the process occurs to even good approximation. The justification for such processes, he suggested instead, "lies completely in the region of the methods used with success in the heat theory since Gibbs."[43] In essence, therefore, the argument was a purely theoretical one. According to the second law, the behavior of a solution was entirely determined by the way in which the volume, the energy, and the entropy depended on the temperature, the pressure, and the number of molecules in the solvent and in the solute. Operating on the *equation* that related these terms, one could move through states of the system that could not be realized in actuality, but whose theoretical validity could be assured, as long as one was rigorous in one's mathematics. An analogous example, provided by Planck, lay in Maxwell and Clausius's manipulation of the state equation relating the dependence of the pressure of a homogenous substance to its temperature and density. This equation represented, it was noted, not only observable states, where the substance was either liquid or gas, but also intermediate states between the two phases, that "could never really be represented," the properties of which "arise only through a mathematical extrapolation and could only arise from such."[44]

The proof of the theoretical pudding, for Planck, lay in its experimental eating. Experiment, that is, was to be the guide for testing the outcome of his theoretical work, not the intermediate steps of its development. Certainly, as his work on cavity radiation would show, a conflict between empirical data and the conclusions of theoretical work would lead Planck to change his theoretical analysis, but the inability to produce in the laboratory the processes assumed by a theory did not, in itself, negate that theory's validity. And in the case of thermochemistry, experimental data and theoretical predictions seemed to be in fine accord. Using his ideal process, Planck had been able to explain the validity of Avogadro's law applied to dilute solutions. He had contributed to the theory of electrolytic dissociation "when it had not yet begun its victory march [*Siegeslauf*] through the continent," and he had explained the deviations from van 't Hoff's laws, so that the experimentally determined depression in the vapor pressure and freezing point of a dilute solution provided a measure of the total number of molecules therein, "independent of their nature."[45] This last point was key, and although Arrhenius would see its statement as proof that Planck had accepted the failings of his theoretical approach, it seems clear that Planck saw it, rather, as a benefit:

It is, admittedly, self-evident, that this result, taken in itself, which has exercised to this point no lasting influence on the viewpoints customary in chemistry, yields, in particular for those expressions only the *number* of the dissolved molecules, [and] in contrast absolutely nothing about their properties, neither chemical nor electric, whereby a visualizable [*anschaulichen*] image for the time being remains withdrawn from every clue [*Anhaltspunkt*]—but yet it offers today, where the images in favor of these expressions have been altered, an incontestable historical proof of the effectiveness of theory.[46]

With the basis of his own work established, Planck turned next to his competitor, who, he acknowledged in a subtle jibe, had published his first works on the dissociation theory "*almost* simultaneously."[47] Arrhenius's work, he claimed, had been based, not on a single general proposition, but on a series of facts of a chemical, electrical and thermal nature. Different (and less general) in its foundations, however, Arrhenius's approach, Planck emphasized, could not be considered purely empirical and hence, perhaps, more valid than his own. "Once the way was paved for the new theory," he declared, "confirmations and extensions followed, as is well known, in a surprising way, arranged to a great extent, through the use of precisely those ideal processes that we have considered here." Increasing the number of those who used such processes, Planck noted that, in addition to Arrhenius and van 't Hoff, "Ostwald and Nernst earned the greatest credit."[48]

In his closing remarks, Planck turned explicitly to the relationship between theory and experiment. In the face of the "undoubted success" that accompanied their utilization, ideal processes had to be recognized as "highly valuable tools for research" and one could expect further discoveries on their basis. "Indeed," Planck declaimed with some vigor, "I do not stand back from holding them to be a particular triumph of the human spirit, with the help of which the connection of the laws of nature can be pursued into areas that are utterly closed off from direct experiment." Nevertheless, such idealizations could not be granted *probative* force. For theoretical work to function as a proof, every phase of its development had to be verified through the action of the senses. Ideal processes, valuable as they were, merely played the role of "pathfinders" for theoretical explanation. At least here Planck was willing to cede ground to those who sacrificed generality for realizability. This concession, however, was soon deemed insufficient.

The year after Planck's lecture at Halle, Arrhenius published a stinging attack "On the Validity of Planck's Proof for van 't Hoff's Law." Planck, he claimed, "not insignificantly over-valued" the methods he had used in his thermodynamic work on the theory of solutions, while simultaneously paying insufficient attention to the other contributing factors involved. Taking umbrage to the suggestion that his own and van 't Hoff's work was less "authoritative" than the theoretician's, Arrhenius proceeded to a systematic critique of Planck's arguments. Several paragraphs followed, which disposed of the priority question by noting that Planck had, in his first publication on the subject, cited one of the papers in which Arrhenius had already (so he claimed) "clearly pronounced" his theory of electrolytic dissociation. Turning to the details of Planck's claim, it was next emphasized that—by not discussing the nature of the dissociated molecules—Planck had produced a theory that simply could not be brought into line with chemical understanding. "Ordinary (non-electrolytic) dissociation" was an assumption that was "wholly untenable," since it could not be reconciled with the facts of diffusion, and since none of the results of the electrolytic dissociation theory could be derived from it. The heart of the short paper, however, dealt with a comparison of van 't Hoff's and Planck's modes of proof for the expressions involving the relative depression of freezing points and vapor pressures of dilute solutions. Van 't Hoff's, Arrhenius claimed, was grounded on "statements of experience" and his own theory of electrolytic dissociation was based on this and yet more experiential facts. "It is probably obvious," he wrote, "that the validity of van 't Hoff's law can in no way be derived *a priori* without other assumptions than both the laws of the mechanical heat theory."[49] Nonetheless, this was precisely what Planck claimed to do, largely by "forgetting" an important assumption with which he had begun, namely the linear dependence of the total energy of a dilute solution on the individual energies of the solvent and solutes (equation 1).[50]

In spite of Planck's emphasis on his "ideal process" as the novelty of his theory, and the point to be defended in most detail, this was not the element that Arrhenius would critique. Leaving entirely aside the question of the validity of a proof founded on non-realizable processes, Arrhenius focused on that element of the proof that could be tested, arguing that the expression Planck wrote down was, strictly speaking, only true for the case of *infinitely* dilute solutions. "It is really very doubtful," he wrote, "whether one can adhere *a priori* to a similar assertion, but if one were to also acknowledge this, it would be highly conceivable that the region in which this relation was valid with sufficient precision possessed such extraordinarily restricted boundaries that it would perhaps be inaccessible to observation."[51] Planck's proof could then only be considered binding after a determination of how closely his analysis approximated the experimental situation from which he drew the data suggesting the deviations associated with the i term. Otherwise, Arrhenius noted, one simply could not be sure of the extent to which deviations from van 't Hoff's laws, in Planck's theory, required the assumption of dissociation, and to what extent such deviations simply followed from the difference between (over-idealized) theory and experiment. Taking this task on himself and using data derived from earlier experimental work by Thomsen, Arrhenius then showed a dramatic difference between Planck's assumption and recorded results. For the solutions used in the laboratory, Planck's statement of linear dependence did not even seem to come close and so his claim for the necessity of the assumption of dissociation could not hold.

Planck's reply was rapid, brief, and to the point. Stressing the theoretical validity of his assumption of linearity, which was "evident" and which he had therefore not emphasized, Planck nonetheless could only weakly defend himself from the substance of Arrhenius's critique. His formula and the deductions that flowed from it, he acknowledged, only possessed validity for suitably dilute solutions, but he was not willing to accept that it might only be sufficiently precise in regions inaccessible to observation. This question, he suggested, "can, in any case, only be decided through experience." A pointwise response to Arrhenius's other claims followed, with equally pointed references to his discussion of precisely these issues in his earlier paper. He had not, he noted, claimed that dissociation followed merely from the two laws; it required, in addition, certain hypotheses regarding ideal processes. He had neither claimed that his theory offered information about the properties of the molecule, nor argued that his method offered more than a "heuristic significance." Indeed, he had made a point of declaring that any divergence between the results offered by his thermodynamic analysis and the facts of experience had to lead to the theory's transformation. Finally, in regard to the question of credit for the theory of dissociation, Planck offered a sensitive rebuttal, declaring his "regret" that Arrhenius had read—unjustly, he implied—certain of his remarks as claiming a priority that was not his due. Leaving open the question of what was, in fact, deserved by either party, he noted simply that he had not been aware of Arrhenius's first paper when he had written his own.[52]

One can understand why Arrhenius may have seen Planck's short paper as a concession. His opponent seemed to give ground on precisely the issues that Arrhenius thought were most important: the question of the nature of the dissociated substances and the heuristic (but not probative) value of the thermodynamic method. Moreover, both men largely left the field after the debate's conclusion.[53] In contrast to the flurry of papers published before the debate with Arrhenius, Planck produced comparatively little new work between 1892 and 1895, the year that he took up the study of Hertzian resonators in preparation for his work on black-body radiation.[54] Yet Arrhenius had misread what Planck took to be the most significant elements of his contributions to thermochemistry. On these issues his opponent had altered his stance very little, as one can see from the handful of papers Planck published on the theory of solutions at the very beginning of the twentieth century. By 1902, Arrhenius's critiques had simply been folded into a more familiar form of analysis. Planck now spoke of solutions in states of "ideal dilution." Under such conditions, as one might expect, the energy and volume depended linearly on the concentration of dissolved molecules. Dealing thus with the ideal, Planck felt no need for rigid determinations of the limits bounding the concept. "One can say no more about how dilute to assume a solution is in order that it satisfies the conditions of ideal dilution," he wrote, "than that it is also impossible to completely generally specify the maximum density which a gas may have in order for it to be in the state of an ideal gas."[55] Arrhenius's experimentally grounded criticism had been circumvented by the expedient of a theoretical definition.

That same year, however, a more fundamental challenge was leveled at Planck's original theory. Matthias Cantor, theoretical physicist at the University of Würzburg, published a paper in *Annalen der Physik* ("On the Foundation of the Theory of Solutions") in which he discussed the justifications that had been offered for the connection between ideal gases and dilute solutions.[56] Planck's investigations, he noted, had seemed to avoid the assumption of a mere analogy between the two states by grounding the link between them on the laws of thermodynamics and a single, general, statement of experience. Planck's proof had then been taken as the "secure foundation" of the theory of solutions. "A more precise treatment shows, however, that it proceeds from an assumption which is grounded neither theoretically nor through direct experience, but is purely hypothetical, and that therefore does not possess the advantage presumed."[57] The problematic assumption was Planck's "ideal process." Cantor noted Planck's claim that it was only necessary that one be able to carry out this gasification "in an ideal sense" and that it was not necessary to assume that such a state was stable, a point he was willing to grant: "but it must be proven that it represents an actually *possible* state of matter, that it would be possible at least as a momentary state of motion. For this, however, a theoretical proof is never adduced and direct experience never authorizes such an assumption."[58] Planck's ideal process seemed, at least in chemical terms, impossible and hence was, for Cantor, illegitimate as the basis for a proof.

Planck was goaded, once more, into defending his methodology and published a reply in the same journal at the beginning of 1903.[59] Defending first—and perhaps unnecessarily—the mathematical validity of his proof, he then turned to the essence of Cantor's criticism: the implication "that a physically genuinely binding proof, can never be derived—not even indirectly—from the consideration of processes that may never be approximately realized in nature."[60] This general objection to the use of ideal processes in thermodynamics was "so self-evident and at the same time, from a purely logical standpoint, so incontrovertible," Planck claimed, that he had used every opportunity to raise and discuss the issue in the past, including most explicitly, his lecture in Halle more than a decade before.[61] For the most part, as he himself noted, the discussion offered in the later paper merely provided similar arguments in a somewhat altered form. Yet the tone of the 1903 paper was different; considerably less defensive, Planck now spoke of the popularity of the "thermodynamic method of proof" and no longer emphasized its purely heuristic efficacy.[62] Indeed, the probative force of the method—if not fully demonstrated—was now to be understood as a matter of personal choice. "Certainly any physicist is at liberty to avoid the use of ideal processes entirely if he does not trust their probative force," Planck wrote equably, "But this standpoint simultaneously signifies an avoidance also of a tool of research that is just as fruitful as it is well tried and well tested."[63] If ideal processes could be shown to lead to numerous results that were not in contradiction with the facts of experience—and were in many cases supported by these facts—and if they were not at odds with the fundamental principles of physics, then this constituted a genuine demonstration of the "theoretical admissibility" of such processes. When making use of the thermodynamic method, realizability—experimental or otherwise—was, for Planck, no longer a necessary criterion for the fundamental elements of a valid theoretical proof.

On Irreversible Radiation Processes

Two of Planck's most detailed defenses of his thermodynamic method—the first in Halle in 1891, the second in the pages of the *Annalen* in 1903—stand as bookends around his work on black-body radiation. Between 1897 and 1899, Planck wrote five papers bearing the title "On Irreversible Radiation Processes." In all of them he sought to do with the continuous processes of electromagnetic theory what could not be done with atomistic mechanics: prove that the second law of thermodynamics held absolutely. Deploying results gained from two papers published in 1896 and early in 1897, Planck made use of a "Hertzian resonator," an idealized linear oscillator small in relation to its characteristic wavelength. Unlike Hertz, however, Planck was not merely interested in mapping the field of the resonator itself. Adding another condition, that the resonator possess a small damping coefficient, Planck hoped to prove that such an object, when excited by incident radiation, would emit radiation of a qualitatively different form to that it had absorbed. This change in the structure and intensity of the electromagnetic field around the resonator, if it could be shown to be unidirectional, would constitute proof that conservative processes could produce irreversibility.

Planck's five papers have been discussed in considerable detail within a literature concerned with retracing his path to the quantum theory.[64] Here the aim is somewhat different, for I am more interested in recovering the argumentative logic of the papers in relation to each other than in the elements of future work that they contain. Rather than looking forward to the quantum, in fact, the discussion below traces continuities backward to the work on chemical thermodynamics, pointing to strong structural similarities in the theoretical methods used in each case. In particular, Planck's introduction of the notion of "natural radiation" in the fourth paper in the series—a move that has been taken to mark a turning point in his thinking toward the methods previously used by Boltzmann—can in fact be shown to be a generalization of ideas introduced almost from the beginning of his analysis. As with his work in thermochemistry, Planck was aware that specific results were difficult to derive from general hypotheses. The unidirectionality that he sought in the behavior of his oscillators was, in fact, an element introduced by an additional restriction on the structure of radiation incident upon them. And as with the introduction of his ideal process, this additional hypothesis was to be justified on theoretical grounds alone.

Central to Planck's analysis in the five papers was his insistence that little needed to be known about the detailed structure and "nature" of the resonator in order to describe its equilibrating action. The aim was to provide as general a solution as possible, one that eschewed specific assumptions about the substances being analyzed. Clearly an element of his overall methodology, Planck had an additional reason to avoid extensive modeling. In later reminiscences, he described what had drawn him to the problem of black-body radiation in the first place, namely the universality of the function relating the distribution of energy across frequencies for a black body at a temperature T. "This so-called normal energy distribution is therefore represented through a universal function of T and v, *dependent upon no substance of any kind*[.] Since, in my opinion, a law of nature sounds all the simpler the more comprehensive it is, the task of seeking for this function appeared to me particularly alluring."[65] Planck reported in 1895 that one of the great advantages of using a Hertzian resonator to model the absorption and emission of electrical waves in a cavity was that one could thereby derive Kirchhoff's law without making any assumptions about the nature of the material of which a black body was composed. The particles of the body, whatever their nature, could be shown to function in this context in a manner analogous to the resonators, which absorbed radiation and then scattered it in all directions: "Kirchhoff's law of the proportionality of absorptivity and emissivity is an immediate consequence of this characteristic. In order to calculate the amplitude of vibration one does not need to go any further into the nature of the emitting particles; it suffices to assume that the dimensions of the vibrational center are small in comparison to the wavelength...."[66]

Their lack of structure was precisely what drew Planck toward the use of Hertz's theoretical object. That lack of specificity had been important for Hertz as well, but in comparing the two researchers' reasons for deploying a common object one gains an insight into the effect of Planck's novel status as a pure theoretician. Certainly, the avoidance of assumptions about the detailed structure of theorized objects was a tenet of Helmholtzian practice upon which both Planck and Hertz would draw.[67] And, for both men, simplifying assumptions made calculation easier; solving field equations close to a linear oscillator of finite length is a fiendishly difficult proposition. If for Planck, however, avoiding the investigation of the detailed nature of his resonators both accorded with Kirchhoff's law and offered a greater generality and security for his solutions, for Hertz the issue was more one that concerned the relevant region of interest around his theorized object. In this particular case, Hertz's resonator was unstructured less from principle, than from experimental practice. From Helmholtz, Buchwald has argued, "Hertz learned to watch for novel interactions between objects in the laboratory without worrying overmuch about the hidden processes that account for the object's effect-producing power. His dipole and detecting resonator evolved out of attempts to investigate interactions of that sort. Neither device required or attracted analysis from Hertz, because he had learned from Helmholtz to probe rather the character of the interaction between the devices than their inherent, perhaps deeply hidden, structure."[68] One finds little, if any, trace of this emphasis on experimental interaction in Planck's use of the resonator. The kind of object-based thinking characteristic of Hertz's (and Helmholtz's) theoretical work was simply not a part of Planck's conceptual apparatus. As was the case for his previous studies, however, it would be wrong to suggest that Planck simply ignored the work of experimentalists.

In his earliest papers on resonance phenomena he made an effort to demonstrate the accord of his theoretical model with data derived from studies of circular resonators, seeking to account for the discrepancies between theory and experiment.[69] The connection to experiment, however, was a matter for the conclusion, not the process of theoretical analysis.

Irreversibility and Experiment
Planck's first paper on irreversible radiation processes considered the case of a resonator at the center of a spherical, mirrored cavity. Making use of results derived from earlier papers, he began by writing down an expression relating the moment of the (lightly damped) resonator, f, to the electric field intensity, Z, in terms of three constants: c (the velocity of light), L, and K:

$$L\frac{d^2f}{dt^2} + \frac{2K}{3c^3L}\frac{df}{dt} + Kf = Z.$$

This expression, Planck had noted previously, closely resembled that for a general vibration, excited by external forces and damped through friction. But the key difference was an important one for Planck: the damping coefficient was independent of the substance making up the resonator.[70] Assuming that the resonator emits spherical waves when excited by an arbitrary field (taken as finite and continuous throughout) and making use of a number of simplifying assumptions,[71] the paper concluded with an expression for the change induced in the primary field by the action of a resonator in terms of its period, given by $\tau_0 = 2\pi\sqrt{L/K}$:

$$Z' = Z - \frac{16\pi^2}{3c^3\tau_0^2}\frac{df}{dt}.$$

It was a result like this that Planck hoped to use in a subsequent paper to demonstrate that the resonator's action was unidirectional and equilibrating.[72] Upon publication, however, this introductory paper was almost immediately critiqued by Boltzmann, who protested that, given the fact that the equations of electromagnetic theory were no less time reversible than those of mechanics, any unidirectionality Planck seemed (or hoped) to find was simply a product of his own assumptions.[73] One merely had to imagine the system Planck had described running in reverse: rather than having a plane wave excite a resonator into producing spherical waves, one could as easily have a spherical wave moving inwards and—reversing every action in the previous case— have the resonator emit a plane wave. Planck countered in a brief second paper by arguing that Boltzmann had misunderstood what he was doing and that the simple time-reversal thought experiment that had been proposed as a counter-example had been expressly disallowed. He claimed to have barred, in advance, any incident spherical wave centered at $r = 0$ on the grounds that the intensity of such a wave would rise to infinity at the origin. Reversing the direction of time to allow a spherical wave that had been emitted from the resonator to be absorbed by it constituted a "singular case," forbidden at the outset:

> As I have shown, the electromagnetic processes in the immediate proximity of the resonator I have investigated are determined through a single function F of the time t and the distance r, which satisfies the familiar equation
>
> $$\frac{\partial^2 F}{\partial t^2} = c^2 \Delta F = \frac{c^2}{r^2} \frac{\partial}{\partial r}\left(r^2 \frac{\partial F}{\partial r}\right).$$
>
> The general integral of this equation is
>
> $$F = \frac{1}{r} f\left(t - \frac{r}{c}\right) + \frac{1}{r} g\left(t + \frac{r}{c}\right),$$
>
> where f and g signify arbitrary functions of a single argument. f thereby signifies a spherical wave propagating outwards, g one propagating inwards, from which it follows that f belongs to the excited, secondary wave, g to the exciting, primary wave. According to the given assumption as to the finiteness and continuity of the exciting wave at the site of the resonator ($r = 0$) the function g is therefore necessarily equal to zero for all values of its argument. *This statement offers, in a certain sense, the kernel of my theory, it cuts in half the number of processes represented by the differential equation above and guarantees thereby the unidirectionality of those left remaining.*[74]

Planck's statement that the removal of certain solutions from his equations offered the "kernel" of his theory comes as somewhat of a surprise. The point is not emphasized strongly in the first paper. The condition that the intensity of the field over the resonator be finite and continuous for all time is not, in fact, explicitly made there, having been noted, rather, in Planck's 1896 paper "Absorption and Emission of Electrical Waves Through Resonance." No statement of comparable clarity outlining what he here argued to be the central point of his analysis appeared in Planck's writings before the second paper. That said, it does seem clear that by late 1897, Planck had realized that the limitation of possible solutions to the equations governing the field around his resonator was crucial to his proof of irreversibility. An analysis of the resonator alone—in spite of the existence of a damping term—could not produce an absolute second law. With this in mind, one can begin to understand the strategy that Planck would then use in the third installment of his writings on irreversible radiation processes—clearly intended to be the immediate successor of the first paper—which relied still further on the exclusion of certain states of the field in order to guarantee the irreversibility of the remaining solutions.

Before proceeding to that paper, however, it is necessary to understand precisely what Planck had done to the ontological status of his resonator by making an argument that depended on the behavior of the field at the origin. If we return to the paper by Hertz that Planck had followed, one notes immediately a peculiarity about the diagrams Hertz had drawn to indicate the field around the resonator. Field lines in those diagrams cannot be found within a small ring drawn around a dumbbell-shaped ideogram, meant to represent the resonator.[75] Hertz described the logic of these diagrams thus:

At the origin is shown, in its correct position and approximately to correct scale, the arrangement which was used in our earlier experiments for exciting the oscillations. The lines of force are not continued right up to this picture, for our formulas assume that the oscillator is infinitely short, and therefore become inadequate in the neighborhood of the finite oscillator.[76]

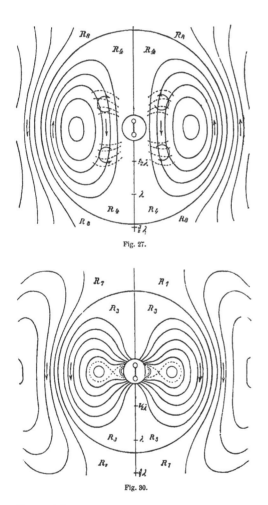

Figure 4.1
Heinrich Hertz's diagrams of the electric field around a dipole.
Source: Hertz, *Untersuchungen über die Ausbreitung der elektrischen Kraft* (1891).

Hertz's diagram involved a fusion of experimental and theoretical analysis, where the resonator that Hertz had himself built maintains a presence in his theorizing. Drawn to scale and correctly positioned, that real resonator then conditions the representation drawn from Hertz's theoretical analysis, which assumed an infinitely short object. The behavior of Hertz's theoretical resonator near the origin was not even considered, since it was precisely in this region that the correspondence between theory and experiment broke down.

Planck's analysis, in following Hertz, had begun with this restriction as well. Introducing the set of equations he used in his 1896 paper, Planck wrote that "The equations which condition the state of the electromagnetic field around [the linear oscillation], *excluding a small space containing the center of the oscillation*, have been completely given by Hertz."[77] In the analysis offered in the long quotation above, however, Planck had made a shift. Both he and Hertz had assumed in their theoretical analysis that the oscillator was infinitely small. Planck, however, now took this approximation to reality (valid for Hertz only within a given region excluding the origin) as a condition on the possibilities for the radiation states due to the resonator. It is true that if the resonator is taken to be infinitely small, spherical waves converging on it cease to be finite, yet as Boltzmann would note, such an assumption is both physically unallowable and (perhaps worse for Planck) does not, in any case, guarantee irreversibility. Planck, having earlier followed Hertz closely in analyzing the match between theoretical analysis and experimental data, continued from this point on a path concerned considerably more with the internal logic implied by his investigation of what was now a theoretical, rather than a theorized object.

Irreversibility and Natural Radiation

Paper three was published in late January 1898 and began by noting the three conditions that would need to be satisfied for a process to be judged irreversible: a direct reversal of the process must be excluded (i.e. simple reversibility of the kind suggested by Boltzmann must not be allowed), "recurrences" must not be possible (i.e. one must prove that, for any future time, the system will not return to a previous state, either exactly or to close approximation), and a stationary state must be shown to exist after some finite, sufficiently large time. Further, once the system achieves such a state, it cannot move away from it.[78] Planck claimed that the analysis of the third paper, using an "extreme" case (that of a resonator at the center of a spherical cavity where the primary and secondary waves were restricted to pure spherical waves) satisfied all three conditions. The first, he argued, was always satisfied. Although he did not state the point here, his derivation made direct use of the "kernel" of his prior theory, the limitation of the function F for the resonator to spherical waves moving outwards. Points two and three functioned by other limitations on the field, expressed now in terms of a Fourier series. Recurrence was barred through the introduction of a distinction between what Planck termed processes "tuned" or "untuned" [*abgestimmt* or *nicht abgestimmt*] to the system. Tuned processes included those where one or several Fourier components were of the same order as the total radiation intensity (i.e. where a small number of frequencies dominate the expansion); where gaps in the series recurred systematically; or where the series broke off after a limited number of terms. In all of these cases, the system would display a degree of periodicity, and so they were excluded in advance. Planck could show, inversely, that processes not tuned to the system—in which the series possessed no or merely unsystematic gaps and in which the Fourier components were each vanishingly small in comparison to the total radiation intensity—the system would not return to any previous state within the long period considered in the analysis.

Having assumed that the radiation was untuned, Planck then turned to his third criterion—the existence and persistence of a stationary state—and introduced a second distinction: between ordered and disordered processes. In an ordered process, many neighboring Fourier components (each individually vanishingly small) interfere constructively due to a certain regularity (or "order") in the values of their phases. As a result, such components together make a measurable contribution to the total radiation intensity. When this regularity ceases, so too do deviations from the total intensity, which is therefore effectively constant. Stationary states, in other words, were "disordered."[79] The final section of Planck's paper was dedicated to demonstrating that any ordered state must ultimately progress toward a disordered one and once there would not move away. Once a process became disordered, that is, it would remain so forever.

Only a single caveat entered into what appeared to be a definitive proof of Planck's initial claims. The final paragraph of the paper flagged the possibility that cases may exist where phase constants possessed *initial* values such that the system could begin in a disordered state and then progress to an ordered one: the intensity might start constant and then exhibit measurable deviations. Much as Boltzmann could find initial states for his gas molecules that seemed to move away and not toward equilibrium, Planck was aware—had been made aware by Boltzmann's criticisms—that analogous states may well exist for the case of electromagnetic radiation. Yet for Planck, such initial conditions represented merely a mathematical possibility, not a physical reality. "According to these results," he wrote,

> I believe that I can make the following statement, that the radiation processes regarded here possess all the essential characteristics of irreversible processes, assuming that one excludes certain special cases. It should not be difficult to show that those ideal assumptions about the initial state which lead to these special cases are never satisfied by the irreversible radiative processes (absorption and emission of heat radiation) which occur in nature.[80]

Planck, no doubt, would have felt some satisfaction with such a conclusion, but it was to be short-lived. Boltzmann quickly objected to the proof offered of strict irreversibility, locating the problem precisely where Planck thought he had found a solution: the exclusion of the reversed field of the resonator in the expression for the "exciting wave." Boltzmann demonstrated that one could, beginning with Planck's own equations, produce fully reversible solutions as long as one wrote for the reversed exciting field, not Planck's expression: $\varphi_u(t) = -\varphi(-t)$, but his own, true for an arbitrary argument ω: $\varphi_u(\omega) = -\varphi_u(-\omega) - f(-\omega)$.[81] In other words, reversibility followed if Planck's expression for g was not set to zero, but rather was written as the time-reversed expression for the field emitted by the resonator. Given that Boltzmann had, in a sense, done little more than restate what Planck had taken to be his basic point, it seems unlikely that this proof of reversibility alone carried much weight. Probably more convincing when coupled with this was Boltzmann's earlier argument, made in his second objection to Planck's papers in 1897, which had noted that, even though it was "in any case, physically inadmissible" to assume that the resonator was "in a mathematical sense infinitely small," the resonator's assumed infinitesimal size still did not rule out reversibility:

> One would, certainly, then have to assume, in order to maintain the overall diffusion of a finite quantity of energy, that in an infinitely small space, yet still very large in relation to the dimensions of the resonator, infinitely strong electrical oscillations occur that still always satisfy Maxwell's equations. Apart from this difficulty, there still always corresponds also in this case to every outgoing wave from a resonator a just as possible ingoing one in the opposite direction and for the general theory of these processes, which does not exclude a number of cases, but rather encompasses all equally, every process is again reversible.[82]

Planck acknowledged this mistake in the final section of the fourth paper, admitting that his earlier proof was invalid. Replacing the earlier restrictions on the state of the field, Planck suggested a new one: "If a resonator at a given time is excited by natural radiation of variable intensity, the entry of the reverse process is absolutely excluded for all later times, as long as the exciting wave maintains the characteristics of natural radiation."[83]

Planck's introduction of the concept of "natural radiation" has led a number of scholars to assume that the fourth paper in the series marks a "new phase" in his thinking about radiative processes. Planck had been previously been quite open in his hostility to the kinds of statistical reasoning developed by Boltzmann in his gas theory. In 1900, however, in a summary of the five papers published in the *Annalen*, he would make explicit the analogy between natural radiation and "molecular disorder," the assumption that gas particles in their initial state were not artificially ordered.[84] The statement that an electromagnetic beam possesses the properties of natural radiation is described there as meaning that "the energy of the radiation is distributed completely *irregularly* over the individual partial vibrations from which the beam can be thought to be composed."[85]

Kuhn's argument, predicated on the similarity between Planck and Boltzmann's respective derivations, reads Planck's conception of this analogy backwards to the 1898 paper and lists three pieces of evidence: (1) that the fourth paper "opens with the announcement that the explanation of irreversible radiation processes requires a special hypothesis, precisely the step his program had initially been undertaken to avoid"; (2) that this hypothesis, natural radiation, "resembles an electromagnetic version of Boltzmann's molecular disorder, and its development demands explicit recourse to averages over resonator bandwidth"; and (3) that Planck then "proceeds at once to seek a function that, like Boltzmann's H, can vary only monotonically approaching a stationary value with time." These parallels, he argued,

> are too close to be plausibly attributed to independent discovery, though that possibility cannot be categorically barred. Instead, they suggest that by mid-winter 1897–98, at the latest, Planck was studying Boltzmann's version of the second law with care, was exploiting suggestions he found there, and had abandoned, or all but abandoned his resistance to Boltzmann's approach. Unfortunately for historians, he did not explicitly acknowledge his change of mind for almost two years....[86]

Absence of evidence does not, of course, constitute evidence of absence, but the argument for a substantial change in Planck's thinking in the fourth paper relies on several related assumptions, a number of which do not hold. First, one may note that this paper does not offer the first instance of Planck's introduction of a "special hypothesis": the third paper had already done this, in those precise terms, and in doing so, Planck was following a logic similar to that used already in the second paper. Perhaps more importantly, there is sizeable evidence *within* papers three and four suggesting that Planck understood these as deeply connected. In other words, Planck did not see himself—or at least, did not portray himself—as diverging significantly from previous arguments. This does not mean that he was not studying Boltzmann's arguments carefully and exploiting relevant similarities—given his interactions with Boltzmann one would expect him to do so—but it does mean that substantial evidence points away from the claim that Planck understood natural radiation as a precise analogue to Boltzmann's conception of molecular disorder in 1898 and hence saw the fourth paper of the series as an "abrupt about-face."[87]

Note the structural continuity between papers three and four. Sections in the later paper follow consecutively on from the earlier one. Planck, in fact, was quite explicit about their relation. "Because the following investigation is tied directly to the results of my previous communication," he wrote at the end of the introduction to paper four, "and builds a totality [*ein Ganzes*] with it, all the assumptions and notations made there are, firstly, retained and also the numbering of the sections, paragraphs and formulas is carried forth in a consecutive series."[88] Certainly, Planck was forced to admit that his proof of the impossibility of strict reversibility had been faulty in the earlier paper, but, even while accepting that a modification was thus necessary, he emphasized that nothing else needed to be altered in his prior analysis. The final paragraph of the paper, which followed immediately upon his claim that the hypothesis of natural radiation guaranteed irreversibility, read: "Accordingly, the statement put forward here is to replace the sentence cited above from section 12 [of the third paper] as well as the place referencing it in the introduction. For the rest, however, no alteration to the presentation is thereby required."[89] In particular, one may note, arguments concerning the recurrence paradox and the nature of stationary states remain unchanged in their wording, in spite of the introduction of the concept of natural radiation.

Most significantly, attention must be paid to the way in which Planck describes the hypothesis of natural radiation when it was first introduced in the main body of the paper. The paper's final section dealt with the question of the relationship between the oscillation of the resonators and the intensity of the radiation surrounding them. Writing out an expression for the energy of such a resonator, Planck quickly noted that the relevant expression includes terms corresponding not only to a measurable quantity—the intensity—but also to sub-measurable quantities, like the amplitudes and phases of individual partial vibrations, about which Planck's unstructured Hertzian resonators could provide little information. "It follows, therefore," he wrote, "that a determinate generally valid relation between the energy of the resonators and the intensity of the exciting wave does not exist at all, or that the energy of the resonator does not depend on the radiation intensity of the incident radiation at a corresponding frequency, but rather depends also on certain other properties of these waves, relating to the individual partial vibrations."[90] This theoretical result, however, did not accord with experience, which indicated that such a relation between energy and intensity does indeed exist, compelling a modification in the assumptions made in Planck's previous analysis:

If one wants, nevertheless, to arrive at a determinate relation between the energy U of the resonators and the corresponding radiative intensity I_0 of the incident wave—and the phenomena of the absorption and emission of heat radiation speak for the fact that such [a relation] really exists in nature—nothing then is left but the introduction of a new restrictive assumption about the character of the individual partial vibrations [*Partialschwingungen*] of the exciting wave.

Planck claimed that one particular assumption was the closest to hand: one merely set the sum of all terms in the energy expression that involved variables corresponding to the amplitude and phase of the partial vibrations to zero, thus leaving an equation involving only the energy and intensity of the radiation and the frequency and damping term of the resonator. Having written out this assumption in the form of five equations, he continued:

We want, henceforth, to regard these equations as valid overall and simultaneously, in order to express the fact that we have introduced a certain restriction of generality with them, to designate each and every [form of] radiation which satisfies these equations as "natural radiation," *and on the other hand to designate any other as "tuned to the resonator" in generalization of the definition put forward for this purpose in Section 6* [paper three].[91]

The terminology used here is significant. Since his earliest papers in thermodynamics, Planck had deployed a strict dichotomy between "natural" and "neutral" processes.[92] The first were irreversible, the entropy of the final state larger than the initial; the second were reversible, the total entropy change equal to zero. "Natural" radiation was thus almost tautologically irreversible and Planck signaled early in his fourth paper the result that "all radiation processes which possess the properties of natural radiation necessarily proceed irreversibly, in that the waves passing over the resonator always display smaller fluctuations in their intensity of radiation than prior to doing so."[93] It is striking, however, that Planck did not use the term "neutral" radiation to describe those processes that did not proceed irreversibly, preferring the terminology of his previous paper, thereby emphasizing the conceptual relations between the two installments.

No further explicit discussion of the physical meaning of the concept of natural radiation was offered in the fourth paper. At best, Planck supplied a defense of the thermodynamic *style* of his analysis, arguing that it was precisely his refusal to elaborate on the structure of his resonators that impelled the introduction of a concept like natural radiation as a means of providing a full solution to the problem. "Precisely in the gap left behind [by this indeterminacy of structure] the hypothesis of natural radiation finds its place; were the gap not to hand, the hypothesis would be either superfluous or impossible, because then the process would also be completely determined without it."[94]

Commentators arguing for an originary and constructive analogy between Planck's understanding of natural radiation and Boltzmann's notion of molecular disorder—in the absence of any explicit mention of such a connection when the term was first introduced—have turned instead to the fifth and final contribution to the series, completed almost a year later. There, mention of "tuned" radiation was dropped. Instead, Planck devoted an entire section of his analysis to explaining the difference between "rapidly varying" and "slowly varying" quantities, a distinction, he noted, that corresponded to one made in acoustics and optics between oscillations and beats. In the former category were to be found the amplitudes and phases of electromagnetic radiation within a cavity; the latter category included its energy and intensity. Only slowly varying quantities were assumed accessible to direct observation. The hypothesis of natural radiation, in this formulation, involved the assumption that terms involving rapidly varying quantities in the expression for radiation intensity could be replaced by their slowly varying averages, which appeared as amplitudes in the physically measurable intensity. Intuitively, as Planck put it, one could grasp the concept as the stipulation that "the deviation of the rapidly varying magnitudes...from their slowly varying average values...shall be small and irregular."[95]

What, then, is the connection between this definition of natural radiation, in terms of averages, and the earlier one, which Planck described as a generalization of his notion of radiation untuned to the system? Functionally, of course, both conceptions serve a similar purpose. One of the important results of the third installment, claimed in the introductory paragraphs of paper four, was the fact that "certain special kinds of waves must be expressly excluded" in order to securely guarantee the irreversibility of the processes examined. Those "special" waves were those designated as "tuned to the system."[96] The notion of natural radiation clarified and extended the category of processes the exclusion of which guaranteed irreversibility. A more specific relation, however, would seem to arise not through the concept of tuned or untuned processes, but through the related definition of a "disordered" state. Planck, it will be recalled, had defined a stationary state as one in which the sum of all terms involving phase constants could be set to zero, the intensity thereby reaching a constant value. The newer formulation of the concept of natural radiation depicted in paper five generalized this logic to describe *any* value of the intensity, not merely that at equilibrium. In effect, each averaged value of the intensity functions as a stationary state, its value constant for a time period long compared to that of the rapidly varying quantities involved. The sum of deviations from this average within the given time period may be set to zero. *All* natural radiation, by this logic, is therefore disordered in the sense in which Planck uses that term in the third paper. To understand Planck's original conception of natural radiation no recourse to Boltzmannian molecular disorder is required; his own papers show the independent development of ideas that would later be deemed analogous.

This argument for a lack of an original connection between natural radiation and molecular disorder in Planck's thinking, based on evidence internal to the papers, also accords implicitly with one of Planck's most detailed later accounts of the events leading up to his black-body theory. Natural radiation was discussed twice in his 1943 paper on the "History of the Discovery of the Physical Quantum of Action," once when dealing with the events of 1898 and again when discussing his turn to combinatorics around 1900.[97] The first time, no mention of the idea of molecular disorder was made, natural radiation being introduced as a dynamical solution to the problem of reversibility raised by Boltzmann. It was not until Planck described his attempt to give physical meaning to the expression for the entropy of a resonator that he had derived in 1900 that he discussed the analogy with the kinetic theory. "I myself had not attended to the connection between entropy and probability up until that point," he wrote, "it therefore held no appeal for me because each law of probability also allows exceptions and because at that time I ascribed exceptionless validity to the second law of thermodynamics. That the proof of irreversibility of the radiation processes that I had considered could only work under the assumption of the hypothesis of "natural radiation," that therefore such a restrictive hypothesis is just as necessary in the theory of radiation and plays there the same role as molecular disorder in gas theory, only became completely clear to me over time."[98] Further, it should be noted that although Planck claimed to accept a similarity between the concepts of natural radiation and molecular disorder, he did not do so on Boltzmann's terms. Rather than accepting the statistical interpretation of the second law contained in Boltzmann's kinetic theory, and assuming that this applied to his own theory of black-body radiation, Planck argued that in any real-world scenario, the kinetic theory would also lead to an absolute increase of entropy. In other words, by incorporating Boltzmann's idea, Planck had not accepted his reasoning, but had rather subverted Boltzmann's ideas to his own position.[99]

It may thus be argued that natural radiation *became* a mirror to the concept of molecular disorder at some point between the middle of May 1899, when Planck completed the final installment of his series and the beginning of November of the same year, when he published his summary of the papers in the *Annalen* and made the connection explicit for the first time. A footnote to the later paper, in the section directly before that dedicated to natural radiation, suggests a means by which this shift in thinking may have occurred. Planck noted that the derivation of a result that had appeared in paper five had been altered following a "verbal remark" made by Boltzmann.[100] It does not seem unlikely that Boltzmann, studying papers four and five carefully, had spotted an analogy that Planck had missed, especially given that he had been pushing the significance of the "order" of initial conditions on Planck for some time. Another "verbal remark" to that effect may well have been the impetus required for Planck to think that connection through systematically.

Irreversibility and Entropy
With the completion of his fourth paper, Planck had achieved—at least for the simplified case of spherical waves in a spherical cavity—almost all of his original aims. Using the concept of natural radiation, he had determined the conditions under which microscopic processes were irreversible. In a final section of the paper he proved the existence of a function, analogous to Clausius's entropy function, which increased when the resonator system underwent any change. The fifth paper offered a considerable generalization and extension of these results, exploring the case of arbitrary electromagnetic waves in all directions for a mirrored cavity of any shape. Deploying methods similar to those used in his previous papers, Planck derived what he termed "the fundamental equation of the developed theory," relating the energy of a resonator at a given frequency to the intensity of incident radiation:

$$U_0 = \frac{3c^3}{32\pi^2 \nu_0^2} I_0.$$

The use of this expression in subsequent sections, Planck noted, was predicated on the continuing application of the conditions of natural radiation.

The paper's second section made use of this expression to demonstrate that the resonator system obeyed the principle of the conservation of energy. Introducing a quantity defined as the "electromagnetic entropy," it showed, in addition, that the function increased monotonically, reaching a maximum at the system's stationary state. Similar in purpose to the calculation performed in his previous paper, the entropy function introduced in paper five was intended to have considerably greater physical significance. In the fourth installment, Planck had noted that the "extremely special character" of the radiative phenomena he had explored meant that a large number of functions existed that might serve as analogues to the thermodynamic entropy. Choosing the simplest of these to represent the entropy of a resonator, $S_0 = \log U$, Planck nonetheless made explicit his unwillingness to ascribe a significance to the function for more general radiative processes. Less reticent in the final paper of the series, he wrote down, without explanation, a different definition for the same quantity,

$$S = -\frac{U}{a\nu} \log \frac{U}{eb\nu},$$

which he subsequently argued was not merely the analogue but the equivalent of the thermodynamic entropy.[101] Describing the *total* electromagnetic entropy, i.e. the sum of all individual resonator entropies and that of the surrounding field, he wrote: "The significance of the preceding definition of the electromagnetic entropy rests on the fact that, with its aid the principle of entropy increase can be proven to be valid for the radiative processes considered here and further that the same definition, via an identification of the electromagnetic with the familiar thermodynamic entropy, leads to a thermodynamic meaning of the electromagnetic radiative processes, as well as to a formulation of the second law of thermodynamics for all the phenomena of heat radiation."[102]

Given both Planck's dedication to finding a monotonically increasing function that could represent the entropy of his system and the logarithmic forms suggested for this function, it seems undeniable that he was closely patterning his approach in this part of his analysis upon Boltzmann's work on the kinetic theory of gases. Yet one may also note, in Planck's efforts to relate his results to those achieved for the case of black-body radiation, strong structural similarities to his own earlier work in thermochemistry. The problem he had faced in the late 1880s had been that of finding thermodynamic expressions relating matter in two different phases: as a dilute solution and as an ideal gas. In the final section of the fifth paper on irreversible natural processes, his aim was to connect—again thermodynamically—two different forms of radiation: thermal and electromagnetic. In each case, previous researchers had offered analyses of the phenomena using the suppositions of Maxwell and Boltzmann's kinetic theory—van 't Hoff for the case of physical chemistry, Wilhelm Wien in his derivation of his distribution law. And in each case, Planck sought to circumvent such appeals to a materialist specificity by invoking the greater generality of his own approach. The fact that the laws of black-body radiation developed first by Kirchhoff and then by Wien were independent of the characteristics of the body, being fully determined by its temperature alone, "urged one to the direct conclusion that also the stationary radiative state of the vacuum treated in the previous section satisfies the conditions of black-body radiation, completely without reference to the question of whether the electromagnetic resonators assumed there exhibit a greater or lesser similarity to the centers of heat radiation in certain real bodies."[103] If thermal and electromagnetic *radiation* could be identified, Planck argued further, then it followed that the same could be done for thermal and electromagnetic *entropy*, offering a parallel again to the thermochemical case, where Planck's ideal process rendered thermodynamically identical the two phases of matter and hence allowed the transfer of known results from one problem to the other.

Although Planck could make the general argument for the identity between the entropy of heat and electromagnetic radiation, he was nonetheless faced with a more specific difficulty. What justified his assumption that the expression he had written down for the total electromagnetic entropy was in fact the unique and correct one? That answer was one he could not give, noting only the effort, without success, that he had put into trying to do so. All that could be definitively demonstrated was that, starting from his definition of resonator entropy, and applying the thermodynamic relation $\partial S/\partial U = 1/T$, it was possible to derive an expression for the energy distribution of radiation of a given wavelength λ at a given temperature T—an expression that was identical to that put forward by Wien and—within certain limits—confirmed by multiple experiments:

$$E_\lambda = \frac{2c^2 b}{\lambda^5} e^{-ac/\lambda T}.$$

This apparent success was somewhat less impressive than it seems. Further remarks by Planck make it clear that he had not arrived at his entropy expression independently, but rather by working backward from Wien's law and then adding the condition that the multiple possible "entropies" arrived at in this way satisfy in addition the law of entropy increase for his system of resonators.[104] What many may have perceived as a circularity in his logic, Planck nonetheless proceeded to portray as a strength of the theory, concluding that both his entropy expression and Wien's law itself were "necessary conclusions" that flowed from applying the entropy law to the theory of electromagnetic radiation, the boundaries of their validity coterminous with that of the second law.

When Planck published a summary of his series of five papers in the *Annalen* early in 1900, he could afford to feel some satisfaction with his work of the last half-decade. Without making assumptions about the material makeup or structure of his electromagnetic resonators, he had proven that the second law was an absolute, as valid and general as the law of energy conservation. He had offered, more specifically, an expression for the value of the total entropy of a system of resonators in interaction with an electromagnetic field, one seemingly confirmed by the available experimental evidence. After a brief attempt to locate the source of irreversibility in the resonators, and in particular in their assumed infinitesimal size, he had subsequently located that property in the nature of radiation itself, suggesting that it was precisely the lack of structure in the resonators that allowed, indeed required, such a move. What had been a methodological and practical emphasis on phenomenology for Hertz—one maintained only within certain theoretical limits—became both essential and general for Planck. The series concluded not only with these fulfillments of a program of research, but with what Planck must have seen as a bonus and confirmatory result. Combining the constants, a and b, which appeared in his definition for the entropy of a resonator, with c, the speed of light and f, the gravitational constant, one could, in Planck's words, "establish units for length, mass, time, and temperature which, independent of specific bodies or substances, necessarily retain their meaning for all time and for all cultures, even extra-terrestrial and non-human ones and which, therefore, may be designated "natural units of measurement.""[105] With his strongly held faith in the absolute, this conclusion provided Planck with the confidence in his derivations that their somewhat shaky provenance may not have fully deserved.[106]

Entropy, Experiment, and the Physics of the Future

Planck's works from 1900 to 1906 (the year he published his *Vorlesungen über die Theorie der Strahlung*, or Lectures on the Theory of Heat Radiation) are now among the most closely studied—and debated—publications in the history of the modern physical sciences.[107] It was, of course, in this period, in pursuit first of a better derivation for Wien's law, and an eventual justification for the expression he suggested as an improvement to it, that he introduced the expression relating the energy of a resonator to its frequency for which he is now most famous: $\varepsilon = h\nu$. Only a few of the details of these developments are germane to the narrative here, however, which is restricted to sketching the means by which Planck introduced Boltzmann's combinatorial understanding of entropy and incorporated it within a set of practices and assumptions to which it had previously been foreign. The move would prove essential, for by 1908, the same year that saw support for his black-body theory begin to swell, Planck could depict the probabilistic conception of entropy as a solution to a problem that had dogged him since his early work on the theory of solutions: the question of the necessity of experimental realizability as a criterion for a valid theoretical proof.

Experimental data appearing to call into question the general validity of Wien's law appeared the same month that Planck submitted the summary of his five-paper series to the *Annalen* in 1899. Yet the close link he had drawn between Wien's expression and the principle of entropy increase was not one he could give up lightly. A paper on "Entropy and Temperature of Radiative Heat," published the following April, devoted itself to providing a systematic derivation of his earlier result. Almost seven months passed before mounting experimental criticism and his own identification of a flaw in his earlier reasoning led Planck to offer a purely mathematical "derivation" for a *new* black-body law. Wien's law, he had shown previously, followed from setting the second derivative of entropy with respect to energy equal to $-\alpha/U$. Since that expression seemed inadequate, the "next simplest" alternative was to write

$$\frac{d^2S}{dU^2} = \frac{-\alpha}{U(\beta+U)}.$$

Integrating twice, making use of the expression $dS/dU = 1/T$ and Wien's displacement law, in the form $S = f(U/v)$,[108] Planck arrived at the "Planck distribution":

$$E = \frac{C\lambda^{-5}}{e^{c/\lambda T}-1}.$$

Experimental confirmations quickly followed, while Planck himself began a search for a theoretical derivation of the empirical law he had found. In his October paper, Planck had flagged the method he would soon deploy, parenthetically remarking that the logarithmic relation between entropy and energy in his expression was suggested by "probability calculations" (*Wahrscheinlichkeitsrechnung*). A lecture delivered to the German Physical Society in December 1900 offered an outline of a proof based on the combinatorial arguments that Boltzmann had used in his kinetic gas theory. "Entropy requires disorder," Planck claimed there, and suggested that the new method deployed for the calculation of resonator entropy required comparatively little change to the conceptual structure of his earlier works, needing only that a "somewhat more advanced version" be given of the notion of "natural radiation." The requisite disorder was to be found in the "irregularity, with which the vibrations of the resonators change their amplitudes and phases, even in completely stationary fields of radiation."[109]

Highly schematic, Planck's analysis introduced as "the most essential point of the whole calculation" the postulate that the energy of N resonators of frequency v was made up of "an entirely determinate number of finite equal parts," the size of which was determined by the "natural constant" h so that the "energy element" ε was equal to $h\nu$. The problem that Planck described was that of determining the number of ways that P such elements ($P = E/\varepsilon$) could be distributed among the N resonators. The equilibrium distribution was then defined as the state in which the sum of such numbers, for all resonators (N' at frequency ν', with total energy E', N'' at frequency ν'' with total energy E'', and so on, such that $E_{\text{total}} = E + E' + E'' + \cdots$) was a maximum, considered over all possible energy distributions. Planck, however, did not perform this maximization procedure, merely noting that he possessed a "more direct" means of deriving his black-body law. That more direct method was submitted to the *Annalen* a little more than three weeks later. In the new paper, Planck simply assumed—much as he had in his previous *Annalen* paper, in which he had derived the Wien distribution—that his system of resonators was already at equilibrium and thus that no maximization was required.[110] In addition, he now considered only resonators at a single frequency, with total energy U_N, assumed equal to N times their average energy U. He then wrote down a single expression for the number of ways, R, that energy elements might be distributed among his resonators:

$$\frac{(N+P-1)!}{(N-1)!P!}.$$

Given Stirling's law, which for large N to first approximation gives $N! = N^N$, this may be written as

$$R = \frac{(N+P)^{N+P}}{N^N P^P}.$$

Making use now of an expression bearing a "close relation to a law of the kinetic gas theory," $S = k \log W + \text{const}$ (where W is the probability that the N resonators together possess a total energy U_N), Planck set $W = R$, choosing the constant such that it vanished, and wrote $P = NU/\varepsilon$ to give the total entropy of all N resonators. Dividing by N to yield the average entropy of each resonator gave

$$S = k\left[\left(1+\frac{U}{\varepsilon}\right)\log\left(1+\frac{U}{\varepsilon}\right) - \frac{U}{\varepsilon}\log\frac{U}{\varepsilon}\right].$$

By inspection, Wien's displacement law, written in the form Planck had used previously ($S = f(U/\nu)$) and which he now formally derived, now requires that the size of the energy element be proportional to the frequency. Planck's relation emerges in this calculation, therefore, not as a postulate, but as a result. Application of the relation $dS/dU = 1/T$ and some substitution finally reproduces Planck's distribution law, with the constant C rewritten in terms of h[111]:

$$E = \frac{8\pi c h}{\lambda^5} \frac{1}{e^{ch/k\lambda T}-1}.$$

Atomism and Ideal Processes

Although he wrote several papers in 1901 summarizing his path to the aforementioned result, Planck wrote little new on the black-body problem until the publication of his *Lectures*. Instead, in addition to a number of papers on electromagnetic theory, his main areas of research were the theory of solutions and further explorations of the meaning of Boltzmann's relations between entropy and probability.[112] A paper ostensibly "On the Distribution of Energy between Ether and Matter," written as part of a Festschrift for the Dutch scientist Johannes Bosscha, brought these last two topics into explicit connection by tracing the historical "extension or generalization of the conditions which we regard as characteristic for the state of a material system."[113] Within the context of this "generalization," states available for observation or measurement were merely to be regarded as special cases among much more numerous and more general possible states that "are not, or not easily, available to observation." Planck's initial examples for such generalized states were those he had considered more than a decade earlier in his discussion of 'ideal processes': for example, the "purely theoretical states" assumed to arise in moving from a dilute solution to an ideal gas without the occurrence of chemical changes. The kinetic theory, once the object of Planck's critique, now offered further examples of states that were inaccessible to observation, but were nonetheless useful, even essential, for theoretical analysis. In order fully to describe the state of a monatomic gas at fixed volume one required knowledge of the position and velocity distribution of the atoms making it up. For the equilibrium state, only one such distribution was observable (a uniform spatial distribution for positions and the Maxwell-Boltzmann distribution for velocities), but one could imagine a multitude of other possibilities, from which, using Boltzmann's relation ($S = k \log W$ + const), one could determine the corresponding entropy. Thermodynamically, the stationary state was that in which the entropy is a maximum, but to determine that maximum through Boltzmann's method required the theoretical consideration of an enormous number of states that could not occur in nature, providing, as Planck put it, "an extension of the definition of entropy beyond the bounds of thermodynamics into that of the kinetic gas theory."[114]

Central to the change in Planck's thinking about the atomic theory was, of course, his reformulation of Boltzmann's statistical understanding of gas theory. Taking "molecular chaos" as an absolute restriction on the possible states for a gas meant that molecularity no longer posed a challenge to the unrestricted validity of the second law. What had not changed, however—indeed, what seemed only to have been strengthened—was Planck's understanding of the limited place of experiment and observation in the practice of theory. His paper for the Bosscha Festschrift offered a strong claim that certain valid theoretical proofs *required* the conceptual utilization of states that could not be observed because they could not occur in nature. Six months later, as we have seen, his response to Cantor's criticisms of his solution theory made explicit the essential converse of this position: that experimental realizability—even to close approximation—was not a requirement for a logically binding proof.

That such a claim was radical may be discerned from the fact that another theorist, writing in the same year, felt the need to defend the use even of "conceptual operations with bodies or processes that are, in reality, only *approximately* realizable," suggesting that these "may appear strange at first sight." Friedrich Pockels, writing about the work of Gustav Kirchhoff, Planck's predecessor at Berlin, argued that such operations were "perfectly admissible as a means of simplifying the argument, for the truths of the facts to be demonstrated cannot depend on the degree of perfection of our artificial instruments." For a strict phenomenologist like the Göttingen theorist Woldemar Voigt, even this concession may have appeared too much. In his definition, the phenomenological view "means that the foundations of the theoretical treatment are taken exclusively from direct observation."[115]

Apart from making clear Planck's significant divergence from what Heilbron has described as a "widespread descriptionist consensus" in the German physical community at the turn of the century, the prevalence of competing claims about the epistemological merits of various theoretical practices helps to explain the insistence and frequency with which Planck would describe his own position.[116] In a two-page letter to the editors of the *Philosophical Magazine* written toward the end of 1904, for example, Planck offered a rebuttal to a claim made by a critic of the English translation of his *Treatise on Thermodynamics*. Professor William McFadden Orr had argued—*contra* one of Planck's definitions of reversibility—against the possibility of a body expanding without this "producing a change of density in some other body." In response, Planck noted that a container with gas inside, with weights upon it acting against the pressure of the gas, will, if placed in a vacuum, expand by lifting the weights, without a density change in any other body. The exchange seems unexceptional, except that Planck immediately added a caveat, suggesting his expectation of a rapid contradiction: "That it is impossible to obtain an absolute vacuum, that absolutely unchangeable weights do not in reality exist, and that there may be other difficulties opposing the realization of this process, does not of course, affect the validity of the proof."[117]

The Unity of the Physical World Picture

The longest and most coherent articulation of Planck's conceptualization of the relationship between theory and experiment came in 1908 in a lecture delivered at the University of Leiden: "On the Unity of the Physical World Picture." This lecture, the site of a stinging attack on the phenomenological philosophy of Ernst Mach, has long been familiar to philosophers of science.[118] In it Planck outlined his faith in a physics the principles of which were not "purely historical, accidental, and conventional," as Mach maintained, but were true representations of the world, providing a "unity with respect to all places and times, unity with respect to all researchers, all nations, all cultures."[119] Rejecting a position he had once claimed to support, Planck now railed with all the vehemence of an ex-convert against the barrenness and inadequacy of Mach's positivism.[120] Although he was convinced, he asserted, that Mach's system, correctly followed through, contained no inner contradictions, he was just as convinced "that its significance, at base, is only a formalistic one, which misses entirely the essence of science." That essence was the search for a *constant* world picture, "independent of the variation of times and peoples."[121] Although Mach might affect to find the principle of economy behind the development of the principles of physics, as sense impressions were brought under the rubric of ever-fewer descriptive statements, such an economic vision was foreign to great physical scientists of the past, such as Copernicus, Kepler, and Faraday. In the end, like all "false prophets," Mach and his system needed to be judged by their fruits, and these, by Planck's reckoning, were few.[122]

Mach's name, as he himself would note in his reply, does not appear until the fourth and final section of Planck's lecture. And, although he was probably right to note that earlier passages contained critiques "clearly intended for me or my kind," a focus on the polemical elements of the piece has tended to obscure some of the insights to be gained concerning Planck's understanding of his own physics at the time, expressed with clarity in the first three parts.[123] The ostensible aim of the piece was to provide a description of the historical advance of physics, to be judged by the way in which its fundamental concepts were defined and used. Utilizing both Mach's historico-critical method and the results obtained in his many books and treatises, Planck noted the original connections between the sciences and human particularities. Geometry arose out of geography or land surveying; mechanics from the theory of machines; acoustics, optics, and heat theory from the corresponding sense impressions. "In short, all of physics, both its definitions and its entire structure, originally bears, in a certain sense, an anthropomorphic character." If that sounded Machian, however, it was not a compliment, for this anthropomorphism was a marker of the physics of the past: "How different from this is the picture which the edifice of teachings of modern theoretical physics offers!"[124] Indeed, the marker of the unity offered by contemporary physics was its gradual exclusion of all anthropomorphic elements and one could hope for further unifications in the future. At present, Planck declared, two great areas stood in opposition to one another, mechanics and electrodynamics, to the extent that some spoke of a "battle between the mechanical and electrodynamical worldviews."[125] Yet the gulf between them was not as large as had been depicted, and many signs pointed to their eventual unification within a more general dynamics. A deeper dichotomy was that between reversible and irreversible processes, expressed in the difference between the first and second laws of thermodynamics. It was toward an elaboration of the de-anthropomorphization of our understandings of these that the rest of

Planck's discussion was turned. As the reader follows the ensuing narrative, it becomes clear that the problem of anthropomorphism was not merely one of the original entrance of sense impressions into science. The majority of his speech on the unity of the physical world picture may, in fact, be read as a sustained defense of the notion that the foundations of theoretical physics could not rest on the question of the feasibility of experiments conducted to test their suppositions. In an argument supporting one made earlier in favor of practically unrealizable "ideal processes," Planck now went one step further, declaring such ideal processes themselves too anthropomorphic, and removing even the notion of a thought experiment from his system of a unified theoretical physics.

In Planck's view, the energy principle had long since sloughed off its original formulation in terms of the impossibility of producing a perpetual motion machine of the first kind. It was now independent of the "refinement of methods that we currently possess of testing the question of the realization of a *perpetuum mobile* experimentally." The generalized principle was, strictly speaking, unprovable, but it was precisely in this that its "emancipation from anthropomorphic elements" resided.[126] The second law, on the other hand, was a more difficult proposition. Derived from a technical problem—the realization by Sadi Carnot that irreversible processes were "less economical" than reversible ones—it maintained much of this earlier connection to the problems of the steam industry. According to Planck, it was precisely this connection to older industrial concerns that now held physicists back from a more adequate mathematical understanding of the law:

We see therefore, that the way that has been adopted in order to grasp the irreversibility of a process mathematically has, in general, not reached its goal, and we see simultaneously also the real reason why this cannot succeed. The mode of phrasing the question is too anthropomorphically colored, it is tailored too much to the needs of mankind, in that it depends primarily on the production of useful work. If one wants to gain from nature a determinate answer, one must approach her from a more general, less economically interested standpoint.[127]

Such a standpoint was provided by expressing the essence of irreversibility as nature's "preference" [*Vorliebe*] for some states over others.[128] A process was impossible if it involved a final state that nature preferred less than the initial one. Entropy, Clausius's term, was then simply a measure of nature's preference for any state, and the second law could be rephrased as the law of entropy increase. "It will," Planck declared to his audience, "now probably appear understandable why I...express my opinion that in the theoretical physics of the future, the most important division of all physical processes will be that into reversible and irreversible processes."[129] Yet, even though the technical and economic elements had been removed from the statement of the principles of physics, a "strong dose of anthropomorphism" remained in the presence of a need for *experiment* in their comprehension. Planck objected to the fact that "in the definition of irreversibility as well as in that for entropy, reference is made to the *practicability* [*Ausführbarkeit*] of certain changes in nature, and that means fundamentally nothing other than that the division of physical processes is made dependent on the capacity of human experimental art, which certainly does not always stay at a particular level, but steadily perfects itself more and more. Therefore, if the difference between reversible and irreversible processes shall have lasting meaning for all time, so must it be still essentially deepened and made independent of all reference to human capacities."[130]

Of course, pure positivism had hardly reigned in thermodynamic theory in the nineteenth century. Clausius had made use of ideal reversible processes in his definition of the second law: "experiments" conducted by ideal physicists, capable of perfect precision. The thought experiments of physical chemistry, once Planck's own field, were particularly striking in this respect, a great deal being achieved with an object—the semi-permeable membrane—practicably impossible in many contexts. Planck did not here name his own "ideal process," but he surely had it in mind. In the 1890s, the fact that he had made use, in a theoretical proof, of a process that was nearly impossible to achieve in reality was an element of his methodology that he felt called upon to defend. Now, however, Planck called into question the very issue of experimental realizability as a criterion. Ideal processes were challenged once more, but this time it was for their too-close connection to an anthropomorphic, if ultimately fictive, experimental practice, rather than their distance from it. Since ideal processes seemed to provide a roundabout, rather than a direct route to the meaning of the principle of entropy increase, Planck proposed their avoidance, and the establishment of the second law on grounds independent of both real and thought experiments.[131]

Such an "emancipation of the concept of entropy from human experimental art" and the elevation of the second law to a "real principle," was, according to Planck, the scientific life work of Ludwig Boltzmann. Whether Boltzmann himself would have agreed with such a profoundly anti-positivistic description of his scientific legacy is perhaps beside the point, for Planck had found in his colleague's probabilistic definition of entropy the de-anthropomorphized foundation he had been seeking.[132] Nature could be understood as "preferring" more probable states over less probable ones, and entropy simply became the measure of this probability, given by Boltzmann's equation: $S = k \log W$. Anthropomorphism is, Planck noted with some satisfaction, "in a word completely eliminated from this definition and the second law is placed therewith on a real basis, just as is so with the first."[133] If the foundation of the physics of the deep past lay in sense impressions and its present lay in an experimental sensibility, the basis of the physics of the future was to be found in principles that transcended any question of "realizability" and were independent "of all arts of human technique."[134] For Planck, at least, that future had already arrived.

Conclusion

As a young man, departing Munich University in order to complete his studies in theoretical physics in Berlin, Planck stopped to take his leave from his former physics professor, Phillip von Jolly. As Planck recalled later, von Jolly sent him away with less than hopeful thoughts about the future:

> Theoretical physics, that is certainly an utterly beautiful subject, although there are no positions in it nowadays. But you will scarcely be able to achieve anything more in it. Because with the discovery of the principle of the conservation of energy the terrain of theoretical physics is probably largely perfected. One can probably still sweep a mote of dust out of one or another corner here or there; but you will not find something new in principle.[135]

Certainly, as with so many statements about the "completeness" of physics in the latter half of the nineteenth century, this (possibly apocryphal) prediction would prove false. But von Jolly was not wrong about the state of theoretical physics when Planck began his career. Not only were there few chairs in the subject, it was also unclear what it actually meant to be a theoretician. Indeed, in Planck's own memory of his time as a young professor in Berlin, his emphasis on theory to the exclusion of all else, including the experimental part of physics, brought a sense of isolation: "I was just then the only theoretician for far and wide, certainly a physicist *sui generis*, which didn't make my debut entirely easy."[136]

The shift away from experimental practice as a necessity for all physicists had more than a social impact on Planck as a theorist. The work that he produced in the last two decades of the nineteenth century mapped out a space for theory that no longer drew explicitly on experiment as its guide. Where Helmholtz had, for example, vaunted the use of variational principles in his theoretical work at least partly on the grounds that variance was of the essence in his analysis of object states in the laboratory, Planck would find the greatest virtue of the variational least action principle solely in its absolute and all-encompassing nature.[137] Where Kirchhoff framed the proof of the black-body law that bears his name in terms of laboratory objects—diaphragms, lenses, prisms, and mirrors—Planck largely stuck to general principles and symmetry considerations to achieve the same aim.[138] And, where Hertz, as we have seen, maintained the presence of a real resonator in his theoretical work—drawing it to scale in his diagrams and restricting the application of his theory to those areas where it reasonably accorded with empirical data—Planck sought to exploit the properties of a theoretical object to limit the solutions available in practice.

What made this new vision of theoretical physics possible was its foundation in a particular view of thermodynamics. It was one that eschewed any study of the material basis of the substances it examined in favor of a more general analysis, one founded on abstract, absolute, and all-encompassing principles. As Planck realized, however, principles alone were not enough to produce theories that described the world in detail. His theoretical practice, therefore, incorporated a logic that dictated the rules governing the use of supplementary hypotheses and it was in justification of these that Planck's most vigorous defenses of his theorizing were launched. No more concerned with substance than were the principles themselves, these hypotheses must be understood as an integral part of the thermodynamic method. If, for Sommerfeld, as we shall see, the problem with thermodynamics was that it produced safe but unsatisfying solutions, ones which provided no insight into mechanism or electrodynamism, then for Planck, this was precisely the method's most striking advantage.

5 The Dynamical and the Statistical: Sommerfeld, Planck, and the Quantum Hypothesis

The careful farmer cultivates new ground
Whereas I have merely a few flowers found.
—Sommerfeld to Planck

What I have plucked, what you have plucked
Let them be, then, combined
We shall exchange them all between ourselves
And beauteous wreaths around them wind.
—Planck to Sommerfeld[1]

Principles and Problems at the Solvay Conference[2]

In the middle of 1911, Sommerfeld and 24 other leading physical researchers received an invitation to attend what would become known as the first Solvay congress on physics, to be held at the end of October. Initiated by the physical chemist Walther Nernst in 1910, the conference was the first meeting to devote itself entirely to the question of "Radiation Theory and Quanta."[3] Held in Brussels and funded by Ernst Solvay, a Belgian industrialist and patron of the sciences, the gathering remains perhaps the best-known scientific conference of the twentieth century. Of the participants in the five-day event, nine were current or future Nobel laureates and the posed photograph depicting them has attained iconic status as a representation of the putative transition from "classical" to "modern" physics.[4]

In the image, Planck stands to the left of the viewer, close to a seated Nernst and in front of a blackboard bearing a single equation. His black-body law seems to emerge without warning or preparation, directly from his thoughts, while the centerpiece of the background is a white board, portraying experimental data on specific heats. Sommerfeld, most likely because of his stature, occupies a row between the seated and standing invitees, slightly to the left of the photograph's geometrical center. Near the heart of the visual representation of the congress and, in spite of a cold, an integral part of many of its recorded discussions, Sommerfeld was

1 Goldschmidt	7 Knudsen	13 Kamerlingh Onnes	19 Lorentz
2 Planck	8 Hasenöhrl	14 Einstein	20 Warburg
3 Rubens	9 Hostelet	15 Langevin	21 Perrin
4 Sommerfeld	10 Herzen	16 Nernst	22 Wien
5 Lindemann	11 Jeans	17 Brillouin	23 Curie
6 de Broglie	12 Rutherford	18 Solvay	24 Poincaré

Figure 5.1
The Solvay Conference, 1911. Photograph by Benjamin Couprie. Courtesy AIP Emilio Segrè Visual Archives. Numbering by MIT Press.

nonetheless not among those Nernst had initially considered as potential contributors. His name appears in neither Nernst's first formal letter to Solvay, written at the end of July 1910, nor in the memorandum written to Solvay in the middle of March 1911 by his scientific collaborator, Edouard Herzen, offering the finalized text of the invitations.[5] This initial exclusion can hardly seem surprising, given Sommerfeld's first skeptical responses to Planck's theory. Although he briefly acknowledged the possible utility of the quantum in a paper written in 1909, it was not until the end of the following year that a flurry of papers began to signal his support for a—suitably modified—version of the quantum hypothesis. Planck may well have had Sommerfeld in mind in June 1910, when he wrote to Nernst suggesting that the proposed congress be postponed a few years, to see "how the gap which begins to open in the theory shall develop, and how finally those who still stand at a distance will be forced to join in."[6] Yet Planck's fears proved groundless. Those who attended the congress and had previously been either poorly informed (like Poincaré) or hostile toward the quantum hypothesis (like Jeans) returned home as enthusiastic converts.[7]

Sommerfeld appears to have been brought around in advance of the meeting, at least in part, by a paper written by Debye on "The Concept of Probability in the Theory of Radiation." The short article, published in the *Annalen* in 1910, derived Planck's distribution law without mention of resonators, the peculiar properties of which Sommerfeld had critiqued in his earlier lectures.[8] A year after the congress, in the Winter semester of 1912–13, he delivered a set of lectures on the quantum theory and in 1913 presented two long papers on the topic that were co-authored by or that presented the results of former students.[9] In 1914, he supervised the first of many quantum-theoretical dissertations.[10]

For many of the participants in Brussels, and for several years afterwards, the most pressing area of research for the incipient quantum theory was no longer the radiation theory with which Planck had begun his own studies, but rather the problem of specific heats and, more generally, the kinetic theory of matter. As Kuhn has noted, this latter topic was a standard subject for physicists and physical chemists, much more so than black-body theory. Nernst's demonstration in 1910 that the specific heat of monatomic substances vanishes as temperatures approach absolute zero (in accord with Einstein's quantum-theoretical prediction and in contradiction to the classical result) expanded significantly the size of the community aware of the puzzles associated with the quantum.[11] Sommerfeld, however, offered a paper neither on the problem of cavity radiation nor on specific heats, choosing instead to emphasize the significance of Planck's relation within existing electromagnetic theories of the electron, particularly those concerned with the production of X and γ radiation. Like his earlier rejection of Planck's theory, Sommerfeld's eventual utilization of it was embedded in his work in support of a program associated with the electromagnetic view of nature.

If Sommerfeld offered a novel articulation of the quantum hypothesis at the Solvay Congress, then so, too, did its originator. Early in 1911, Planck put forward what came to be known as his "second theory," which outlined an asymmetry between the processes of absorption and emission for his resonators. Discontinuity, he proposed, would enter only in the latter process; absorption would proceed classically. The response to both men's papers from their elite audience was detailed and extensive. Perhaps most striking, however, was the suggestion, made by several colleagues, that Sommerfeld and Planck had articulated two largely antithetical approaches, which between them represented the most promising future methodologies within the field. As Marcel Brillouin famously represented the issue in the discussion following Planck's paper, the acceptance of the reality of a "quantum discontinuity" required new and radical theories:

> It seems certain that we must from now on include among our physical and chemical conceptions that of discontinuity, of an element varying by jumps of which we had no notion a few years ago. How is it to be included? That is what I see less well. Will it be in the first form

proposed by M. Planck, despite the difficulties which that raises, or will it be in the second? Will it be in the form proposed by M. Sommerfeld, or in some other still to be sought?[12]

Paul Langevin was more explicit in delineating where he saw the differences to lie:

It appears to me to be difficult to reduce the two principles to one other and to prove their equal validity. One, stemming from Planck, is a statistical statement. It says that one may not, in the calculation of the number of complexions corresponding to a specific energy of a system, regard as different two complexions that differ in the values of the parameters p and q by only a little from each other. The points, which represent two states of a system, must find themselves in two different finite elements of phase space, if the states are to be regarded as different....That is the qualitative content of Planck's statement: it is of an essentially statistical nature.

In opposition to this, the statement of Sommerfeld [defining h in terms of the principle of least action] is of a purely dynamical nature. Perhaps one will be able to prove its equality with Planck's statement, or to show that both are contained within a more general principle, yet I still see no way to such an end.

In summary, I would like to remark: Sommerfeld's statement leads to a finite element of action, Planck's statement to a finite element of phase space; both things seem to be fundamentally different to me; one is of a dynamical, the other of a statistical nature.[13]

The opposition between "dynamical" and "statistical" approaches to the quantum, an opposition that Planck, at least, would soon come to accept, was characteristic of Sommerfeld and Planck's approaches to theoretical physics more generally. Their Solvay papers are near-emblematic examples of their broader methodologies—and the differences between them—brought to bear on the "same" question: the meaning of the quantum. Interested in further delineating what Sommerfeld's style was, this chapter also, therefore, takes up the question of what it quite explicitly was not.

During and after the meeting, both men sought to articulate what was distinct and valuable in their separate approaches. Two years after Solvay, at a conference held in Göttingen, Planck made use of Langevin's distinction—and its personifications—himself. The "statistical approach" was the "safer" of the two options, he claimed, since it stuck closely to the most trustworthy elements of physical knowledge: the principles of thermodynamics. Sommerfeld agreed with the characterization, but not with the value that Planck accorded each side. To the contrary, he argued on a number of occasions that the best route forward for theoretical physics was not the "safe" one of the thermodynamic method, which avoided the detailed consideration of microphysical phenomena, but the "bold" approach epitomized by the electron theory. Deploying hypotheses deeply rooted in his work on the electron-theoretical explanation of the production of X-rays, Sommerfeld explored the details of the electromagnetic processes involved in individual microphysical events. In so doing, he claimed to have achieved a fundamental victory for the electromagnetic view of nature, the point of view from which he had originally critiqued Planck's theory of radiation. Where others saw Maxwell's theory and the quantum as irreconcilable, for Sommerfeld the two were complementary, each serving for the solution of problems upon which the other gave no insight.

The multiple papers that Sommerfeld delivered on the quantum hypothesis in 1911 provided an effective introduction to a small but growing group of quantum theorists. The equations for which he is now most famous—the so-called Bohr-Sommerfeld quantization conditions—were derived only a few years after this first entrée into the field. That derivation, eventually published in 1915, utilized arguments similar to those Planck had employed at Solvay. In Brussels, Sommerfeld had distinguished his new approach from previous ones by noting its ability to deal with aperiodic processes. Faced with the periodic phenomena exhibited by Bohr's planetary model of the atom, Sommerfeld turned back toward Planck's phase-space representation of the quantum condition. Yet, in adopting Planck's statistical definition, Sommerfeld did not thereby adopt the thermodynamic reasoning advocated by his colleague. Although he began with statistics, he rapidly moved toward the pursuit of dynamical problems. And, as Sommerfeld made ever more detailed calculations—with ever-closer attention to a growing wealth of experimentally obtained spectroscopic data—Planck continued to explore a more general and abstract means of understanding the issue, studying idealized exemplars in order to understand the physical structure of phase space. By the middle of the war, the Sommerfeld School had developed a novel program for the study of the quantum theory of spectral lines, one that no longer contrasted its approach to Planck but instead looked increasingly to Copenhagen and the work of Niels Bohr.

Energetics vs. X-Rays: Sommerfeld's Route to the Quantum

Two events, each occurring in the same year at the beginning of Sommerfeld's career, were to shape profoundly his approach and attitude toward Planck's quantum hypothesis. One was a fiery debate over the appropriate material and epistemological foundations for physics, to which Sommerfeld, as a student, was witness, conducted at the annual meeting of German doctors and natural scientists in Lübeck in 1895. The other was the completion of his Habilitationsschrift, on the mathematical theory of diffraction. The Lübeck debate concerned the possibility of replacing mechanical understandings of nature with "energetic" ones, all processes to be explained not in terms of matter in motion, but rather in terms of the transfer and transformation of energy.[14] Finding support for their pointed critiques of the atomic hypothesis in the phenomenological philosophy of Ernst Mach, the mathematician Georg Helm, the physical chemist Wilhelm Ostwald, and other "energeticists" put forward perhaps the first coherent alternative to the mechanical view of nature. At least for a short time in the early 1890s, they found a number of sympathetic ears, including Berlin's new professor for theoretical physics. Planck, however, like many, was soon to decide that the promise of the new approach was not supported by its products and he was among those rising to speak against the members of the energetics camp at Lübeck. The following year saw the publication of an essay in the *Annalen* "Against the Recent Energetics" in which he chided proponents of the worldview for their inflated claims and warned, rather priggishly, of the effect on students that such "dilettantish speculation" may have if it were allowed to continue unchecked.[15]

Sommerfeld's recollection of the events at Lübeck—written nearly 50 years later—remains, perhaps, the most colorful firsthand description:

The paper on "Energetik" was given by Helm from Dresden; behind him stood Wilhelm Ostwald, behind both the philosophy of Ernst Mach, who was not present. The opponent was Boltzmann, seconded by Felix Klein. Both externally and internally, the battle between Boltzmann and Ostwald resembled the battle of the bull with the supple fighter. However, this time the bull was victorious over the torero in spite of the latter's artful combat. The arguments of Boltzmann carried the day. We, the young mathematicians of the time, were all on the side of Boltzmann; it was entirely obvious to us that one could not possibly deduce the equation of motion for even a single mass point—let alone for a system with many degrees of freedom—from the single energy equation....[16]

The critique remembered here is a technical one, but that was not the aspect of energetics that Sommerfeld found most objectionable in the early twentieth century.[17] In 1901, he opened a paper on X-ray diffraction with one of the clearest methodological pronouncements he would make in print. Many supporters of energetics, he claimed, saw it as a veritable "evil" [*Uebel*] to consider specific representations of physical phenomena, recommending instead that experimental researchers merely record the facts of experience and the theoretician content himself with organizing these into a mathematical system. Proponents of such a descriptionist view, he continued, protected themselves from taking any steps backward, but they also abandoned any hopes of advancing knowledge through the appropriate construction of hypotheses about the natural world. This anti-phenomenological criticism was repeated in his lectures on Planck's theory of radiation in the Summer semester of 1907. "Thermodynamics," he noted there, "is the most secure but least satisfying foundation [for the theory]. In opposition to energetics, one requires an understanding of mechanism or electrodynamism."[18] It was this logic, it will be remembered, that led Sommerfeld to prefer the Jeans-Lorentz electron-theoretical method to Planck's.

Whether Sommerfeld's representation of the energeticist position was a valid one or not is less significant here than the fact that he followed his criticism in 1901 with an elaboration of a more acceptable alternative:

By contrast, the most fruitful path for theoretical physics appears to be this: to lay down as specific and determinate [*specielle und bestimmte*] hypotheses as possible, to develop their consequences exactly, and to compare these with experiment: if this shows no contradiction with experience, good, then our hypotheses were valid and may be retained until later; if, however, an opposition emerges, then all the better; then our hypothesis is displayed as invalid and we have gained a definitive, if also negative, finding.[19]

An example given for such a hypothesis—one deployed in the paper—was the electron-theoretical impulse theory put forward by Emil Wiechert and George Gabriel Stokes, whereby X-rays were assumed to be produced when electrons in a cathode-ray tube suddenly decelerated upon striking the anode, emitting high-energy electromagnetic radiation. For Sommerfeld, then, electron theories, and the electromagnetic program more generally, offered not merely an alternative *ontology*—opposed both to a mechanical and an energetic view of nature—but also an alternative *methodology*, one that preferred "specific and determinate hypotheses" about "mechanism or electrodynamism" to the economical summaries of experience of Machians or the generalities of Planck's thermodynamic method.[20] Sommerfeld was hardly alone in this. As Elizabeth Garber has persuasively argued, one of the main reasons for the apparent initial neglect of Planck's black-body work after 1900 was the *form* of his solution to a pressing problem in radiation theory. While his distribution law was rapidly incorporated into handbooks, his theoretical analysis was deemed much more problematic, precisely because it avoided any clear statement as to the mechanism of interaction between matter and the ether. Lorentz and others working on radiation theory "knew that a direct consideration of the actual mechanism of exchange of energy was necessary for a full analysis of the problem. This was at the heart of the problem because such a mechanism would imply a particular molecular and ether structure to account for that particular distribution of energy and interaction mechanism. This was the method Wien used in his derivation of his distribution law and this was precisely the issue Planck avoided in his first quantum papers of 1900 and 1901."[21] Keeping this element of the electromagnetic program in mind goes a long way toward explaining the kinds of theoretical practices—ones based on the solution of specific problems, rather than a focus on general thermodynamic principles—that Sommerfeld would employ at the Solvay conference.

Understanding the detailed content of Sommerfeld's Solvay paper, on the other hand, requires our return to his first major publication in mathematical physics, his habilitation thesis on diffraction.[22] Earlier studies like Gustav Kirchhoff's, Sommerfeld had claimed with a youthful hubris, lacked mathematical precision, serving only as (dubious) approximations and owing their "relatively good" agreement with experience merely to the fact that the wavelength of visible light was a very small quantity.[23] For the cases of Hertzian oscillations and sound waves such approximations did not hold and under certain conditions the older diffraction theories even failed for optical wavelengths. In contrast, Sommerfeld's more general solution considered the optical case merely as one part of a "mathematically precise treatment, based solely on the differential equations and boundary conditions securely established through the electromagnetic theory." Expressed in physical terms, the problem he would solve was that of diffraction. Mathematically, this meant carrying out the integral of a partial differential equation ($\Delta u = k^2 u = 0$) on a Riemannian surface, using particular boundary conditions.[24] That difficult mathematical analysis took up the lion's share of his thesis, but the result was a significant one, for he was able to re-derive Kirchhoff's expressions, noting their accordance with his own to first approximation. "We therefore come to the remarkable result," he concluded, "that we can confirm the results of the older diffraction theory, while we must declare as completely incorrect the methods through which they were derived."[25] His new theory provided with numerical precision the (limited) region of validity for Kirchhoff's formula and for regions beyond that stood in accordance with results derived by Poincaré, using a different method. "As our theory casts each as suitable approximations of the exact formula," he wrote with some pride, "it forms a bridge between the two theories and apportions to both of them their restricted regions of validity."[26]

Sommerfeld's diffraction theory remains one of the most significant treatments of the subject, but its importance within late-nineteenth-century physics may well have been even more striking in light of two new experimental results: Röntgen's 1895 discovery of X-rays, and the growing acceptance of the existence of the electron after several measurements of the ratio of its charge to its mass in 1897. As early as 1896, British researchers had proposed what would become known as the "impulse theory" of X-rays. By the end of the century a number of prominent researchers, including Stokes, J. J. Thompson, and William Thomson (Lord Kelvin) could be counted as supporters of the theory. On the continent, where many scholars favored a wave-theoretical explanation of the nature of cathode rays, the acceptance of the impulse hypothesis was slower, shaped by the gradual acceptance of the experimental evidence in favor of the electron and by the theoretical successes of Lorentz's electron theory.[27]

It was an electron theorist and one of the earliest proponents of the electromagnetic worldview, Emil Wiechert, who first put forward the impulse theory in Germany.[28] As Paul Volkmann's assistant in Königsberg, Wiechert had worked closely with Sommerfeld in the early 1890s, and the two remained in close contact thereafter.[29] When Sommerfeld published his article on the diffraction of Röntgen rays in the *Zeitschrift für Mathematik und Physik* in 1901, he credited Wiechert for repeatedly drawing his attention to the problem.[30] The new element brought to the diffraction problem was the introduction of the impulse theory. Sommerfeld described the nature of Röntgen's rays as consisting of an "*impulsive* (i.e., short and strong) disturbance of the equilibrium of the ether, which propagates spatially and temporally according to Maxwell's equations." The cause of this disturbance, he argued, could be understood in terms of the "nowadays securely established [*wohl sicher gestellte*] conception of the nature of cathode rays" as rapidly moving, charged particles of low inertia. Such particles, striking a sheet of platinum or the walls of a cathode-ray tube, came to a halt in a very short period of time, producing the "impulsive disturbance of the ether" or *Aetherstoss* detected by Röntgen.[31] In other words, rather than treating X-rays as a periodic phenomenon, they were treated as a rapidly produced pulse, propagating through the ether.

Sommerfeld's success in relating his theory to the best available experimental evidence was to have two effects. On the one hand, as Wheaton has argued, the diffraction theory contributed strongly to the general acceptance of the impulse explanation of X-rays in Germany and elsewhere by the early 1900s. "[B]y 1903," Wheaton notes, "the impulse hypothesis had virtually everywhere supplanted the view that X-rays are periodic waves."[32] On the other hand, as Sommerfeld himself was to point out, the explanation of the origin of the new rays, along with a series of other theoretical results, considerably extended the range of validity of the electron theory. "When Lorentz developed the theory," he wrote in 1904,

his gaze was essentially directed toward the old optical problems of aberration, the Fresnel transport etc. Since then an abundance of recent most wonderful facts has been encouraged by experimental research into light, the Zeeman phenomenon, Röntgen rays, Becquerel rays. And with each of these discoveries, which appear to lie far from the original fiefdom of the electron, theory has been in a position to speak a redemptory word; it has either illuminated in its general contours that which was mysterious in the phenomena as is the case with Röntgen rays, or it has supplied totally determinate numerical evidence for the evaluation of phenomena, as is the case with the Zeeman phenomenon and radium rays.[33]

In his lecture notes on Maxwell's theory and the electron theory, he took cathode rays to provide the "existence proof" for electrons, while in Röntgen radiation one might perceive their "most elegant action."[34]

Sommerfeld's first published mention of the quantum hypothesis came in 1909, in the context of a further elaboration of his X-ray studies. His attitude, however, could hardly have been described as enthusiastic. Responding to a paper on the topic by Johannes Stark, he took up the question of the angular distribution of polarization for Röntgen-rays.[35] Stark, one of the earliest converts to Einstein's light quantum hypothesis, had applied the theory to experimental data on the angular variation of X-ray intensity and penetrating power, arguing that the asymmetry that he had observed could not be explained using previous approaches.[36] Sommerfeld replied quickly and publicly, suggesting that Stark had not properly understood those approaches in the first place, and demonstrated that his own theory fit the experimental results perfectly well, without recourse to the light quantum being necessary.[37] He did, however, give ground on one point, to do with evidence suggesting that X-ray radiation was, in fact, made up of two distinct components. One part, polarized and spreading out in all directions with varying degrees of intensity, he associated with the electron's rapid deceleration. The other part, which was unpolarized and spread out equally in all directions, he associated with the absorption and emission of energy within the atoms making up the anticathode after they had been struck by the cathode rays. This he referred to as the "fluorescent component" of the radiation and suggested: "It is very possible that Planck's quantum of action hereby plays a role." For the "braking component," in contrast, the quantum hypothesis "appears to me to have nothing to do.... One should seek to understand the properties of this component through the pure electromagnetic theory."[38]

Two years later, while still skeptical about the light quantum, Sommerfeld had found a place for the quantum of action within his impulse theory. A footnote to a paper published in January 1911, "On the Structure of γ-Rays," explicitly took back his earlier objection. "I stood at that time," he wrote, "under the probably generally widespread impression that the quantum of action could come into question only for processes of a determinate frequency, v. I have first seen now that our braking duration, τ, can take the place of Planck's $1/v$."[39] One could be excused, however, for finding the mode of the introduction of h somewhat arbitrary. The paper derived a relativistic expression for the ratio of the energy of a β particle given off during radioactive decay to the energy of the γ radiation accompanying the emission.[40] All of the terms in this expression were known, except the path length l of the β particle's acceleration, assumed to be a fraction of the radius of the molecular sphere of action of the radioactive substance from which it was ejected. Noting that the considerations that had led to this expression were already highly "hypothetical," Sommerfeld nonetheless proposed taking a further step "by way of experiment" in order to obtain the ratio of energies as a function of velocity alone. Assuming a quantum of action, h, was released in each particle emission, one could set the action of the process equal to the total energy multiplied by the total acceleration time, τ (i.e., $h = E_{\text{total}} \times \tau$). Since Sommerfeld had already derived relativistic expressions for the total energy, and for the relationship between τ and the path length, the latter term could be determined immediately and the expression manipulated to give a numerical value for the ratio of energies:

$$\frac{E_\beta}{E_\gamma} = \frac{6\pi ch}{e^2}\frac{1-\beta^2}{\beta^2}\left(1-\sqrt{1-\beta^2}\right).$$

Comparison with experimental data, especially that from work on X-rays, confirmed the accuracy of the theoretical expressions, at least within an order of magnitude. If this was perhaps not entirely convincing for the reader, Sommerfeld himself was clearly converted to the side of the quantum theorists. Asked to speak on the topic of relativity at the annual meeting for the Society for German Scientists and Doctors in September 1911, he declined in favor of a presentation on "Planck's Quantum of Action and its General Significance for Molecular Physics." Where Planck, delivering a plenary lecture to the same group the year before, had spoken of a "crisis" within physics induced by debates over the relativity principle, the characteristically more optimistic Sommerfeld saw those debates as settled. Einstein's theory, he claimed "scarcely belongs any longer to the real current questions of physics. Although only 6 years old...it appears already to have passed over into the secure possession of physics." Quite different was the case of the quantum theory, where "fundamental concepts are still in flux and problems are innumerable."[41] Nonetheless, Sommerfeld was now a partisan participant and took as his aim the description of the successes of the theory, offering a detailed summary of previous approaches before turning to a modified version of his quantum hypothesis applied to the impulse theory and the photoelectric effect.[42] He would present a lengthier and more detailed version the following month at the Solvay conference.

The Electromagnetic Worldview and the Quantum of Action

Sommerfeld began his Solvay paper with a criticism of work on the energy quantum up to that time, comparing it to a mechanics that only examined periodic circular motion. [43] One could certainly go a long way with such an approach, dealing with most of the problems of astronomy and many of those of mechanical engineering. Yet it would give only a very incomplete view of the general laws of mechanics. Such, he suggested, was the state of quantum studies, which only concerned themselves with periodic phenomena. This worked well for the problems of radiation theory and for specific heats, but failed in the light of further questions of physics. Precisely because of its "special development," the theory of the energy quantum was now somewhat puzzling and appeared difficult to reconcile with the "very secure foundations of the electromagnetic field."[44] One needed a more general standpoint from which to view a wider range of problems. Such a standpoint could be found, Sommerfeld claimed, by examining more closely the meaning of Planck's quantum. Rather than seeing this as a restriction on the way *energy* could be parceled out (a quantum of energy), one could take the dimensions of the constant as reflective of a deeper meaning. The quantum of *action*, which had the dimensions of energy × time, could be seen as a limitation on the product, rather than on either component. "Phrased completely generally, a large quantity of energy in a shorter time, a smaller in a longer time is taken up and given out by matter, so that the product of energy and time, or (closer to the definition) the time integral of the energy is determined through the magnitude of h."[45] Sommerfeld gave two, now familiar, examples as a means of motivating a relation that he had earlier simply stated. The first, concerning energy absorption, arose from studies of the production of X-rays: high-velocity cathode rays produced strong X-rays, and lower-velocity cathode rays produced weaker ones. The strength of an X-ray, however, was taken as an inverse measure of the duration of the braking time for the electrons impinging on the anode. So, high-energy X-rays went with short braking times, low energy with longer ones. The product, Sommerfeld inferred, was a constant, dependent on h. A similar result could be gleaned from the inverse case, that of energy emission. Here Sommerfeld drew on results from work done on radioactivity: large quantities of energy are emitted by radioactive substances in a short time; smaller quantities take a longer time. It is a rule that contradicts everyday experience, since a bullet, say, traveling very rapidly takes longer to slow down when it hits a wall than a slow-moving one. Sommerfeld suggested that the peculiarity of the quantum of *action* might explain this anomalous behavior for particles considerably smaller than those considered in ballistics problems.[46] Expressing his result mathematically—and in a different form to that used in January—Sommerfeld connected the quantum of action with the least action principle, which one "with Helmholtz-Planck sees as the highest foundation [*Grundsatz*] of mechanics and physics," and offered the "following fundamental hypothesis about the general meaning of h: with every purely molecular process a fixed, universal amount of action is taken up or given out from the atom."[47] That amount was given by

$$\int_0^\tau H dt = \frac{h}{2\pi}. \tag{1}$$

Presumably it was this equation that Langevin had in mind as Sommerfeld's dynamical "statement," to be contrasted with Planck's "statistical" definition of h in terms of areas in phase space,

$$\iint dq \cdot dp = h.$$

We have seen now how rooted Sommerfeld's expression was in areas of research to which he had been devoted for more than a decade, so that it is hard to agree with a perspective that depicts his 1911 papers as "consist[ing] in his embodying Planck's recent ideas by performing detailed calculations on individual processes," as Nisio has claimed.[48] Indeed, reversing this suggestion might be more meaningful, noting that it was only within the context of a program rooted in specific problems that h came to be significant for Sommerfeld's research to begin with.

In a close analysis of Sommerfeld's use of his "dynamical" expression, one may also discern other characteristic elements of his theoretical practice. Note, for example, that in this treatment, equation 1 is not presented as the deduction of an expression from a set of general axioms. Nor is it simply definitional. Rather, it follows from the consideration of specific problems: X-radiation and radioactivity. We will also see below the centrality of experiment within Sommerfeld's analysis. Not merely deploying experimental data as a final point of comparison for theoretical results, Sommerfeld made use of this data *during the process* of constructing his mathematical expressions, experimentally derived results replacing theoretical analysis as he avoided any modeling of intra-atomic processes. Finally, Sommerfeld's Solvay paper illuminates the shift in his adherence to the electromagnetic view of nature. In chapter 1, I tracked Sommerfeld's gradual acceptance of the failure of this worldview as a universalizing ontology. In Brussels, the traces of his old allegiance remained strong, even when coupled with modifications and limitations required by the hypothesis of action quanta. He found a great deal of satisfaction in the apparently necessary "symbiosis" between his preferred electromagnetism and the apparently essential quantum.

Following a proof that his new formulation of the quantum hypothesis was relativistically invariant—a calculation that, as he acknowledged, did indeed owe a great deal to Planck[49]—Sommerfeld concluded the opening section of his Solvay paper with the proviso that he held his analysis to be "in many points hypothetical and incomplete." No doubt realizing the company he was in, he made an effort to avoid ruffling too many feathers through his iconoclastic approach:

Regarding the general comparison of the energy quantum and the action quantum.... I would admittedly like to emphasize [*festhalten*] the greater consequences of the action quantum, but at the same time would only like to place my view with all caution alongside the view of other researchers, who have concerned themselves so much longer and more fundamentally with these questions, and have up until now exclusively supported the standpoint of the energy quantum with such great success.[50]

The requisite modesty having been displayed, the "greater consequences of the action quantum" were then pursued in the following sections, each of which dealt with aperiodic phenomena in terms of specific problems: the production of X-rays and γ-rays, the photoelectric effect, and ionization. The first of these followed the pattern of his paper on γ-rays, where he had divided X-radiation into two components: one polarized and due to the deceleration of electrons, the other unpolarized and due to the vibrations of atoms within the anticathode. The sum of the energies from these two processes gave the total energy ($E_{pol} + E_{unpol} = E_r$). Sommerfeld's theory only treated the first of these terms. "While the mechanism of the *Eigenstrahlung* is unknown and touches on intra-atomic processes, we can pursue the electromechanism of the braking radiation according to the Impulse theory."[51] True to the programmatic aims of the electromagnetic worldview, the intra-atomic mechanism was not considered (indeed, was calculated out) in the discussion. Sommerfeld's result (equation 9 below) was derived entirely through electromagnetic considerations. To secure it, Sommerfeld began by noting that, since the unpolarized radiation spreads out uniformly in all directions, a simple expression connects its energy (E_{unpol}) and intensity (S_{unpol}):

$$E_{unpol} = 4\pi r^2 S_{unpol}. \tag{2}$$

He then obtained two initial equations, for the energy and the intensity of the polarized part of the radiation (E_{pol} and S_{pol} respectively) through a purely electrodynamic argument, derived by Abraham in his *Theorie der Elektrizität*.[52] Sommerfeld had derived this in his earlier paper on γ-rays, and merely wrote the results down for the Solvay report[53]:

$$E_{pol} = \frac{e^2 \dot{v}}{6\pi c^2} \frac{\beta}{\sqrt{1-\beta^2}}, \qquad (3)$$

$$S_{pol} = \frac{e^2 \dot{v}}{16\pi^2 c^2 r^2} \int_0^\beta (1-\beta^2)^{3/2} d\beta$$

$$= \frac{e^2 \dot{v}}{16\pi^2 c^2 r^2} \beta \left(1 - \frac{1}{2}\beta^2 + \frac{3}{40}\beta^4 + \cdots\right). \qquad (4)$$

(Here e = the elementary charge on the cathode-ray electrons, v = their velocity, β = v/c, and $\dot{v}$ the deceleration during the braking). Rewriting equation 3 in terms of equation 4, Sommerfeld obtained

$$E_{pol} = \frac{2}{3} 4\pi r^2 S_{pol} \left(1 + \beta^2 + \frac{11}{20}\beta^4 \cdots\right). \qquad (5)$$

Dividing equation 5 by equation 2, one obtains

$$\frac{E_{pol}}{E_{unpol}} = \frac{2}{3} \frac{S_{pol}}{S_{unpol}} \left(1 + \beta^2 + \frac{11}{20}\beta^4 \cdots\right). \qquad (6)$$

The ratio of the polarized to the unpolarized intensity had been determined experimentally by one of Röntgen's students in Munich, Walter Friedrich,[54] as

$$\frac{S_{pol}}{S_{unpol}} = \frac{3}{7}. \qquad (7)$$

Sommerfeld assumed a value for β of 0.4, to accord with Friedrich's data, and inserted equation 7 into equation 6 to obtain a numerical result for the ratio of polarized to unpolarized energy:

$$\frac{E_{pol}}{E_{unpol}} = \frac{2}{7} \times 1.17. \qquad (8)$$

If one now writes the above in terms of the total radiation energy ($E_r = E_{pol} + E_{unpol}$), one can remove the term representing the energy of the unpolarized radiation—and hence the question of the intra-atomic mechanism—from the calculation

$$\frac{E_{pol}}{E_r} \approx \frac{1}{4}. \qquad (9)$$

Sommerfeld's final step was to express equation 9 in terms of the ratio of the polarized energy to the energy of electrons at the cathode (E_k). Wien and one of his students at Würzburg had measured E_r/E_k.[55] That measurement was taken at a higher voltage than the corresponding data that Friedrich had obtained, so Sommerfeld scaled the former (the ratio being proportional to voltage) and multiplied the two results, to obtain the value of the ratio of the polarized energy to the cathode energy:

$$\frac{E_r}{E_k}\frac{E_{pol}}{E_r} = \frac{E_{pol}}{E_k} = \frac{1}{6} \times 10^{-3}. \tag{10}$$

The quantum hypothesis could now be brought to bear on an expression for E_k. Assuming that the potential energy of the electrons (U) was vanishingly small compared to their kinetic energy (T), one can set $H = T - U$ equal to T. The hypothesis concerning the quantum of action then becomes

$$\int_0^\tau T dt = \frac{h}{2\pi}. \tag{11}$$

If the deceleration of the electrons, $dv/dt = \dot{v}$, can be taken as constant during their braking, then the braking time, τ, is given by $\tau = v_i/\dot{v}$ (where v_i is the initial, i.e. pre-braking, velocity).[56] $\dot{v}$ can then be rewritten as $\dot{v} = \beta c/\tau$. Introducing the variable v' to represent the instantaneous velocity of the braking electrons, one can write

$$dt = -\frac{dv'}{\dot{v}},$$

$$T = \frac{m}{2}v'^2. \tag{12}$$

One thus gets

$$\int_0^\tau T dt = \frac{m}{2\dot{v}}\int_0^{v_i} v'^2 dv' = \frac{1}{3}\frac{m}{2}v_i^2\frac{v_i}{\dot{v}} = \frac{h}{2\pi}. \tag{13}$$

If one now writes $E_k = (m/2)v_i^2$, then equation 13 implies that

$$E_k \tau = \frac{3h}{2\pi}. \tag{14}$$

Manipulating equation 3 to obtain E_{pol} in terms of τ gave

$$E_{pol} = \frac{e^2}{6\pi c\tau}\frac{\beta^2}{\sqrt{1-\beta^2}}. \tag{15}$$

Sommerfeld divided both sides by E_k and inserted the expression for τ derived in equation 14 ($\tau = 3h/2\pi E_k$) on the right-hand side of equation 15:

$$\frac{E_{pol}}{E_k} = \frac{e^2}{9hc}\frac{\beta^2}{\sqrt{1-\beta^2}}. \tag{16}$$

Using known values for h, c, and e, and assuming $\beta = 0.4$, Sommerfeld thus obtained another numerical value for the ratio of the polarized energy to that of the cathode electrons:

$$\frac{E_{\text{pol}}}{E_k} = 2.7 \times 10^{-4}. \tag{17}$$

Equation 17 agreed well with the value given in equation 10, especially in view of the approximating assumptions made along the way.

The basis of Sommerfeld's result should be clear. On the one hand, he used electrodynamics and experimental data to calculate the ratio of the polarized energy to that of the total radiation energy. He avoided the energy contribution of the unpolarized radiation and hence the mechanics of atoms. On the other hand, he appealed to Planck's quantum theory to provide a value for the braking time of the electrons, τ, which could not be obtained through the electrodynamic theory. It was precisely this inter-reliance that seemed to appeal to Sommerfeld. Where Wien had attempted both approaches, the electromagnetic in 1905, the radiation-theoretic in 1907,[57] and had assumed that the two were essentially antithetical, Sommerfeld emphasized that each needed the other in order to solve the radiation problem:

For Wien no connection existed between his electromagnetic and radiation-theoretical calculation of λ; both appear to exclude each other. On the contrary, we have seen that both standpoints are well unifiable with one another. For the calculation of the energy of the polarized Röntgen radiation we proceed from the *purely electromagnetic formula* (4) [equation 3 here; emphasis added]. In this a quantity remains undetermined, $\dot{v}$, the deceleration corresponding to the co-dependent braking time τ. About this the electromagnetic theory can teach us nothing, because it is a function of the braking molecule. In its place the concept of the quantum of action now intervenes....It is not until the intervention of the radiation theory that the electromagnetic theory of Röntgen radiation is completely determined.[58]

In fact, Sommerfeld saw the main benefit of his formulation of the quantum concept over that of Planck as lying in the easy connectability of his version to electromagnetic theory: "The opposition between our application of the quantum of action and Planck's method of the energy quantum has already greatly occupied us. Both depictions are foreign to classical electrodynamics and mechanics. But while our version is reconcilable with electrodynamics, the original depiction by Planck stands in an unmistakable [*unverkennbaren*] opposition to it."[59] The radiation theory may have put an end to the dream of an all-encompassing electromagnetic worldview, but that did not mean that this latter was now completely defunct. It would, in Sommerfeld's hands, rise again as the partner of its would-be assassin: "Radiation theory and Electromagnetism do not exclude, but rather complement one another."[60] The lenses of the electromagnetic view of nature, through which Sommerfeld had seen and criticized Planck's theory in 1906, now compelled a new compromise: "If as can scarcely be doubted, our physics needs a new fundamental hypothesis, which is to be added as [something] new and strange to the electromagnetic world picture, then it appears to me that the hypothesis of the quantum of action is called upon before any others [*vor anderen berufen*].[61]

This fruitful fusion could then be applied to a range of problems that otherwise threatened electromagnetic theory. The photoelectric effect, for example, "doubtless posed one of the greatest difficulties for customary electrodynamics." Again, Sommerfeld's quantum of action could save the situation: "the difficulties seem to disappear as soon as one depicts the freeing of the electrons from the atomic bond as a Planckian action process and apply to this our fundamental hypothesis."[62] The same logic applied to the problem of ionization. Each time, the problem could be divided into two sections, one of which was electrodynamic, the other of which was ruled by the quantum. The division, however, was not an arbitrary one. The strength of Sommerfeld's quantum theory was not that it offered different, but perhaps more empirically viable answers to electrodynamic problems, but that it answered precisely those questions about which electrodynamics had no answers. Unlike either Planck's theory or Einstein's light quantum hypothesis, which posited quantum explanations in place of those based solely on Maxwell's equations and the electron theory, Sommerfeld's fundamental hypothesis, in his eyes, offered no contradiction to electrodynamics at all: "in fact, it supplements this with regard to the course of such processes about which electrodynamics, in and of itself, knows nothing."[63] The dichotomy, that is, was not (as it would eventually become) one between "classical" and "modern" physics, but one between quantum and electrodynamic theory. And, unlike the former dichotomy, this was explicitly and necessarily not an either/or. In Sommerfeld's vision, the quantum and the electromagnetic field were *both* required in order to understand the physical world.

Sommerfeld's paper provoked a good deal of discussion. Henri Poincaré wanted an explanation of the fact that Newton's third law (that of action and reaction) appeared not to apply in Sommerfeld's hypothesis. Curie and Rutherford engaged in an extended debate over whether it had actually been experimentally verified that γ-rays always accompanied decelerating electrons (Rutherford insisted that it was). Most welcomed, at least in principle, the extension of Planck's theory to non-periodic phenomena, and a number of later treatments over the following years mentioned Sommerfeld's analysis with approval.[64]

Yet, although it could hardly have been predicted at the time, the focus on periodicity would soon return. After the introduction of Bohr's atomic model in 1913, Sommerfeld himself turned away from his own dynamical and aperiodic formulation of the quantum condition, beginning his analysis of the Keplerian motion of an electron around the nucleus with Planck's statistical statement. His sense of the intimate and necessary connection between the electromagnetic worldview and the "new and strange" world of the quantum, however, remained with him. Part 1 of his *Atombau und Spektrallinien*, a book that became known after its publication in 1919 as the "Bible" for quantum spectroscopists, dealt with "Introductory facts" and began with a "retrospect of the development of electrodynamics." Neither thermodynamics nor mechanics merit general discussion at all. From its beginning as a series of "disconnected laws" that functioned merely as an analogue to Newton's laws of gravitation, Sommerfeld described a new view in the second half of the nineteenth century that opposed field theory to action at a distance. "The greatest triumph of this view," he wrote,

> occurred when Hertz succeeded in connecting *light*, the phenomena of physical nature with which we are most familiar, with *electromagnetism*, which was at that time the most perplexing phenomenon. From [Hertz's electric waves] an almost unbroken chain of phenomena leads by way of heat rays and infrared rays to the true light rays, whose wavelengths are no more than fractions of μ [micrometers]. The greatest link in this chain came later as a result of Hertz's experiments, namely the waves of wireless telegraphy, whose wavelengths have to be measured in kilometers...the smallest and most delicate link is added at the other end of the chain, as we shall see, in the form of Röntgen rays, and the still shorter γ-rays, which are of a similar nature.[65]

The Sommerfeld School had studied almost every link of this chain and in *Atombau*, electrodynamics was the gateway to, and foundation for, the quantum theory. Where many would see the electron theory as having been superseded by the quantum and relativity theories, for Sommerfeld this most fruitful construct of the electromagnetic worldview possessed a lasting and independent existence:

> The original theory of Maxwell which had been perfected by Hertz retained its significance for phenomena on a large scale, such as in electrotechnics and wireless telegraphy.... But to render possible deeper research leading to a knowledge of elementary phenomena, a deepened view became necessary. Maxwell's Electrodynamics had to give way to Lorentz's Dynamics of the Electron; the theory of the continuous field became replaced by the discontinuous theory, that of the atomicity of electricity. So the theory of action at a distance and the theory of action through fields were succeeded by the atomistic view of electromagnetism, the theory of electrons, which still holds today.[66]

For Planck, in contrast, the Solvay conference would mark the beginning of a gradual elision of the electromagnetic origins of his quantum hypothesis, as his statistical statement came to describe, not the quantization of energy within a resonator, but the finite divisions of phase space, independent of a substantivist basis.

Planck's Paper: Entropy, Probability, and the "Second Theory"

Planck, as befitting a man usually portrayed as a reluctant revolutionary, had been slow to accept the most radical implications of his own work. It was almost a decade before he acknowledged the necessity of a quantum discontinuity, and even in 1908 he spoke of such a thing with distaste. In a letter to Lorentz written in 1910, Planck ruefully accepted that "the discontinuity must enter somehow." Nonetheless, he balked at embracing it completely and "located the discontinuity at the place where it can do the least harm, at the excitation of the oscillator."[67] In essence, this introduced an asymmetry, as continuity and the electromagnetic theory governed the process of emission, and discontinuity was introduced through the quantized energy involved in the absorption of radiation by an oscillator. Early in 1911, he would stick to his "damage control," but now reversed the asymmetry, making emission, rather than absorption, the site of discontinuity. This idea was presented again in Brussels as the Solvay conference provided another site for the discussion of what came to be known as "Planck's second theory." The description of this theory, however, takes up only a relatively small portion of Planck's Solvay paper. More significant, in Planck's eyes, was the reformulation of the quantum hypothesis that the paper put forward, one that subordinated the question of the detailed *mechanism* of discontinuity to its expression in thermodynamic terms. Beginning, as so often, with a historical overview, he quickly arrived at the fundamental difference between the familiar equations derived theoretically by Lorentz and Jeans for the black-body law and Planck's own. The former was the "classical" theory, and if one wished to evade the empirically false conclusions to which it led, then a deep modification to previous understandings was required. The locus of that modification, for Planck, was to be found in the relationship between temperature and energy, a relationship that could only be understood through considerations of probability. Equating the statistical definition of entropy with the thermodynamic for a system at equilibrium,[68] Planck then wrote down "the general definition of temperature,"

$$\frac{1}{T} = k\frac{1}{W}\frac{dW}{dE}.$$

"The laws of heat radiation," he concluded, "are therewith led back to the calculation of the probability W of a given quantity E of the radiant energy, and with that we reach the central point of the whole problem."[69] Without mention of oscillators, electrons, or even the electromagnetic field, the problem of radiation was now, most fundamentally, to be solved in terms of a statistical thermodynamics.

Planck's next steps now proceeded via an analogy with classical mechanics. Liouville's theorem, a result that follows from Hamiltonian mechanics, states that areas of equal size in phase space are equiprobable. Planck defined an infinitesimal "elementary area of probability" as $dp \cdot dq$, and noted that if one uses this to calculate the probabilities (and hence energy distribution) of heat radiation, one arrives at the (incorrect) classical equation. One thus needs a quantum equivalent for Liouville's theorem, a "special physical hypothesis" that will allow one to arrive at a black-body equation in accord with experiment. Planck suggested the following modification:

$$\iint dq \cdot dp = h. \tag{18}$$

That is, instead of infinitely small areas of probability, one must now deal with finite areas, of size equal to the quantum of action. This is the statistical "statement" to which Langevin would refer.

Neither the use of phase space nor the emphasis on h as a quantum of action was novel here.[70] Both were elements of Planck's systematic espousal of his theory in the first edition of his lectures on heat radiation, published in 1906. What *was* novel was the detailed analysis of the quantum of action independent both of a conception of the quantization of energy (i.e., defined by the relation $\varepsilon = h\nu$) and of the electrodynamic mechanism that had rendered meaningful the phase-space analysis in the first place. In his *Wärmestrahlung*, as in his derivation from 1901, h had first entered as the constant ratio between energy and frequency that emerged from the combinatorial determination of the entropy of a system of resonators.[71] However, the same expression could be derived, Planck had noted in 1906, not by an analysis of the energy of a given state, but by direct reference to the electromagnetic state of the individual resonators. From the resonator equation,

$$U = \frac{1}{2}Kf^2 + \frac{1}{2}\frac{g^2}{L}.$$

it is obvious that the curve for which U (the energy) is constant is an ellipse.[72] Its area is $2\pi U\sqrt{L/K} = U/\nu$, where ν is the frequency of oscillation. The area between two ellipses, of energies U and $U + \Delta U$ is then a constant, given by $\Delta U/\nu = h$. The relation between an element of energy and oscillator frequency, in other words, may be obtained without making use of expressions from radiation theory, like Wien's displacement law. "At the same time," Planck noted, "the elementary quantum of action h shows itself to us here in a new light, namely as the size of an elementary area in the state surface of a resonator, valid for resonators of completely arbitrary periods of oscillation."[73] This was, indeed, a new take on the meaning of h, but the Solvay paper offered an additional novelty. There, the oscillator enters only as a specific example to which is applied a far more general relation, true independent of any electromagnetic mechanism.[74] Whereas in 1906 h was explicitly tied to the electromagnetic state of a resonator possessing frequency ν, and was derived from calculations on resonator energy, in Brussels the logic was reversed, and the quantization of energy emerges as a *result* of a quantized probability.[75]

The turn toward a greater emphasis on the independent significance of h may well have been spurred by Planck's reading of Sommerfeld's use of the quantum hypothesis in his paper on γ-rays. Planck wrote enthusiastically to his colleague in April, commending the "significant advance" involved in extending the meaning of h to aperiodic processes. In his previous work, he noted, he had only regarded oscillators of a determinate frequency, and had hence only considered energy elements of size $h\nu$. "But if one takes an oscillator that possesses no distinct period, then no determinate energy element exists for it as well and one must return to the primary meaning of *h*."⁷⁶ He had alluded, very briefly, to that "primary meaning" in a paper published in 1908, which he now recommended to Sommerfeld, not least for its presentation of the relativistic integral form of the principle of least action, which Sommerfeld would adopt for his papers in Karlsruhe and Brussels. In a section of the paper that demonstrated that the principle of least action was relativistically invariant, Planck concluded, somewhat enigmatically, with the claim that "to every change in nature there corresponds a determinate number of action elements, independent of the choice of reference systems. It will be understood that through this statement the significance of the principle of least action is extended in a new direction."⁷⁷ As Planck himself admitted in his later letter, however, he had not been "able to come up with anything worthwhile" following this insight.⁷⁸ No elaboration of the point follows in the 1908 or subsequent papers until after the correspondence with Sommerfeld, intervening publications continuing with the use of energy elements of size $h\nu$. One can thus begin to see why both Solvay presentations, in spite of their many differences, nonetheless offer a similar reinterpretation of the meaning of the quantum, not of energy, but of action.

Having laid out this reinterpretation, Planck took up the question of the "physical nature of the constant *h*," offering two alternatives, depending on whether one understood the quantum of action as being tied to the laws governing the propagation of radiant energy in a vacuum or to processes of emission and absorption alone. The former, clearly offering a total contradiction to Maxwell's laws, involved the hypothesis of light quanta, which Planck (along with most others, to Einstein's chagrin) declined as too radical.⁷⁹ He turned instead to the more familiar questions of his earlier work. It was in this context that he now repeated, with a significant alteration, the phase-space calculation of *Wärmestrahlung*. Writing down, once more, expressions for the energy and frequency of a simple oscillator, he now calculated "the size of that energy ε *which corresponds to an elementary area of the probability*, therefore to the size of the quantum of action *h*."⁸⁰ Remarks at the conclusion of the paper made clear the significance he attributed to this inversion and the importance of Sommerfeld's paper in shaping it:

Above all things it is to be emphasized, that, at least in my opinion, the quantum hypothesis is no energy hypothesis, but is an action hypothesis. The first element of the hypothesis is the size of the elementary area of the probability, the quantum of action *h*. The energy or light quantum *h*ν is first derived from this and only possesses significance for processes, for which a determinate frequency ν exists. There can be no doubt, however, that, insofar as a deeper meaning inheres in the quantum hypothesis in general, the quantum of action *h* must possess a fundamental significance for non-periodic and non-stationary processes as well. This has already been shown directly in a few cases by Sommerfeld.⁸¹

The original basis of the quantum hypothesis, in other words, based on the periodic behavior demonstrated by an oscillator, possessed a limitation overcome by the new definition, which also corresponded more closely with the de-anthropomorphized understanding of the entropy law that Planck now took as fundamental.

With an expression for the quantization of energy in an oscillator in hand, Planck then (but only then) turned to a consideration of more specific questions. "It is now further to be asked," he wrote "how these energy elements are to be interpreted physically, or in other words: what *dynamical* law the oscillator frequencies must, at base, be governed by so that precisely the probability law found above results."[82] With his probability law as guide, Planck analyzed several physically permissible assumptions, beginning with his own, and considering in turn suggestions previously offered by Einstein, Lorentz, and Nernst. All, he proved, led to the same result, namely to the Planck distribution equation,

$$E_N = N \cdot h\nu / (e^{h\nu/kT} - 1).$$

This conformity of results from such drastically different approaches provided, Planck felt, a certain guarantee of the correctness of his basic presentation.[83] Yet all of these dynamical representations suffered from the same problem. In combining electrodynamics and the hypothesis of the quantum, one introduced a fundamental opposition. Maxwell's equations assumed that energy was radiated continuously, the quantum theory necessitated discontinuity.

Even if one accepted this opposition as a fact, and introduced special physical assumptions to cope with it, more problems bedeviled contemporary quantum theory. Low temperature physics presented particular difficulties. Near absolute zero the intensity of radiation was far too low to raise an oscillator's energy by the minimum requisite amount of one quantum. As the temperature dropped, the absorption time grows incredibly rapidly, so that, as Planck put it, one "can scarcely speak of a "sudden" absorption of a whole quantum of energy."[84] For non-stationary fields, the situation was even worse. Given the finite time it would take an oscillator to absorb a quantum, it would seem that the oscillator would need to know in advance that the process would last long enough for it to absorb a whole quantum.[85]

Einstein's solution to many of these problems was his light-quantum hypothesis. Planck's alternative, introduced more than three-fourths of the way through the paper, was his second theory, which divorced the process of absorption from that of emission. Absorption would, in his new theory, proceed continuously, with energy given by $E = n\varepsilon + \rho$, with $\rho < \varepsilon$. Emission would proceed discontinuously, in whole quanta of size ε, with a probability that would depend solely on n, the number of entire quanta. When n was zero, no emission would occur. An oscillator could thus effectively "bide its time" until it absorbed at least one quantum of energy. Then, but only then, would it be governed by probabilistic laws governing discontinuous emission.[86]

Nonetheless, and Planck made a point of emphasizing this aspect, the second theory was still compatible with the "fundamental thought of the quantum hypothesis," for the "elementary areas of equal probability are still represented through the finite quantum of action."[87]

The shift in Planck's thinking about the quantum hypothesis was enshrined in the second edition of his *Vorlesungen über die Theorie der Wärmestrahlung*, completed at the end of 1912. Considerably shorter than the first edition, the new theory of heat radiation de-emphasized the electromagnetic route to the quantum in favor of the statistical one, removing most of his derivations of formulas describing the actions of emission and absorption and reversing the order of presentation, so that the electrodynamic analysis now followed a discussion of "Entropy and Probability."[88] The preface to the book played down the significance of any dynamical explanations offered within. "For anyone who would make his attitude concerning the hypothesis of quanta depend on whether the significance of the quantum of action for the elementary physical processes is made clear in every respect or may be demonstrated by some simple dynamical model, misunderstands, I believe, the character and the meaning of the hypothesis of quanta. It is impossible to express a really new principle in terms of a model following old laws."[89] The quantum hypothesis, instead, is framed statistically, in terms of areas of phase space, as at Solvay.[90]

A few months later, in 1913, at a conference organized by the mathematician David Hilbert and held in Göttingen, Planck made explicit both the reasoning behind his emphasis on the statistical approach and, at the same time, who he associated with the alternative. Discussing the quantum hypothesis, he wrote:

We seek at first to formulate its content as fully as possible and simultaneously as conclusively as possible. Only a purely dynamical formulation would probably be completely satisfying, one which laid out a certain elementary law, which all actions among atoms, electrons [and] electrodynamic waves obeyed. That is the direction in which the various works by A. Sommerfeld tend. Until the definitive statement of such a dynamical law is achieved [however] it appears to me safer to restrict oneself to a statistical formulation of the quantum hypothesis, and this will join most suitably to the two laws of thermodynamics, particularly the second law, with which the problem of temperature and the quantum hypothesis is bound most tightly.[91]

Planck's suggestion that Sommerfeld's dynamical approach might be less "safe" but more "satisfying" than his own statistical-thermodynamic formulation offers a direct echo, albeit with inverted valence, of Sommerfeld's comments on Planck's original theory. And the tension between the demands of specificity and generality in physical theory was one that both men would raise in their discussions of the difference between their two approaches to the quantum. In his letter written in April 1911, Planck had noted that Sommerfeld only considered individual acts of emission, over time periods comparable to τ, whereas he considered a number of emissions, over much longer time periods. "You go right into the details of an act of emission, whereas for my consideration an individual emission is a singular process of vanishingly short duration, about the precise details of which my hypothesis reveals nothing at all."[92] For Sommerfeld, who clearly agreed with this characterization, the difference was one of method. Knowledge about individual processes was unnecessary to "statistical considerations," while his own approach was much more dependent on "risky [*gewagte*] hypotheses" and "risky individual assumptions."[93] Both men, in other words, agreed that an approach utilizing statistics and the principles of thermodynamics was safer, in that it avoided detailed dynamical modeling; both agreed that this was less satisfactory than one that sought an explicit representation of micro-physical phenomena. Yet each took from this opposite—and oppositional—conclusions: Sommerfeld with his embrace of the specificity of problem-based dynamics, Planck with the pursuit of the generality of a principles-based statistics.

The Bohr-Sommerfeld Quantization Conditions

At the end of 1915, Sommerfeld delivered a lecture before the Bavarian Academy of Sciences titled "On the Theory of the Balmer Series."[94] It was research that drew upon work that Niels Bohr had published in early 1913. Bohr's now-famous atomic model had fused a "planetary" structure for the atom—with negatively charged electrons revolving around a small positively charged nucleus—with what even Bohr referred to as other "horrid assumptions."[95] Among these was the notion that electrons revolved in particular circular orbits of radius a, the value of which was determined by a modification of Planck's relation $\varepsilon = h\nu$, given in terms of angular momentum $p_\theta = ma^2\omega = nh/2\pi$. Only those orbits that obeyed this quantum condition were allowed, but for these the motion of an electron was decreed stable. The usual loss of energy through radiation that accompanies charged accelerating bodies was assumed not to exist for these "stationary states." Only in moving between such states, dropping with the emission of energy from a higher state to a lower, or absorbing energy and climbing to a higher energy state, was radiation given out or taken up.

Bohr's model had been strikingly successful in explaining the existence and positions of spectral lines in basic elements, particularly hydrogen. It was with a comment on this success—and its effective limitation to "hydrogenic" spectra—that Sommerfeld began his talk:

> The theory of the Balmer hydrogen spectrum appears at first glance to have been brought to a conclusion through the wonderful investigations of N. Bohr. Bohr could explain not only the general form of the series law, but even the numerical value of the constants detailed therein, and their refinement when the motion of the nucleus is taken into consideration. One might even say that the capability of the Bohr theory is for the time being limited to this hydrogen series and to the hydrogenic series (ionized Helium, X-ray spectra, series ends of the visible spectra).[96]

Sommerfeld, in spite of the praise, wanted to point to a sizeable gap in the present theory of the Balmer series: Bohr's model only considered circular orbits. Extending Bohr's analysis to include elliptical orbits would, Sommerfeld wrote, not only fill this gap via a "deepening" of the quantum hypothesis, but it would also illuminate one of the peculiarities of the hydrogen spectrum.[97] While other elements displayed a number of different spectral series, hydrogen had remarkably few. The very simplicity of the hydrogen atom, Sommerfeld explained, with only one electron orbiting the nucleus, meant that several series effectively overlapped. His suggestion was that one could explore this by splitting the lines in an electric field (Stark effect). Since the field would affect different quantum numbers differently, one would expect that a single line could be resolved into several non-degenerate lines. In general, quantizing the different degrees of freedom of the system independently of one another held out the promise of explaining what was known as the "fine structure" of spectral lines.

As with his original entry into the study of the quantum, Sommerfeld's path to the quantum theory of spectral lines began with the electron theory. One of the theory's early successes had been the explanation of the "Zeeman effect," the splitting of spectral lines in a magnetic field, discovered by the Dutch physicist Pieter Zeeman in 1896. In his theoretical model of the phenomenon, published in 1897, Lorentz had assumed that it was caused by the oscillations of a quasi-elastically bound "ion." In an external field, he showed, such a particle should exhibit three spectral lines, two circularly polarized (in opposite directions) in the direction of the field, one linearly polarized perpendicular to it. The distance between two lines, he demonstrated further, should be proportional to the applied field.[98] A remarkable success at the time, Lorentz' enthusiasm would be short-lived, as further experiments showed the existence of what was later termed the "anomalous" Zeeman effect, which exhibits many more lines, and different distances between them, than Lorentz's theory predicted. If confusing, the data nonetheless presented evidence of regularities. In 1898, Thomas Preston claimed that within a given spectral series, lines for any substance exhibit the same splitting pattern; more significantly, perhaps, he showed that for different elements, analogous spectral lines (those in the same position within a series) show the same splitting. In 1907, Carl Runge, after years of collaboration with Friedrich Paschen, put forward what would be termed "Runge's rule," which posited that the separations between lines were integral fractions of the separations predicted by the original Lorentz theory.[99] As more and more researchers became interested in the fascinating complexity of spectral lines, results in the field multiplied rapidly. The same year that Runge formulated his rule, Sommerfeld pressed him for an article on spectral analysis for the *Encyclopedia of the Mathematical Sciences*. "I well understand," he noted however, "that the moment for the formulation of a classic article on spectral analysis is not auspicious, since here as in other parts of physics the foundations [*Grundveste*] appear to wobble. But what can one do?"[100] As it stood, Runge's article did not appear until the mid 1920s.

Sommerfeld had touched upon spectral analysis in a major work on the electron theory, published in 1904, which he suggested to Runge "throws a tentative light on spectral lines," but the discussion of the topic was brief.[101] His first substantial engagement appeared in 1913,[102] after Paschen and his student Ernst Back published results showing that for sufficiently strong magnetic fields, the multiple lines of the anomalous Zeeman effect reduced to the triplet first predicted by Lorentz. Sommerfeld's paper dedicated itself to giving a "theoretical interpretation, or better expressed, a theoretical analogue to the beautiful and simple results of Herrs Paschen and Back" by offering a simple generalization of Lorentz's elementary theory.[103] Instead of assuming that the electron was isotropically bound (i.e., that its frequencies of vibration were the same in all directions), he assumed that the frequency corresponding to each perpendicular axis differed slightly. This meant that even in the absence of a magnetic field, one would observe a triplet. For weak fields, the triplet would increase in size, but for a sufficiently strong field, the small frequency differences would become unimportant and Sommerfeld's model would approach Lorentz's. If a good model of the Paschen-Back effect, however, Sommerfeld's simple representation could do little to provide even an analogue of the complexity of lines in a weak field, leaving as he acknowledged in a later letter "the larger half of the problem by the wayside."[104] A more complete treatment followed the next year, in large part as a response to complaints Sommerfeld had received from the Göttingen theoretician Woldemar Voigt, who testily noted that the model of the anisotropically bound electron was one that he had suggested more than a decade ago before dismissing it as "untenable." He pointedly recalled informing Sommerfeld about it in person some time around 1900.[105] Sommerfeld quickly sought to pour oil on the waters, adding a note to his paper in publication registering Voigt's priority and then devoting an article to a simplified treatment of "Voigt's Theory of the Zeeman Effect."[106]

It is not surprising that in Sommerfeld's immediate response to Bohr's theory, expressed in a letter written in September 1913, the first question about further work concerned the application of the model to the Zeeman effect: "I would like to occupy myself with that."[107] A letter to Langevin a few months later made the connection, in Sommerfeld's thinking, between his and Bohr's spectral work even more explicit. "In a few days I will send a short note about the Zeeman effect. If it contains nothing much new compared to Voigt, it nonetheless shows that in the atom a number-theoretical symmetry and harmony appears to rule, as from another side Bohr has shown. Clearly a great deal is true in Bohr's model and yet I think that it must be fundamentally reinterpreted in order to satisfy."[108]

The re-interpretation would be more on Sommerfeld's side than Bohr's. In the Winter semester of 1914–15 he lectured on "The Zeeman Effect and Spectral Lines," and in January he gave a paper in the Munich colloquium on the Stark effect (discovered in 1913) in which he derived preliminary results on the basis of Bohr's theory.[109]

By the time he published a paper that year on "The General Dispersion Theory according to Bohr's Model," the sheer weight and precision of the data—spectroscopic and otherwise—that seemed to confirm its assumptions had removed most of his doubts.[110] In the past, as we have seen, he had sought systematically to avoid a consideration of intra-atomic processes and had expressed his "skepticism" about "atomic models in general" in his September letter to Bohr.[111] Now he concluded that the numerical value for e/m "is so good that a doubt about the exact correctness of Bohr's hydrogen model is not really possible." All the more, he continued, since more recent values for the ratio brought it even closer to the theoretical prediction.[112] The transformation was a radical one, for the model offered a stark contradiction to the electromagnetic theory, with its rotating but not radiating electrons, but by the beginning of 1916 Sommerfeld was unequivocal.[113] Experimental evidence showed that the quantized electron paths he had derived using his generalized model "correspond exactly to reality" and possess "real existence." "Therewith is opened to us an insight into the details of processes not only for hydrogen and hydrogenic atoms, but also—via a corresponding extension—into the atomic fields of the other elements through an exploitation of the huge [amount of] material piled up in the spectroscopic data."[114]

The acceptance of Bohr's stationary states may well have brought with it a more minor but still significant change in Sommerfeld's approach to the quantum. In light of the previous discussion, the derivation of the results that would soon be known as the "Bohr-Sommerfeld quantization conditions" draws our attention in particular, for at first glance it goes against the "dynamical" methodology for which Sommerfeld was seen to stand. In fact, in deriving his result, Sommerfeld drew upon precisely that expression that Langevin had labeled Planck's "statistical" statement at the Solvay conference in 1911. Section 1, titled "The Quantum Hypothesis for Periodic Paths," began with words that Planck could have spoken:

Before every application of probability considerations one has to ask oneself a question about equi-probable cases (about the correctness of the dice to be used). In the area of statistical mechanics, Liouville's theorem provides here the only starting point. This says, as is known, that equal-sized elements in "phase space" (q, p) are equally probable, insofar as and because they are temporally transferred into each other. q are the position coordinates, p the corresponding impulse coordinates

$$p = \frac{\partial T}{\partial \dot{q}}.$$

T is the kinetic energy, and one has as many coordinates q and p as the system has degrees of freedom. If one considers the element of the phase space, $\Pi(dqdp)$, one operates from the beginning with continuous probabilities. The quantum theory replaces these with discrete probabilities and considers instead of the phase elements $dqdp$ as the elementary area of probability, the finite phase integral $\int dqdp = h$.[115]

Proceeding next in explicit analogy with Planck, Sommerfeld reminded the reader of "the famous ellipse figure for the harmonic linear resonator" to which Planck had often referred, but had never actually drawn.[116] These ellipses marked out the allowed motions, in phase space, for a resonator. The curves were those of constant energy, and the area between each ellipse was a constant equal to Planck's constant h. Sommerfeld noted that others had already extended the scope of Planck's analysis. Debye had considered the case of arbitrary periodic motion of one degree of freedom, and Ehrenfest had applied Planck's reasoning to the case of simple rotation, with radius a. While the angular position ($q = \vartheta$) in this case varied smoothly between $+\pi$ and $-\pi$, the angular momentum ($p = ma^2\dot\vartheta$) remained constant, producing a set of rectangles in phase space (which Sommerfeld *did* draw), of length 2π and height $h/2\pi$. The allowed quantum states are thus those determined through the relation $p = nh/2\pi$. In other words, comparing Planck's case of the resonator with Bohr's case of orbiting electrons, "in place of the discrete energy elements for the oscillating point mass, Bohr's hypothesis of the discrete impulse element for the rotating point mass enters here."[117]

From these "introductory examples" Sommerfeld now proceeded to his general formulation of the quantum hypothesis. For the case of a rotating mass point, one selects only those paths for which the energy between the nth and the $n - 1$th curve is equal to h. Denoting the ordinates of these curves by $p_0, p_1, p_2, \ldots$, Planck's equation for the

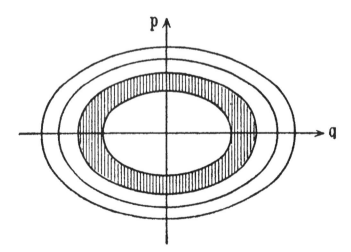

Figure 5.2
Phase-space representation of allowed orbits. Area of shaded region = h. Source: Sommerfeld, *Atomic Structure and Spectral Lines* (1923).

division of phase space becomes

$$\iint dqdp = \int p_n dq - \int p_{n-1} = h.$$

Assuming that $\int p_0 dq = 0$ and summing successive terms gives what Sommerfeld termed the "phase integral":

$$\int p_n dq = nh.$$

Why had Sommerfeld adopted Planck's statistical statement of the quantum hypothesis? His own answer came at the end of his paper, where a brief comment noted that the reader might recognize certain similarities between expressions derived through the newer definition and those achieved in Sommerfeld's studies of γ-rays and Röntgen-rays in 1911. The two were to be distinguished, however, by the fact that Sommerfeld now restricted himself to the analysis of periodic processes.[118] Although he had been critical of the earlier quantum theory's over-emphasis on periodic motion, within the context of such a restriction he had found Planck's phase-space representation a "deeper" means of deriving the radiation law than one derived solely from radiation theory.[119] Given the periodicity of Bohr's model—and the fact that the quantization of *momentum* that is central to it emerges immediately from Planck's definition—Sommerfeld clearly saw the new representation as the more useful. At the same time, one may add, one of the most appealing features of the theory he had presented at Solvay was now in tatters. Sommerfeld had vaunted his explanation on the grounds that it complemented, but did not contradict, electromagnetic theory. Accepting Bohr's model meant that this could no longer be true.

The shift to the statistical statement was, nonetheless, a pragmatic one, and in making it Sommerfeld took along none of the theoretical or methodological meaning that Planck associated with the statistical definition. Thermodynamics was not mentioned at any point in the paper on the Balmer series; Boltzmann's probabilistic definition of entropy was not cited. Planck, independently developing his own extension of the quantum condition to systems of multiple degrees of freedom, began by stating both. The contrast between the two approaches is once more striking and illuminating. Sommerfeld, having analyzed some examples as in his Solvay paper, quickly wrote down an expression for his quantum condition and then turned to a detailed investigation of Bohr's model. Planck spent much longer on the general problem of determining "the physical structure of phase space."[120] Having arrived at a result, his applications were, as usual, to generalized mechanisms: a simply periodic oscillator, a molecule rotating around a single fixed axis, a freely turning rod. Although he did take up the question of the production of spectral lines for the case of an electron bound by a Coulombic force, and thereby rederived Bohr's expression for line frequency, he did not accept Bohr's basic assumptions. The notion that radiation was emitted as electrons moved between stationary states was replaced by the more abstract claim that emission for a charged particle in an electromagnetic field occurred at the borders of elementary areas of phase space.[121] As Sommerfeld put it in a letter to Schwarzschild: "The exact coincidence with Planck's structure theory of phase space has also brought me much joy. Exactly the same result from such different starting points and so different modes of thinking (Planck careful and abstract, me somewhat aggressive [*draufgängerisch*] and headed directly toward observations)."[122]

Sommerfeld's analysis depended completely on the exploration and extension of Bohr's model in all its specificity. Having laid down "as special and determinate hypotheses as possible," in the words of his methodological statement of 1901, he now proceeded "to develop their consequences exactly." He began with a calculation of the energy of the elliptical orbit described by an electron moving around a charged center. Adding to this familiar expression the condition that the angular momentum be quantized ($p_\varphi = nh/2\pi$) and then setting the energy difference between two successive orbits equal to $h\nu$, he arrived at an equation for the frequency of the spectroscopic line produced by an electron moving from the nth to the mth orbit (where ε defined the eccentricity of each orbit and N was the Rydberg constant):

$$\nu = N\left(\frac{1-\varepsilon^2_n}{n^2} - \frac{1-\varepsilon^2_m}{m^2}\right).$$

Were the eccentricity allowed to be a continuous variable, Sommerfeld now noted, the frequencies so defined would not produce a discrete series, but a "washed-out band." It thus seemed "irrefutable" that the eccentricity would also need to be quantized, which he proceeded to do by applying his quantum condition to the radial momentum, p_r. The outcome was that the energy of an orbit seemed to depend solely on the quantum numbers of the two degrees of freedom ($W = -Nh/(n + n')^2$) and the frequencies derived corresponded once again to the sharp lines of the Balmer series. "The result is surprising in the highest degree and of striking certainty," Sommerfeld concluded, taking it as proof of the model's and his method's correctness. "It appears impossible to me that such a precise and momentous result could be subscribed to an algebraic chance; I see in it much more a convincing justification for the extension of the quantum hypothesis…to both degrees of freedom of our problem."[123] Successive sections took into account the fact that the charged nucleus was not fixed (and that therefore both it and the electron rotated around a shared center of mass) and sought to provide an answer for the question raised at the beginning of the paper: how many degenerate lines were collapsed into one another in the Balmer series? According to the model, the number corresponded to the possible orbits that could be associated with each sum of quantum numbers. A comparison to Stark's data on the splitting of those spectral lines in an electric field showed, if not precise agreement, at least a "general parallelism," and Sommerfeld held out hope that further experimental work might reveal the existence of weaker components, thus bringing the data into closer alignment with theory.

$m + m' = 3$, drei Möglichkeiten.

$m' = 0, \quad m = 3, \quad \varepsilon = 0, \quad b = a,$

$m' = 1, \quad m = 2, \quad \varepsilon = \frac{\sqrt{5}}{3}, \quad b = \frac{3}{2}a,$

$m' = 2, \quad m = 1, \quad \varepsilon = \frac{\sqrt{8}}{3}, \quad b = \frac{1}{3}a.$

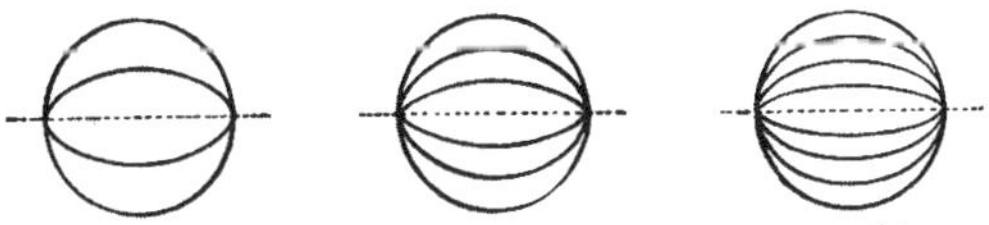

Fig. 3a.   Fig. 3b.   Fig. 3c.

$m + m' = 4$, vier Möglichkeiten.

$m' = 0, \quad m = 4, \quad \varepsilon = 0. \quad b = a,$

$m' = 1, \quad m = 3, \quad \varepsilon = \frac{\sqrt{7}}{4}, \quad b = \frac{3}{4}a,$

$m' = 2, \quad m = 2, \quad \varepsilon = \frac{\sqrt{12}}{4}, \quad b = \frac{2}{4}a,$

$m' = 3, \quad m = 1, \quad \varepsilon = \frac{\sqrt{15}}{4}, \quad b = \frac{1}{4}a$

Figure 5.3
Sommerfeld's representations of the shapes of atomic orbits for different values of their azimuthal and radial quantum numbers. Source: Sommerfeld, "Zur Quantentheorie der Spektrallinien," *Annalen der Physik* 51 (1915): 1–94 and 125–67.

If promising, the new theory was also limited. For the case of Keplerian motion, Sommerfeld's solution contained no extra spectroscopic information on fine structure. The multiple degeneracy of the lines in the hydrogen spectrum meant that the measurable results, in terms of line frequencies, were the same as those for the single degree of freedom case treated by Bohr.[124] Sommerfeld's predictions led to different shapes of orbits, but to no extra spectral lines. The treatment of the Kepler case also raised a more general concern. In solving the problem, one can make use of a variety of different coordinate variables (usually either parabolic or polar) This, of course, makes no difference to the classical solution, but it does result in a different quantization and hence different quantized orbits when one restricts the range of possible motions using Sommerfeld's quantum conditions. The question of which coordinate systems would result in the correct quantization (and why) would not be answered in full until several years had passed.

Sommerfeld's next paper, presented in January 1916, was more successful and added the requisite third component to Sommerfeld's method: quantitative comparison with experiment. Introducing a relativistic correction, he was able to obtain values for the energy difference between new lines that corresponded to the precession of the perihelion of the electron's elliptical orbit, introducing for the first time what is now termed the "fine-structure constant." Later sections of the paper extended the analysis to include series other than the Balmer series, including those at X-ray frequencies, and elements other than hydrogen. The results were well worth Bohr's enthusiastic letter from Copenhagen a few months later. "I do not believe," he wrote, "that I have ever read something that gave me so much joy, and I need not say that not only I, but all here have had the greatest interest in your significant and beautiful results."[125]

Crucial to the positive reception of the theory was its close experimental confirmation, gained through an intense collaboration between Sommerfeld and Paschen in particular. In 1916, a letter arrived from the experimentalist on average once every ten days. In his fine-structure paper, Sommerfeld commented: "It may be remarked that observation and theory have proceeded independently of one another and have only been brought into conjunction with one another through a correspondence." But the relationship was even closer.[126] Sommerfeld's theory was not developed independently of experiment, nor was Paschen's "experimental art" sufficient on its own to confirm the theoretical results. The theory gave no information about line intensities, a profound interpretative problem, since the data could not be understood without them. Sommerfeld noted in 1917 that he had previously offered a "suspicion" about the relative intensities of components in the fine structure of hydrogen-like elements, which was "more in order to have a determinate rule for later diagrams than in order to advance quantitative physical claims with them." If it now seemed that recent results may have proven these guesses justified, he nonetheless emphasized that Paschen's observations "can give to theory's hand valuable clues for the evaluation of intensity questions."[127] In turn, as Kragh has shown, those results "were far from purely observational but depended crucially on Sommerfeld's theory."[128]

Nonetheless, for all except committed anti-relativists, who balked at the claim that spectroscopy had proven the worth of Einstein's special theory, the precision of the experimental confirmation was taken for granted. Indeed, in 1917 one of Sommerfeld's students, Karl Glitscher, completed a dissertation that used the fine-structure theory as applied to ionized Helium to decide between the "rival mass laws" contained in the Abraham and Lorentz-Einstein theories of the electron. In his report to the Munich faculty, Sommerfeld recorded Glitscher's conclusion, that Abraham's theory consistently gave a value $\frac{1}{5}$ smaller than the relativity theory. "This is excluded according to Paschen," Sommerfeld wrote. "Abraham is therefore definitively quashed."[129] The result provided the concluding lines for *Atombau*: "the theory of relativity, just as it has remodeled all our physical thoughts and ideas, has also been able to help forward spectroscopy in a decisive manner. Conversely, we note that, in return, spectroscopy is in a position to lend support to one of the main pillars of the theory of relativity, and to decide in its favor the question of the variability of mass of the electron."[130]

For theorists, Sommerfeld's systematic extension of Bohr's model provided a clear program, one that Bohr himself articulated in a lecture to the Chemical Society in Copenhagen in 1919. Experiments, he claimed, "give the most direct experimental evidence imaginable for the existence of stationary states of atoms, assumed originally purely on theoretical grounds to account for the peculiar formulas holding for the frequencies of the spectra of the elements." He continued:

> In recent years it has been possible to advance fairly far in this way, in that it has become possible to set up conditions holding for the stationary states not only of such simple systems as the hydrogen atom but also of more complex systems in which the motion is not of the simple periodic type. The work in this field has been due mainly to Sommerfeld, but his ideas have been applied and adopted by a large number of eminent German physicists, such as [Paul] Epstein, Schwarzschild, and Debye.[131]

Schwarzschild and Epstein (one of Sommerfeld's former students) had introduced powerful new techniques for dealing with complex atomic models.[132] Their use and explanation of the methods of Hamilton and Jacobi allowed a rapid extension of the kinds of problems with which quantum theorists could deal. For motion that was "not of the simple periodic type" but was nonetheless "conditionally periodic," the Hamilton-Jacobi equation describing the motion could be separated into as many components as there were degrees of freedom. Quantization rules could then be applied easily to the momentum corresponding to each independent variable. In this way, motion that was not strictly periodic—but in which, for a certain coordinate system, each component was periodic in phase space—could be analyzed with facility. Of equal importance, one now saw which of the many possible coordinate systems was the correct one: precisely the one in which the problem could be solved through the method of separation. In his *Atombau*, Sommerfeld would refer to the Hamilton-Jacobi method as the "royal road for quantum problems," and the text is filled with the elaborate models analyzed using it.[133] Deep within the atom Sommerfeld seemed to have found, for the moment, proof of the superiority of the dynamical method.

Conclusion

Writing of Sommerfeld's work in the early twentieth century, one of his earlier biographers had a critique to offer. "Sommerfeld often undertook detailed calculations on problems that required further clarification of fundamentals," Armin Hermann has stated, "such as his papers on electrons moving at speeds faster than light and particularly in his later work on the Bohr-Sommerfeld atom. To solve the principal problems of physics was simply not within his capabilities; rather, he was best suited to provide the mathematical development of a theory and its actual solution."[134] Leaving aside the fact that such a criticism requires a certain degree of hindsight in order to judge which areas were still not clear in their fundamentals, a deeper problem is that it does not seek to understand what value Sommerfeld himself accorded to the detailed calculations that were so integral a part of his physics of problems. In part, the level of detail—even where certainty about an issue was lacking—was clearly tied to a pedagogical purpose. Thus, in writing to Carl Runge about his article on spectral lines, Sommerfeld noted that "It seems to me that in the Enc. [yclopedia] all *Ansätze*, the unsuccessful ones as well—and that's probably all of them—should be discussed at length, excluding, of course, the mathematically false ones.... You want to undertake all that and Much More, and represent it all *really elaborately*, so that future workers on the problems have it easier than present ones."[135] In addition, as his students would note, Sommerfeld was not content with analytic solutions to problems, insisting on their numerical evaluation, a fact that does, indeed, cause his papers to balloon in size. As he wrote in 1901, quoting Fourier, as he often did: "Setting up the general analytic expression only offers the first step toward the solution of a mathematical-physical task. The second, just as important [step] consists in numerical evaluation. 'As long as this is missing, the solution remains incomplete or useless; because the truth that we want to uncover, lies no less deeply hidden in the analytic formulas than in the physical problem itself.'"[136]

The comparison to Planck offered in this chapter, however, makes it clear that there was more to Sommerfeld's emphasis on detail than merely either mathematical virtuosity or pedagogical preference. Detail was of the essence for Sommerfeld's approach to theoretical physics in general, defining the kind of objects that populated his theory and the kinds of solutions deemed most useful for the further development of his field. Mathematically elaborate solutions were to be expected from a program that saw its aim as "developing exactly" the consequences of "specific and determinate assumptions." As Sommerfeld made clear at Solvay, "risky" assumptions that went into the detail of individual physical processes—as opposed to the statistical method associated with the energy quantum, which could avoid such detail—produced calculations that were much more "labored" and dependent upon individual hypotheses.[137] But such hypotheses were, for Sommerfeld, precisely the means by which physics advanced upon the most fruitful path, whether they were ultimately successful or not. Although he noted that statistical methods might allow the generalization of his approach by eschewing such detail, it was not clear that the general conveyed the same high value for Sommerfeld as for Planck. Thus, in 1916—five years after the Solvay conference—Sommerfeld compared the quantum condition used in his theory of spectral lines to Planck's. His, he claimed, could describe more facts of experience, including the problems of fine structure and of X-ray lines. But the question remained: "Can my more direct quantum statement for the phase integral replace Planck's general considerations?" His method of deciding the matter was characteristic, involving the calculation of a single problem, a spatially harmonic oscillator, previously studied by Planck. Aided by a calculational correction suggested by Planck, he showed equivalence with his analysis. "The impression is fortified thereby, that our special and determinate [*speziell und bestimmt*] quantum formulation contains the general solution of the problem posed by Planck."[138] Insofar as the aim was solving specific problems, the general was less significant than the direct.

The comparison to Planck proves useful as a means of illuminating elements of Sommerfeld's practices even outside the realm of their specific interactions over the quantum. Tied to Planck's thermodynamic method were two elements flowing from the "de-anthropomorphized" abstraction of its principles: their removal from the realm of the "economic"; and their separation from experiment. Sommerfeld's understanding of thermodynamics, unsurprisingly given the place of praxis within his theoretical work, did not concur with Planck's. "Thermodynamics," Sommerfeld noted in a lecture, "is the master example of a science built up from axioms," but his own treatment would not be so axiomatic. That the Carnot-Clausius proof of the second law "operates with technical conceptions we see more as an advantage than a disadvantage. Thermodynamics, after all, originally arose from the requirements of steam-engine builders."[139]

Perhaps most striking, however, were their different relationships to experiment. Planck saw the practices of theory and experiment as distinct. The principles of theory were to be understood independent of any question of their realizability, experiment introducing too great an element of "anthropomorphism." For theories involving more than the principles of thermodynamics, experiment was to enter the practice of theory at the end, only in order to check outcomes. The question of the experimental realizability of a theoretical object was not one that Planck found necessary to answer—hence the infinitely small resonators used in his radiation theory or the "ideal" processes of his thermochemistry. In Sommerfeld's dynamical method, on the other hand, theoretical objects—electrons, Bohr's atoms—represented objects of real existence, so that experimental critiques must hold weight. Paschen's data served as a regular check on Sommerfeld's theory at all levels of its construction, and data could substitute where theory was inadequate, as in Sommerfeld's calculation of intra-atomic energy in the production of X-rays. This element would only become stronger as a part of Sommerfeld's practice in succeeding years. In 1924, Max Born, in his own textbook, complained about the amount of experimental detail to be found in *Atombau*.[140] By that point, as we shall see, even the place of models in theorizing had been downplayed, as Sommerfeld and his students adopted an even more "direct" methodology than ever, seeking to read spectroscopic laws directly from the data.

III

6 *Prinzipienfuchser* and *Virtuosen*: Theoretical Physics after World War I

[I]t would be extremely fortunate, a saving deed, if the bearer of the crown voluntarily renounced his rights. But in the word "voluntarily" already lies the impossibility for me to work in this direction; for first I think of the oath I took, and second I feel something that you admittedly will not be able to understand at all...namely a reverence for and an unshatterable solidarity with the state to which I belong, about which I am proud—and especially so in its misfortune—and which is embodied in the person of the monarch.
—Max Planck to Albert Einstein, October 26, 1918[1]

Theory and Experiment in the Aftermath of the War

By the end of September 1918, Erich von Ludendorff, head of the German armed forces, could foresee the imminent collapse of German armies on the western front. In an act that many historians have marked as one of high cynicism, his response was to insist on the immediate formation of a new government made up from those parties then in the majority in parliament. It was thus to be the Social Democrats and liberals who would bear the shame of defeat and against whom the right wing would circulate the myth of the *Dolchstoss*, the stab in the back inflicted by the weak left wing at home on gallant German soldiers in the field. As the military and political situation deteriorated further, tensions that only the hostilities had kept in check erupted into social violence. The thought that, in spite of the apparent hopelessness of Germany's position, the high command might nonetheless attempt a final attack against the allies led to a sailors' insurrection in Kiel in the first days of November. Within a week, Planck's beloved Kaiser had fled to the Netherlands, Berlin found itself under the rule of a new revolutionary government, and the nation entered what has been termed "the period of decisions," the time in which the foundations for the German republic were laid.[2]

On December 3, Sommerfeld wrote to Einstein asking help on a question of the simplest way of explaining the foundations of quantum statistics to the audience of what was described as a popular book about atomic models. That book, of course, was to be Sommerfeld's *Atombau und Spektrallinien*, which soon after its publication in 1919 would become the "Bible" of quantum spectroscopists. Following the query about science, however, Sommerfeld added a note of incredulity. He had heard, he said, that Einstein was supportive of the new political situation. "God preserve you in your beliefs!" wrote Sommerfeld, but it was clear that his sympathies lay elsewhere. "I find everything unspeakably dire and imbecilic. Our enemies are the biggest liars and scoundrels, we the biggest morons. Not God but money rules the world."[3]

Sommerfeld had not heard incorrectly regarding Einstein's position. Indeed, Einstein seemed to be reveling in the reversal of status that the revolution had brought about. Writing to his old friend Michel Besso, he claimed that he could "not be dissuaded from my optimism. I am enjoying the reputation of an irreproachable Socialist; as a consequence, yesterday's heroes are coming fawningly to me in the opinion that I could break their fall into emptiness. Funny world!"[4] His reply to Sommerfeld two days after this letter was more restrained, perhaps, but no less positive about the future:

It is true that I have hopes for these times, despite the many ugly things it is bringing forth in the details. I see the political and economic organization of our planet advancing. If England and America are sensible enough to agree, wars of some consequence will not be able to occur at all anymore. The military economy so repugnant to me will also virtually disappear. If the period of transition becomes quite grievous for us, in particular, in my opinion it is—frankly speaking—not entirely undeserved. However, I am of the firm conviction that culture-loving Germans will soon again be able to be as proud as ever of their fatherland—with more reason than *before* 1914. I do not believe that the current disorganization will leave permanent damage.[5]

As before and during the war, Sommerfeld serves as a far more representative example than Einstein of the majority opinion among German physicists. There are good reasons for Peter Gay's decision to include Einstein as one of the "outsiders" that became "insiders" during the heady years of the Weimar republic (and exiles afterward),[6] but it must be realized that the "inside" was far emptier in the years after 1918 than before. The reason there was space there for those like Einstein was because the previous occupants of an inner space had abandoned it. In fact, perhaps a better characterization of the period, one that covers a much greater part of the population, is to note that "insiders" during the Kaiserreich, those who had participated in political life, pledging a very public support for a state "embodied in the person of the monarch," made a self-conscious decision to remove themselves from the overt political sphere after the war. German nationalism remained, especially a sense of the wrongs committed against the fatherland by grasping and vindictive foreign powers, but few of the politically conservative physicists that made up the majority of the field could bring themselves to a support of the republic. While a comparatively small number confessed themselves to be fundamentally inimical to the republican experiment, most, like Sommerfeld, contented themselves with a move that bore them to the outside on a wave of weariness, disgust, and rage. Such emotions can be easily understood given the social and economic circumstances that followed Germany's acceptance of its defeat. To guarantee that their enemy would be willing to negotiate on their terms, the British maintained their blockade for eight months after the war's end, resulting in the food shortages that had characterized the last years of the conflict. The Weimar government's frankly irresponsible fiscal practices following the attempt to meet heavy reparations payments meant that academics, like other members of the Bildungsbürgertum, could only watch their savings dwindle into meaninglessness as the value of the mark dropped. Yet the German professoriate did poorly in the half decade that followed 1918 even in comparison to those members of the middle and upper classes to which they could most easily be compared, namely the civil service on the one hand, and the free professions (medicine and law, in particular) on the other. A professor working in Berlin before the war could count on an average base salary of about 7,200 marks, plus teaching and examination fees of 8,700 marks, making a total of 15,900. Those at Prussian provincial universities might expect about 4,500 marks less. After 1914, as student numbers plummeted, teaching and examinations fees followed, while basal cost-of-living adjustments failed to keep pace with wartime inflation.[7]

The Civil Service Pay Reform Act of 1920 made this difficult situation worse. Student numbers rocketed back up as soldiers returned home. While almost 72,000 students attended a German university or technische Hochschule in 1914, this number had increased to almost 109,000 in 1922. As numbers in existing courses thus dramatically rose, universities also introduced an extra intermediate semester [Zwischensemester] to help veterans make up for lost time. While more students and more classes should have led to more money, however, the act of 1920 led to the peculiar situation in which academics might receive 1,750 marks less than previously, even as they taught twice as many students. The increases that followed in 1921 and 1922 failed to address the more fundamental issue, namely that in comparison to other civil servants academic salaries were plunging. Approximately equal to one another in 1920, German professors could expect their pay to be 32,000 marks less than a senior civil servant in 1922.[8]

German theoretical physicists suffered with their colleagues in other disciplines under such conditions. Correspondence from the period makes clear the commonality of the exhaustion that Max Born spoke of feeling as the result of increased and excessive labor coupled with insufficient and insufficiently nourishing food. Yet in contrast to their grim personal situations in the months and years immediately after the ceasefire, theoretical physics as a discipline was looking stronger than ever. Wilhelm Wien had spoken of the "now-mighty" theoretical physics in 1915,[9] but it was really after the war that a sense of the strength and independence of the field becomes most apparent. Intellectually, interest in relativity theory exploded after the experimental confirmation of Einstein's predictions concerning the gravitational bending of light in 1919. The mid 1920s would see the other of modern physics' "twin pillars," the quantum theory, enter a new era, as two different forms of quantum mechanics were proposed, elucidated, and contested. Institutionally, the period can be characterized as one of consolidation. By 1918 Max Born and Max von Laue occupied two new full professorships for theoretical physics in Frankfurt and Berlin, respectively. By 1926, all but two of the universities and technische Hochschulen in which extraordinary professorships for theoretical physics had previously existed had converted these to full (although sometimes only honorary) ordinary chairs.[10]

The power that the new sub-discipline was *perceived* to have acquired after 1918 can be discerned in the fact that it rapidly accrued enemies. Sniping about the abstract nature, and the "undeserved" success and popularity of Einstein's general theory of relativity became public in 1920, when a group calling itself the Arbeitsgemeinschaft deutscher Naturforscher (Syndicate of German Scientists) mounted a campaign denouncing the scientific validity of the theory and personally impugning Einstein himself. Einstein issued his reply two days later, accusing his opponents of incompetence and anti-Semitism.[11] A more scientific forum than the pages of the *Berliner Tageblatt*, the 86th Assembly of Scientists in Bad Nauheim, saw a more decorous debate between Einstein and the Nobel Prize-winning experimentalist Phillip Lenard (soon to be a supporter of Adolf Hitler).[12] Few discussions of the theory by Lenard, Stark, Gehrcke, and other opponents would retain any sense of the balance of Nauheim in succeeding years.

The personal attack on an individual theoretician and his work was incorporated into a much more general critique of the nature of modern theoretical physics in Johannes Stark's 1922 pamphlet on "The Present-Day Crisis in German Physics." This crisis, Stark claimed, was being brought about by the struggle within physics of two antithetical approaches. In his view, a dangerous and deluded theoretical physics had not only gained equality with experimental physics, it was now seeking to exert its dominance:

Through the propaganda for Einstein's general theory of relativity and through the dogmatism of the quantum theory a theoretical approach has won a dominant influence, which has begun fundamentally to damage physical science, since it seeks its sources more in intellectual construction than in experience and wants to make this into a handmaiden of The Equation. The experimental approach, by contrast, finds itself on the defensive; it sees the source of physics in observation and measurement and sees in theory a heuristic and systematic aid for the winning and representation of physical knowledge. There can be no doubt which approach will finally win the upper hand.[13]

Stark, increasingly bitter and inclined to perceive enemies where merely differences existed, probably was not entirely unjustified in his suggestion that experimental physics was suffering in the Weimar era. Theoreticians acknowledged the same point, although with substantially different emphasis. Von Laue's review of Stark's book was almost entirely critical, but did agree on one issue: "We would like to agree with the author, however, that Germany should try to increase the amount of experimental research at the expense of theoretical work. But surely we all know that the current disproportion is mainly due to the predicament which our entire nation is in and that physicists are unfortunately not in the position to change it much."[14] Sommerfeld predicted that such a state of affairs could only continue in coming years. "In my country, I am afraid" he said in a lecture delivered in 1922 or 1923 during a tour of the United States, "the experimental work in the next years will be almost impossible because of our extreme poverty. On the other hand, in my country we have developed a very ~~strong and~~ successfull scool of mathematical physicist."[15]

In sensing a threat to experimental physics in the success of theoreticians, Stark may also have picked up on what has proven to be a substantial shift in the *persona* of the theoretical physicist, the origin of which can be dated to the Weimar period. While Sommerfeld had earlier emphasized the importance of experimental work within his institute for theoretical physics, members of the generation he trained were, in the years after the war, already beginning to suggest that, for their own purposes, a knowledge of experimental practice could be abandoned altogether.[16] A series of anecdotes about the theoretician Wolfgang Pauli concern what has been termed the "Pauli effect." Not a description of a new phenomenon in physics, nor to be confused with the Pauli (exclusion) principle, the Pauli effect is said to arise as a result of the "well-known" fact that theoreticians are incapable of handling experimental equipment. Pauli was such a good theorist that his powers in disrupting experiments became uncanny. He and his colleagues would delight in relating stories of experimentalists refusing to allow him in their laboratories, Otto Stern apparently insisting that any conversations be held with Pauli on the other side of a closed door. When a complicated piece of experimental apparatus inexplicably stopped working in the laboratory of J. Franck in Göttingen one day, he wrote to Pauli about it in Zurich, only to receive a reply postmarked from Copenhagen. Pauli, it appeared, had been on a train to visit Niels Bohr, and his train had halted in Göttingen at precisely the time of Franck's experimental failure, a story that suggests that Stern's precautions may have been inadequate.[17]

The anthropologist Sharon Traweek has argued that stories like these make up an integral part of the socialization process of future participants in a scientific discipline.[18] This is to speak about myths and shared narratives in general, but one should also note that there is a great historical specificity connected with descriptions of the Pauli effect. Pauli was part of the first generation for whom the telling of such stories—of not merely an infelicity with experiment, but of a supernatural ineptitude—would not have constituted a form of professional suicide. Around 1900, comments that a candidate for a position in physics lacked experience or skill in laboratory work counted as a devastating blow against them. Sommerfeld made much of the importance of experimental work even within his theoretical institute in the first few years after 1906. And if he had been able to win a full professorship in spite of his lack of experience, this was nonetheless nothing to emphasize, certainly nothing to brag about. As he put it candidly to his mother in 1894, outlining one of his reasons for turning down a position as Voigt's assistant in Göttingen: "I understand nothing about experiments and I have told this to Voigt. I am afraid to make a fool of myself."[19]

Things had changed by the time Pauli arrived as Sommerfeld's student in Munich, and the early emphasis on the importance for theoreticians of a solid experimental foundation seems to have evaporated. Sommerfeld still kept very close ties to experimentalists, especially Friedrich Paschen, whose results were crucial to theoretical studies of quantum spectroscopy, but the practice of physics had now become a partnership between the theoretician and the experimentalist. The idea that training should endeavor to produce a hybrid of both kinds of physicists had dropped away, and the postwar stories of the performance of Sommerfeld's theoretically gifted students in their laboratory work repeat the themes of a lack of interest and ability. Hans Bethe, who began studies in Munich in 1924, later related his lack of enthusiasm for the experimental physics seminar in Wien's institute, calling them "Quaker meetings" where the two extraordinary professors "sat there silently and contemplated the puzzle of some strange fact which had to do with some stop-cock grease problem which I found extremely uninteresting." At least he managed to perform better in the advanced laboratory, however, where he may have been the only theoretical physicist to whom Wien gave an A.[20] This formed a stark contrast to the performance of Werner Heisenberg a few years earlier. Heisenberg, perhaps attempting to emulate the behavior of the senior student Pauli, paid almost no attention to his work with Wien. If his later tales are to be believed (and of course, the power of the Pauli effect discourse must be taken into account when judging the veracity of such non-contemporaneous stories), the depth of his ignorance was profound. As he described it, he was "engaged in theoretical work even during the time when I was in my room at the physical institute." During his examination, Heisenberg was unable not merely to answer questions related to actual experimental practice, but even to discuss theoretical issues concerning experimental equipment. Wien grew increasingly exasperated as Heisenberg fluffed question after question about the resolving power of various instruments and positively annoyed when it was revealed that Heisenberg didn't understand the theory behind the operation of a storage battery. "Then in some way he felt that I had too little interest in experimental physics....He thought that I should have a very poor mark; I don't know whether he wanted me not to pass the examination. I suspect that there must have been quite a discussion afterwards between him and Sommerfeld about the problem."[21] Heisenberg suggested that behind Wien's insistent questioning may have been a larger difference of opinion between him and Sommerfeld. "Wien felt that Sommerfeld forgot about the normal and decent part of experimental physics which everyone must know and should know, and he wanted to insist that these things must be known to every student, even in Sommerfeld's laboratory." Heisenberg's claim that Sommerfeld "always felt that theoretical physicists do some kind of 'highbrow' physics" may well have been true for the time during and following Heisenberg's own years in Munich. It should not, however—as earlier chapters have indicated—be taken as genuinely descriptive of the state of affairs during the first dozen years of the Sommerfeld School (1906–1918). Indeed, perhaps there is no better sign not merely of the general good health of postwar theoretical physics in comparison to the fin-de-siècle, but also of its specifically changed relation to experiment—from subordination to growing independence, even cockiness—than the "discovery" of the Pauli effect.

Undergoing shifts in its relationship with other fields of physics and expanding its representation within tertiary institutions, theoretical physics in the Weimar period also saw a significant change in its intellectual leadership. In 1911, the theologian and President of the Kaiser Wilhelm Gesellschaft, Adolf von Harnack, had cited Planck and Einstein as twin refutations of the charge that the present generation had no philosophers. Seven years later Max Born suggested that both were deserving of the Nobel Prize, noting their shared efforts to deepen to foundations of physics. "Besides an immense experimental progress," he wrote, "contemporary physics shows a clear attempt toward deepening foundations, to acquire universal knowledge [*universeller Erkenntnisse*]. The two main representatives of this direction, which could be described as philosophical, are Albert Einstein and Max Planck. Einstein's theory of relativity and Planck's quantum theory signify radical changes in the domain of science that have rarely taken place in a more colossal and promising fashion."[22] Continuing with the modest claim that he could not speak to the question of which of the two discoveries was the greater or the more important, Born nonetheless promoted Planck as the more deserving for that year, "as the older of the two researchers and as the one whose results have been of the greatest importance for the progress in experiments."[23]

Born was not wrong to emphasize Planck's age. He had been 62 when he received the delayed 1918 Nobel Prize in 1920, and his years of greatest productivity were behind him. In 1926 he announced his retirement, the faculty nominating Sommerfeld as his successor. After Sommerfeld declined, Erwin Schrödinger, second on the list, accepted the call. Yet it might be said that even before his official retirement, Planck had already been succeeded. The years after the war saw him maintain, even cement his socio-disciplinary position as the public face of German physics. In terms of his role in the development of the quantum theory, however, it was Bohr who came more and more to stand as the principal alternative to what remained a powerful

Munich School under Sommerfeld. Looking back roughly 30 years later, Pauli mentioned only those two approaches to the problems offered by the quantum of action—effectively effacing Planck's contribution—and it is telling to note the manner in which Born's understanding of the field of theoretical physics changed in the decade after 1918.[24] Maintaining his belief in the existence of a "philosophical" direction in the development of modern physics in a speech delivered in 1928 on the occasion of Sommerfeld's sixtieth birthday, Born now contrasted this approach directly with Sommerfeld's own. At the same time, one of Born's twin exemplars of philosopher-physicists had been replaced. "Physics as a field of application for philosophical principles," he wrote, "as Einstein and Bohr promote [*treiben*] it, is foreign to Sommerfeld's basic nature; if in spite of this several of his most recent and most significant students have achieved great things in the study of foundations, this must, insofar as, in general, an external impetus should be sought, be due to later influences, particularly to contact with Bohr."[25]

The remainder of this chapter is concerned with delineating the most essential elements of this physics so "foreign to Sommerfeld's basic nature"; a physics of principles deemed both philosophical and foundational. For Einstein, much of this work has already been completed. A great deal of scholarship has traced the importance of thinking in terms of generalized principles for not only the more obvious examples in his research—his use of the principle of relativity for his "special" theory; the centrality of the equivalence principle for general relativity—but also for his work on Brownian motion, the photoelectric effect, fluctuation phenomena, and the quantum theory.[26] Two areas where Einstein's emphasis on principled reasoning have been less closely examined have been, first, the socio-disciplinary meaning of the practice of principles and, second, the role that the principled *form* of his theories of special and general relativity played in their reception, particularly in Britain.

Those members of the physics community who focused on principles rather than on calculational virtuosity, Einstein claimed in the 1920s, marked themselves out as different—and by implication, better—than their colleagues. Unlike Planck's vision of principles as a means of promoting disciplinary unity, in other words, Einstein saw a focus on principles as a marker of intra-disciplinary distinction. Discussing theories, rather than people, Einstein offered another alternative to thinking in terms of generalized principles. In 1919 he offered a clear, widely reported and repeated distinction between "constructive theories" and "theories of principle." The former were "synthetic," building a picture of a complex world from a simpler, materialist schema; the latter were "analytic," using general, empirically grounded conceptual elements from which one derived real-world phenomena. The spectacular wave of publicity that greeted Einstein's every utterance on the topic of relativity after the confirmation of the gravitational bending of light by Arthur Stanley Eddington and Frank Dyson in 1919 makes his postwar writings an obvious starting point for an understanding of the mechanisms by which a principles language might spread. We will see that the distinction Einstein put forward in the London *Times* profoundly shaped, for critics and supporters alike, the reception and subsequent re-description of his general theory of relativity in Britain, particularly within the pages of almost certainly the single most important work on the general theory of relativity for English audiences in the early 1920s, Eddington's *Space, Time and Gravitation*.[27]

Bohr's example enables the sort of direct comparison of theoretical practices that was possible for the case of Planck and Sommerfeld at the Solvay conference. This issue is taken up in chapter 7, where I compare Bohr and Sommerfeld's responses to the problem of reconciling Maxwellian electromagnetic theory with Bohr's quantized planetary atomic model. In the present chapter, Bohr's case will serve to illustrate the dramatic range of meanings that could be covered by the notion of physical principles. For, unlike Planck and (as we shall see) Einstein, Bohr did not see principles as absolutes. For Bohr principles were only ever meant to be approximations to the truth. His principles of correspondence and mechanical transformability (or adiabatic invariance, as it was earlier termed) were the two most important heuristic tools of the early quantum theory. They were, however, like the philosophical principles of his mentor Harald Høffding, intended merely to function as bridges between two incommensurable worlds: of consciousness and unconsciousness, reason and emotion, continuity and discontinuity. If we take at face value Born's judgments of Einstein, Planck, and Bohr as kindred spirits in their philosophical emphasis on foundational principles, we must nonetheless acknowledge that significant differences existed among all of them. The strength and robustness of a language of principles, in fact, did not arise from a consensus over the meaning or content of central terms, but rather from the capacity of the word 'principle' to canvass at times dramatically different—indeed diametrically opposed—philosophical positions.

Prinzipienfuchser and *Virtuosen*

It was with no small joy that Paul Ehrenfest wrote to Albert Einstein in September 1925 to report of a surprise visit from Niels and Margarete Bohr. Niels, Ehrenfest reported, was wrestling prodigiously with quantum problems and longed to speak with Einstein more than anyone else. "I know that no living man," wrote Ehrenfest, "has looked so deeply into the real abysses of the quantum theory as you two and that no one apart from you both really sees how necessary are completely radical new conceptions."[28] It was perhaps this commonality between the two men that Ehrenfest saw as responsible for their effect upon him, for a little further along in the same letter he wrote: "It is always so peculiar for me, that the contact with you and Bohr gives me the courage to push on with my foolery, while contact with by far the majority of other theoretical physicists totally depresses me. This is very paradoxical."

Einstein's reply came with remarkable rapidity, but he failed to see anything peculiar in the situation Ehrenfest described. "This matter, what you have said about Bohr and me in contrast to other theoreticians in relation to your work, doesn't astonish me. There exist Prinzipienfuchser and Virtuosi. We belong, all three of us, to the first category and have (at least certainly we two) little virtuosic ability. Therefore, Effect of an encounter with noted Virtuosi (Born or Debye): depression. It works, by the way, similarly the other way round."[29] The meaning of the terminology used is subtle. While to be called a virtuoso is not, on the face of it, insulting, the implication of the term is not necessarily complimentary either. In German, as in English, the possession of virtuosity can also suggest the lack of genuine insight. Grimm's dictionary notes that in modern German usage, the virtuoso is often opposed to the artist, in that the former possesses all manner of technical ability, but does not have either the judgment or the genius of the truly artistic. In this way, while "Prinzipienfuchser" has the character of a self-deprecating description—one who is obsessed by principles—it also, through its opposition to "virtuoso", implies that these obsessives were also artists, not mere technicians. That this was indeed the implication intended may be inferred from the examples given of those in the "virtuoso" category. Both Debye (trained in Aachen and Munich under Sommerfeld) and Born (trained in Göttingen under David Hilbert) were known for their exceptional mathematical talents. Einstein's use of the term may well have also represented an inside joke. Ehrenfest had spoken of "integration virtuosity" at the "Bohr festival" in 1922. The final footnote of his paper read: "The treatment that Bohr has developed here gives us hope that the quantum system possesses a special preference and aptitude for slipping away from mathematical complications. May Bohr be right here once again: this would provide a deliverance of the quantum theory from the hegemony of integration virtuosity!"[30] At least according to his student, Georg Uhlenbeck, however, Ehrenfest alternated between fear and disgust when confronted with examples of mathematical virtuosity. Ehrenfest, he claimed, "was a man who always had to get it out of his toes. He had, somehow, no technique. Nothing was in his fingers. He always had to think it out completely from the beginning. Although he knew mathematics it was not simple for him. He was not a computer. He could not compute."[31]

In contrast, Sommerfeld and Born would always take up a new idea and begin to calculate. Ehrenfest "was a little bit frightened and also disgusted by them. Always 'diesen komplexen Integrale.'" The fear, it appeared, was mutual. "In several of my visits with Ehrenfest to Göttingen the contrast between these two was very striking. Born, on the other hand, was slightly afraid of Ehrenfest. Born was afraid he'd make the wrong calculations, and then Ehrenfest would, by simple models, by simple logical algebra, show him that it was wrong." That emphasis on simple logical algebra came to the fore in Leiden's colloquia and seminars:

Here Ehrenfest always wanted to have the simple point, "what is the point." Can one say that, so to say, in a few words or with a very simplified model so one could see what the point was. Often he was unable to do it, usually unable. And then Ehrenfest...jumped around. It was always so. It was disgust, really, all the time, all the time. And he never stopped. In a certain sense it was also his tragedy to be interested, because you see when he got older it got more and more difficult for him to get it out of his toes, to learn it all. People like Sommerfeld and also Lorentz, and surely Born, could always fall back on the technique. They could make long calculations. Maybe in the fundamentals it was not very new but it was always interesting.[32]

One should, of course, be wary of accepting Ehrenfest's self-deprecating opinion of his mathematical talents—here refracted through his student's eyes. And certainly few would deny that Einstein exhibited an impressive mathematical ability in many of his papers. Yet perhaps even more interesting than the validity or even the particular implication of the distinction was the very fact that Einstein sought to make it at all. That is, that he sought to demarcate physicists of principle from other kinds of theoretical physicists. The point is strikingly different to that made by Planck, who emphasized the importance of principles within theoretical physics no less strongly than Einstein, but who saw principles as agents of social and intellectual unity, rather than markers of intra-disciplinary distinction. A standard trope in the historical comments that began so many of Planck's papers was his description of the difference between a current age riven by philosophical and methodological debate and a golden age in the past, when a unified perspective had seemed guaranteed. Yet the solution to contemporary problems was clear: regardless of one's particular worldview—whether energetic, mechanical, or electromagnetic—the physics of principles as Planck portrayed it offered a disciplinary commonality, a basis of shared agreement.[33] All physicists could, at one level, be physicists of principles. Indeed, given the conception of the historical development of physics that Planck sketched, there was scarcely another choice. As physics became gradually more and more de-anthropomorphized, a physics of principles naturally came into being, or at least would do so as long as physicists were not blinded by the dubious charms of a Machian phenomenology. In contrast to Planck's unifying inclusivity, Einstein's self-definitional distinction between kinds of physicists who induce mutual discomfort appears both elitist and exclusive.[34]

To be sure, the dichotomy Einstein would set out in 1919 between constructive and principle theories (especially when expressed, as we shall see, by terms including "completeness" and "security") sounds very close to that between "satisfactory" and "secure" physical explanations, the language used by both Sommerfeld and Planck. The difference lay in the presumed connection between practices and people. Certainly, Planck and Sommerfeld associated particular practices with particular researchers. Planck noted that Sommerfeld's many works had tended in the direction of more "satisfying" dynamical approaches. Sommerfeld, in turn, invited to speak at Planck's birthday celebrations by Einstein, repeated the observation he had made in 1911, that "Planck's scientific personality was rooted in Thermodynamics," the epitome of a secure and safe science. And yet, while Sommerfeld and Planck may well have spoken of theories of principle, and people who used them, they did not speak of *physicists* of principle. That is, what was merely *characteristic* for Planck and Sommerfeld was *essential* for Einstein. This facet of Einstein's thought is illuminated most clearly in a speech delivered in celebration of Planck's sixtieth birthday in 1918, one that makes clear that Einstein's elucidation of the schism between various kinds of theoretical physicists in his letter to Ehrenfest was not merely a private joke. Einstein described what he termed the "temple of science":

In the temple of Science are many mansions, and various indeed are they that dwell therein and the motives that have led them thither. Many take to science out of a joyful sense of superior intellectual power: science is their own special sport to which they look for vivid experience and the satisfaction of ambition: many others are to be found in the temple who have offered the product of their brains on this altar for purely utilitarian purposes. Were an angel of the Lord to come and drive all the people belonging to these two categories out of the temple, it would be noticeably emptier, but there would still be some men, of both present and past times, left inside. Our Planck is one of them, and that is why we love him.[35]

This, indeed, was an even greater distinction than that between physicists of principles and those with virtuosic skills. In Einstein's eyes, it was only those men who were like Planck that really counted as "true" physicists at all. If the first two types were the only kinds of scientists, "the temple would never have existed, any more than one can have a wood consisting of nothing but creepers."[36] Such comparatively lowly, creeper-like men, however talented, could have been anything: "For these people any sphere of human activity will do, if it comes to a point; whether they become officers, tradesman or scientists depends on circumstances."[37] By implication, the elite the angel is willing to spare are those called to physics, and the reader will be unsurprised to learn that such men are physicists of principle, willing and able to carry out "the supreme task of the physicist": "to arrive at those universal elementary laws from which the cosmos can be built up by pure deduction."[38]

Einstein clearly saw much of himself in his description of Planck. They shared not only a dedication to developing theories of principle, but also an understanding of the epistemological status of such principles, one shaped by prior support for Machian phenomenology. Each had eventually been struck by the need to reconcile the relativism of Mach's historicism with their own beliefs in the ability to arrive at scientific truth.[39] Where Planck attempted to ground the validity of his founding principles on their ability to transcend their historical origins, Einstein's position was more overtly Machian, yet led to a similar conclusion. In his address he noted that there was no logical path to the acquisition of fundamental principles and thus that one might assume the relativity of systems of theoretical physics, each of which might explain phenomena equally well. However, he argued, clearly following Mach and the principle of biologico-economy, one could apply an *evolutionary* principle to this process and therefore could also assume that at any one moment there existed a system that was *best suited* to its environment. Einstein's next step, however, was assuredly not Machian, for he went on merely to assert the equality of the subjective and objective world. "Nobody who has really gone deeply into the matter," he stated, "will deny that in practice the world of phenomena uniquely determines the theoretical system, in spite of the fact that there is no logical bridge between phenomena and their theoretical principle; this is what Leibniz described so happily as a 'pre-established harmony.'" Einstein opined that epistemologists' failure to pay sufficient attention to this "harmony" was at the root of the "controversy carried on some years ago between Mach and Planck."[40] This rather emphatic vote in favor of Planck stands in strong contrast to Einstein's earlier description of the Planck-Mach dispute, written in a letter to Mach in 1909. Then he had spoken in more positive terms of the epistemologist's role in physics. "You have had such a strong influence upon the epistemological conceptions of the younger generation of physicists that even your opponents today, such as Planck, undoubtedly would have been called Mach followers by physicists of the kind that was typical a few decades ago."[41] Einstein's "gradual apostasy"—that is to say, his movement away from a Machian position (especially after 1918)—has been well documented in the literature.[42] It is worth noting, in addition, however, that a concomitant shift occurs in the notion of a principle between 1905 and the end of the World War. While the postulates of thermodynamics and relativity could be deemed "facts of experience" by Einstein in 1918, in 1907 he proved far more circumspect in his judgment. In line with a Machian position, he argued in response to a note published by Ehrenfest, that "the principle of relativity, or more, exactly, the principle of relativity together with the principle of the constancy of velocity of light, is not to be conceived as a "complete system," in fact, not a system at all, but merely as a *heuristic principle* which, when considered by itself, contains only statements about rigid bodies, clocks and light signals."[43] No absolute statement, nor anything guaranteed by a pre-established harmony, the principle of relativity here is simply a tool which provides new information not by or in itself, but "only by requiring relations between otherwise seemingly unrelated laws."[44] A plurality of principles and their meanings, one can see, proliferated not only among different physicists, but even within the lifetime of a single individual.

Theories of Principle

According to one of his biographers, one can date the birth of the Einstein legend from November 7, 1919, the day the London *Times* broke the news of the verification of the gravitational bending of light.[45] One might, with as much justification, note that the same day saw the birth of the Einstein industry. Within a year, the number of titles of books to do with relativity skyrocketed. In January of 1920, as Max Born reported to Sommerfeld, a lecture series on the subject had netted him 6,000 Marks, with which Born intended to refit his institute.[46] Sommerfeld's own lectures the next summer were repeated, attended each time by about 1,200 avid listeners,[47] and the controversy induced by attacks on Einstein's science and his character by conservative and often anti-Semitic opponents only fanned the flames of publicity. By 1923, in a lecture delivered in Madison, Wisconsin, Sommerfeld could talk of the "general, somewhat sensational and epidemical interest in the theory of relativity."[48]

Among Einstein's first contributions to the epidemic—other, of course, than his scientific works—came in an article commissioned by the *Times*, published on November 28, 1919.[49] Naturally, Einstein took the opportunity to lay out the implications of his theory in as simple a fashion as possible. What is striking about the manner in which he did so, however, is the fact that, after a brief vote of thanks to the English astronomers and physicists who had overcome the hatreds inspired by the war to test his theory, he began by laying out perhaps his single most cogent definition of what it meant to understand physics in terms of principles. The effect was to indelibly imprint the principled *form* of the theory, as well as its shocking results, on the minds of all the article's readers.

The theories of physics, Einstein explained, came in two different kinds. Most are "constructive" in that they "attempt to build up a picture of the more complex phenomena out of the materials of a relatively simple scheme from which they start out."[50] Examples include the kinetic theory of gases. One can only speak of one's "understanding" of a natural process when one has found a constructive theory that covers it. In contrast, there are also "principle theories": "These employ the analytic, not the synthetic method. The elements which form their basis and starting point are not hypothetically constructed but empirically discovered ones, general characteristics of natural processes, principles that give rise to mathematically formulated criteria which separate processes or the theoretical representations of them have to satisfy."[51] Relativity and the laws of thermodynamics fell into the category of such theories of principle and had the advantage of "logical perfection" and the security of their foundations. Constructive theories, on the other hand, offered a greater level of "completeness, adaptability and clearness."

Recent scholarship in the history of science has placed particular emphasis on the importance of Arthur Eddington's technical and popular activities for the reception of Einstein's general relativity theory in England. Eddington was the earliest and most prominent British general relativist, beginning an exploration of Einstein's theory during the war years, and it was Eddington, of course, together with Frank Watson Dyson who reported the success of the expedition to measure the gravitational bending of light, leading to the *Times*'s dramatic headline "Revolution in Science." Just as (if not more) important were Eddington's and Dyson's efforts to prime both their colleagues and the wider public for a favorable reception of the observational results.⁵² Eddington was successful in creating the impression that the only possible results to come from the eclipse expedition could be expressed through a trichotomy: either there would be no result, Newton's result, or Einstein's result. The effect was to build public excitement about the expedition, even before the announcement of its results, to implicitly reject alternative interpretations of results as due to earth or solar atmospheric effects, and to help cultivate the impression that any result larger than Newton's value was a point in favor of Einstein's theory. "Clearly," Alistair Sponsel has argued, "the experiment did not stand on its own: the edifice of the crucial experiment had to be constructed by Eddington and his associates...the JPEC's [Joint Permanent Eclipse Committee's] success must be explained in terms of the credibility that Eddington established and the manner in which audiences were primed by information that was disseminated to the *Times* and elsewhere."⁵³

Einstein's *special* theory of relativity was not met with the same enthusiasm in Britain when it was published in 1905. Largely ignored at first, it was then reinterpreted by Cambridge scholars in such a way that most of its results were understood as already following from an extant electron theory of matter. In particular, to the extent that the British were aware of Einstein's analysis, they saw his postulational approach—deriving results from the principles of the constancy of the speed of light in all reference frames and the principle of relativity—as both illegitimate and superfluous. Einstein's results, it was claimed, were merely "philosophically based" rederivations of relationships already in use as part of electrodynamic theory at Cambridge. "They believed, moreover," Warwick notes, "that the new principles were empirically unfounded and led to no new experimentally testable results."⁵⁴

The question for the case of Eddington, then, is why he—an otherwise typical product of the Cambridge pedagogical system—should have become the foremost promoter of the general theory in Britain. Warwick's answer deals in turn with four major points. First, with Eddington's training by a wrangler who was uncharacteristically interested and practiced in the methods of differential geometry that would prove crucial to the understanding and application of the general theory (and crucially obstructive to those attempting to learn the theory on their own). Second, Eddington, as an astronomer rather than a mathematical physicist, approached the theory with an understanding of its gravitational applications, rather than with an understanding of its apparent electrodynamic failings. Third, as a conscientious objector during World War I, Eddington had time to study the relevant material denied to those of his colleagues who were in military service. Fourth, and finally, Eddington learned the theory through correspondence with the Dutch master of relativity, Willem de Sitter, who not only could draw on the resources of his colleagues at Leiden in preparing a lucid account of the theory in English, but who also could aid Eddington when he encountered difficulties.⁵⁵

The comparison between the reception of the special and the general theories also helps to explain the particular presentational *form* of the latter. If British scholars had failed to notice or to accept the axiomatic structure of the special theory, it would be precisely this aspect that Eddington would recommend as the most significant attribute of the general theory. The novelty of the form of theory would not go unnoticed this time. In 1918, emphasizing the beauty of the theory's mathematical structure drew attention away from the relative paucity of its experimental confirmations.[56] Thus, in the first complete analysis of the general theory published in English, the *Report on the Relativity Theory of Gravitation*, Eddington wrote:

Whether the theory ultimately proves to be correct or not, it claims attention as one of the most beautiful examples of the power of general mathematical reasoning. The nearest parallel to it is found in the applications of the second law of thermodynamics, in which remarkable conclusions are deduced from a single principle without any inquiry into the mechanism of the phenomena; similarly, if the principle of equivalence is accepted, it is possible to stride over the difficulties due to ignorance of the nature of gravitation and arrive directly at physical results.[57]

In his 1920 text *Space, Time and Gravitation*, and hence after the experimental confirmation offered by the eclipse expedition, Eddington praised the axiomatic character of Einstein's theory as its greatest attribute. The final chapter of Eddington's text, titled (one assumes with a certain degree of tongue-in-cheek humor) "On the Nature of Things," begins with this passage:

The constructive results of the theory of relativity are based on two principles which have been enunciated—the restricted principle of relativity and the principle of equivalence. These may be summed up by the statement that uniform motion and fields of force are purely relative. In their more formal enunciations they are experimental generalizations, which can be admitted or denied; if admitted, all the observational results obtained by us can be deduced mathematically without any reference to the views of space, time, and force described in this book. In many respects this is the most attractive aspect of Einstein's work; it deduces a great number of remarkable phenomena solely from two general principles, aided by a mathematical calculus of great power; and it leaves aside as irrelevant all questions of mechanism.[58]

Space, Time and Gravitation was written after Einstein's article for the London *Times* was published, the article in which he opposed "constructive" theories to "principle theories." The reader may thus be a trifle confused by Eddington's usage of the term "constructive" in the first line of his enthusiastic appraisal of the relativity theory as a theory of principle. This peculiarity is, in fact, indicative of the importance ascribed by others—whether favorably inclined or not—to the principled character of Einstein's work. Pointing to the peculiar nature of the theory had its own drawbacks, for those who objected to the general theory now couched part of this objection in the language of construction vs. principle, even as they tried to invert the meaning ascribed to this dichotomy. In early February 1920, at a time when Eddington was in the process of delivering the lectures that would later make up *Space, Time and Gravitation*, the Royal Society hosted a "Discussion on the Theory of Relativity" between some of the most significant personages of British physical science. The meeting, again, had been held after the publication of Einstein's essay, and H. F. Newall drew attention to Einstein's division of types of theory as a means of subtly questioning the validity of general relativity. "Einstein," Newall noted, "has said that the merits of theories of principle is their logical perfection and the security of their foundations. We are here dealing with a theory of principle. We have to be very sure of the foundations before we accept it."[59]

Far more dangerous for an Einsteinian such as Eddington were the comments of James Jeans, who opened the discussion, giving an authoritative description of the theory's merits—and its failings.[60] Perhaps giving a measure of Einstein and Eddington's success in getting the theory acknowledged as different in kind to most others, Jeans began by comparing the principle of relativity to the principles of energy conservation and entropy increase. The three principles, Jeans observed, "have in common that they do not explain how or why events happen; they merely limit the types of events which can happen.... All three principles deal with events and not with the mechanism of events." Given that this came from a man whose work usually dealt with mechanisms, whether literally mechanical or electromagnetic, this may have struck listeners as slightly ominous, yet Jeans boldly and appreciatively continued by claiming that the theory's main interest is not that it offers a new principle, but rather that it "discloses a new universe." This positive judgment, however, was almost immediately undermined by his inversion (or, some might say, logical completion) of Einstein's dichotomy: "New and mysterious continents appear for science to explore, but it is not for the theory of Relativity to explore them. The methods of the theory are destructive rather than constructive, and, when the theory predicts a positive result, it is invariably for the same reason, namely; that a process of exhaustion shows that any other result would be impossible."[61] One can see why Eddington would want to emphasize the "constructive" results of Einstein's theory of principle, if the alternative was to allow it to be described as fundamentally "destructive."

In the end, both Eddington's judgment of Einstein's theory and his formulation of it would win out over most opponents. By the late 1920s, questions on general relativity had begun to appear in examinations for the Mathematical Tripos at Cambridge, and talk of the "fundamental principles of relativity" had replaced talk of systems "traveling through the Aether." Eddington would personally train a significant portion of the best theoretical physicists to come out of Cambridge in the postwar era, including Paul Dirac, whose *Principles of Quantum Mechanics* even today provides a model of the axiomatic, non-materialist method applied to the physical world. The preface to the first edition of that text, written in 1930, began by noting the "vast change" that methods in theoretical physics have undergone in the preceding three decades. Physics had previously been concerned with exploring the mechanisms and forces that connect observable objects. "It has become increasingly evident in recent times, however, that nature works on a different plan." Now, according to Dirac, one looked not for mechanisms, but for the invariants of transformations. "The growth of the use of transformation theory," Dirac wrote in his characteristically clear and dry style, "as applied to first relativity and later to the quantum theory, is the essence of the new method in theoretical physics. Further progress lies in the direction of making our equations invariant under wider and still wider transformations."[62]

The principled thinking of Einstein and Planck had been largely absent in Britain before Eddington's advocacy of Einstein's theory.[63] Though much use had, of course, been made of particular principles (least action, energy conservation, the second law of thermodynamics), they had largely been *used* in the context of material theories of one kind or another, either broadly supportive of mechanical perspectives or of the electron theory of matter. This kind of usage, for example, coupled with detailed mechanical modeling, was at the heart of the problems in which students were drilled in the second half of the nineteenth century. A theory founded on principles—particularly new ones—that were not themselves based on some understanding of mechanism was, although obviously not impossible, nonetheless largely alien. If Dirac overestimated the level of general acceptance of the "essence of the new method of theoretical physics," its novelty could not be contested.

Niels Bohr: The Principles of Irrationality

On the last day of April 1922, Bohr wrote to Sommerfeld thanking him profusely for the copy of the third edition of *Atombau*, which he had just received. Offering his best wishes and admiration for the book, Bohr also expressed his gratitude for the "friendly manner in which you have commented upon the works of my colleagues and me."[64] That friendliness seemed to stand in stark opposition to the ways in which Bohr felt his most recent efforts had been received. "In the last few years I have often felt very lonely scientifically, under the impression that my efforts, to develop systematically to the best of my abilities the principles of quantum theory, have been met with very little understanding. It is not a matter for me of didactic trifles, but of the genuine attempt to win such an inner coherence that one could hope to create a secure foundation for further development."[65]

Bohr's attempt to secure the field's foundations through principles dated back at least to 1916, the year of Sommerfeld's extension of the planetary atomic model to systems of more than one degree of freedom. At the time that he received word of Sommerfeld's successes, Bohr had been working on the page proofs of a paper that was to have been published in the *Philosophical Magazine* in April. "On the Application of the Quantum Theory to Periodic Systems," was, in part, an attempt to explain the stability of stationary states under the variation of external conditions. This assumed stability lay at the heart of what Bohr termed his "fundamental assumption," labeled "Assumption A" in his 1916 paper:

> According to assumption A, an atomic system can exist permanently in a certain number of stationary states. As this necessitates the absence of energy radiation in these states, assumption A implies that the ordinary laws of electrodynamics cannot be applied to the stationary states. This restriction, however, does not hold for the ordinary laws of mechanics, at any rate if these laws lead to periodic orbits; and in this section we shall see that it seems possible to obtain a consistent theory for the stationary states of periodic systems on the assumption that the dynamical equilibrium in these states is determined by ordinary mechanics.[66]

After 1913, Bohr had come to acknowledge that the application of a field to the atomic system would change the energy of the states. One could thus theoretically manipulate the conditions external to the atom in such a way as to cause the system to emit or absorb energy without making a transition between stationary states and hence directly contradict the principle laid out above. To exclude this possibility, Bohr appealed to a recent result put forward by Ehrenfest. This "principle of adiabatic invariance," put simply, meant that for periodic systems certain variables exist that remain constant during a suitably slow variation of external parameters.[67] For Bohr's model, it meant that the ratio of the average kinetic energy to the frequency of rotation (i.e., T/ω) will remain constant (and hence equal to $h\nu/2$) as long as any applied electromagnetic field is introduced slowly enough. Thus, instead of the system emitting or absorbing energy without a transition, the elliptical electron orbits will smoothly expand or contract in size until they reach a value that corresponds to the new energy level.

At the time of writing "On the Application of the Quantum Theory to Periodic Systems," as can be guessed from its title, Bohr had assumed that his analysis could only be applied to simply periodic motions. Sommerfeld's paper dispelled this mistaken belief, and Bohr withdrew his article in order to revise it. It was never, in fact, to be published. For almost two years, Bohr worked on a single project, the aim of which was to provide a uniform description for the major results that had been arrived at since 1913, and by April 1918 he had the first section of what was intended to be a four-part opus titled "On the Quantum Theory of Line Spectra." As with the unpublished work of 1916, Ehrenfest's principle (now renamed by Bohr the "principle of mechanical transformability" to remove the suggestion of a necessary thermodynamic or statistical connection) played a central role. For Bohr, the principle provided not only stability, but also the requisite connection between the classical (i.e. continuous) and quantum (i.e. discrete) aspects of his model of atomic structure. Indeed, Bohr wrote,

> the principle allows us to overcome a fundamental difficulty. In fact we have assumed that the direct transition between two such states cannot be described by ordinary mechanics, while on the other hand we possess no means of defining an energy difference between two states if there

exists no possibility for a continuous mechanical connection between them. It is clear, however, that such a connection is afforded by Ehrenfest's principle which allows us to transform mechanically the stationary states of a given system into those of another....[68]

There was one final aspect of Ehrenfest's analysis that Bohr was to use. Since the second round of Planck's papers in 1906, Ehrenfest had continued to work on the extension of statistical mechanics to the quantum realm. In 1911 he had shown that Planck's use of the Boltzmann relation ($S = k \log W$) was valid precisely because Planck had quantized the ratio of an oscillator's energy to its frequency (E/ν); in other words, the system's adiabatic invariant. In addition, Ehrenfest proved more generally, and more crucially, that it was only if the statistical weighting of each division in phase space depended solely on this ratio that the validity of the oscillator's statistical thermodynamics could be guaranteed.[69] In spite of its relative obscurity at the time, Bohr had seized on this result in 1916 and had made it central to his discussion in the paper on the quantum theory of line spectra. There it was employed as a means of getting at values for transition probabilities between stationary states, the argument being that "if the *a priori* probability of the stationary states of a given system is known, it is possible at once to deduce the probabilities of the stationary states of any other system to which the first system can be transformed continuously without passing through a system of degeneration."[70] In other words, once the probabilities for the stationary states were known for one system, they were, in effect, known for all systems into which the first could be adiabatically transformed.[71]

The principle of adiabatic invariance thus resolved two outstanding problems of the quantum theory: first, that of determining the correct method of dividing phase space for a quantum system, and second—and perhaps more heuristically useful—providing a justification for the application of specific arguments from mechanics to what were, by definition, non-mechanical systems. The correspondence principle, so named in 1920, offered, similarly, a means of applying arguments from classical electrodynamics to systems defined by their failure to accord with Maxwell's equations. Bohr's first fundamental postulate, as noted above, defined non-mechanical stationary states for an atom. His second defined the energy given off or absorbed in a transition between such states in terms of a frequency, ν, by $E = h\nu$. Unlike a classically orbiting electron, however, this "optical" frequency bore no relation to the "mechanical" frequency of rotation. While one could determine the frequency, the polarity, and the intensity of the emitted or absorbed radiation from the equation of motion for the electron in the classical case, for the quantum, Bohr's second postulate meant that one could determine the frequency alone. The "correspondence" of the correspondence principle was posited by Bohr as that between individual quantum transitions and components of the classical frequency spectrum. Each one of the infinitely many possible quantum leaps was related to one of the elements in the infinite Fourier series of the classical motion, and the intensity and polarization calculated from each of these elements then "corresponded" to that for the quantum case. In the limit of infinitely slow rotations (infinitely high quantum numbers) this correspondence became equality. In the next chapter we will discuss this principle—and Sommerfeld's reaction and alternative to it—in detail.

Bohr's route toward and understanding of both these principles was strongly shaped by his readings of the Danish philosopher Harald Høffding.[72] Bohr attended Høffding's lectures at the University of Copenhagen as a young man, and benefited from Høffding's close friendship with Niels's father, Christian Bohr, professor of physiology at the same institution. According to Høffding, all the problems of philosophy—of psychology, ethics, epistemology, and ontology—could be reduced to a single problem, that of the relation between continuity and discontinuity:

> This relationship involves the deepest interests of personality as well as of science. In both realms there is, as already noted, a striving after unity and connectedness, and, in so far, the discontinuous appears as an insurmountable obstacle. On the other hand, it is discontinuity (distinction of time, of degree, of place, difference of quality, of individuality) which more than anything else brings new content, releases locked powers, and opens up the greatest tasks in the realm of life no less than in the realm of science. Thus it would appear that neither of the two elements is the only accredited one.[73]

The principle of mechanical transformability can be seen as an instantiation of Høffding's idea that rational understanding could be achieved only by continuously connecting the discontinuous sense data of experience. Of the problem of consciousness, Høffding wrote, "the proof of a continuity in the processes under investigation is necessary to a true understanding." He continued: "We should reach the ideal of psychology not only if we could secure such a complete description of states of consciousness that each state would stand as a proper member of the whole psychical process, but also if we could reduce the differences of the changing states to such simple forms that any succeeding state would appear as the continuation or as the transformation of the preceding state."[74] Such a complete and ideal understanding is, in principle, impossible, since experience imposes discontinuities (in this case in the form of unconscious intervals between conscious states—e.g., swoons, dreamless sleep, etc.). Continuous connection nonetheless remains the aim of the analyst. The principle of mechanical transformability, which allowed of a continuous transformation between discontinuous stationary states, provided an exactly analogous relation for atomic physics.[75]

The correspondence principle, similarly, can be read as Bohr's attempt to follow Høffding's program of describing irreducibly individual and discontinuous phenomena (quantum jumps, in Bohr's case) in terms of general laws. Again, it must not be supposed that such laws—representing continuity and consistency—could "cover" completely the "confused multiplicity of phenomena."[76] Continuity and discontinuity, rationality and irrationality, community and individuality, and whole and part are antinomies that cannot be reduced to one another. A general law can never fully represent individual phenomena. Nonetheless, we require such general laws and principles for our own understanding. With this in mind, one can begin to fathom Bohr's repeated insistence that the correspondence principle was "a law of the quantum theory" alone, and not a principle that could be understood or judged in classical or familiar terms.[77] The principle made sense only in relation to the discrete phenomena of the quantum theory; it was the continuous counterpart to the discontinuities of quantum jumps.

The connection between certain of Høffding's ideas and the *content* of the principles Bohr would employ in his efforts to provide a consistent foundation for what came to be known as the "older quantum theory" seems clear. Yet there is also a more general perspective from which one can study the relationship between Bohr and Høffding. And it is from this perspective that we can come to an understanding of one of the most striking aspects of Bohr's theoretical methodology and philosophy: his repeated emphasis on the *flexibility* of principles. As we shall see in the next chapter, one of Bohr's critiques of Sommerfeld's methods was that it displayed a "tendency of considering the quantum theory as a set of formal rules," rather than the fluid adaptability and ambiguity that Bohr himself favored. This characteristic has been emphasized most recently by Darrigol, who has noted that both Bohr and his followers repeatedly made a virtue of the near-artistry required in applying the correspondence principle successfully. Ehrenfest, for example, speaking at the Solvay conference in 1921, spoke of the correspondence principle approvingly as "variable and groping" and claimed that it was "not desirable" to cast it in a rigid form.[78] Heisenberg made a similar point many years later, during an interview with Thomas Kuhn. "I always liked Bohr's Correspondence Principle," he said, "just because it gave that kind of lack of rigidity, that flexibility in the picture, which could lead to real mathematical schemes."[79] This appears to be a very peculiar idea of a principle. Even if one understands principles in Einstein's original sense, as heuristics, rather than fixed laws or Planckian commandments, there is little scope for seeing them as malleable in the way suggested above. Mach, of course, saw principles as ever-changing, becoming slowly adapted to an external world in an evolutionary sense. Yet this is a *passive* adaptation, and seems insufficient as a basis for the active flexibility of application that use of the correspondence principle commanded. William James placed Høffding in the same camp as Mach, Avenarius, and others, suggesting that he "aligns himself with the 'economical' school of scientific logicians." But Høffding went a step farther than Mach, and it is by tracing that step that we can come to the basis of the kind of principled thinking Bohr espoused.

In discussing the "problem of knowledge" and specifically the question of the relationship between the world and our understanding of the world, Høffding listed four principle approaches: approaches that considered the principles of knowledge as intuitions (Plato), postulates (Hobbes, Fichte, Kierkegaard), generalizations (Mill, Spencer), and economic tools of thought (Mach, Avenarius). Høffding's preference was for the last, for he had a great deal of sympathy both for the parsimony of the economic approach (which posited no more principles than could be proved, and hence sought to avoid metaphysics or dogmatism) and for that approach's emphasis on the importance of seeing principles in terms of their utility. The economic theory, Høffding argued in a historicist vein, arises in part "from experiences in the history of science which show how principles and hypotheses may for a certain period be valid and fruitful, but later must be displaced by others." Giving a specific example (one of which Mach would have approved), he continued, "the discussions carried on of late as to the validity of the mechanical conception of nature have directed attention to this second [historicist] class of motives."[80] A few pages later, Høffding was even more explicitly instrumentalist and utilitarian: "The significance of principles is that they may lead us to reach a rational understanding in our work. Their truth consists in their *valid application*; and this consists in their *working value*. That a principle is true, signifies that one can work with it."[81]

If one considers the problem of knowledge in relation to the other problems of philosophy, however, the economic theory, according to Høffding, runs into certain difficulties. The approach of Mach and Avenarius fails to take into account the nature of a principle as a *psychological* construct. "When strict induction from previous experiences leads to contradictory results," Høffding notes, "we prefer to assume that the experiences are incomplete rather than that Being contradicts itself."[82] Yet if we were guided merely by notions of parsimony and fitness, this dedication to our principles in the face of refutatory evidence would make no sense. Our behavior in this context can be understood by seeing principles as kinds of *intellectual habits*. "Every comprehensive principle—psychologically considered—is essentially the expression of such a habit, which may be more or less deeply imbedded in the nature of consciousness, i.e. which sometimes has the nature of an instinct, but sometimes seems more like the influence of custom."[83] In this way, "What appears as a hypothesis from the purely empirical view, becomes, empirically considered, a principle, a regulative thought, under whose leadership consciousness may satisfy in the empirical world its demand for continuity and union."[84]

Also considering the problem from a psychological perspective, it becomes clear that principles cannot be completely arbitrary constructs. If, according to the economic theory, a man can create "whatever concepts and principles he may need in order to make himself master of phenomena," this freedom is not accorded the subject of Høffding's analysis.[85] On the contrary, principles must be "psychologically possible," that is, "in accordance with the general laws of conscious life." And so we arrive at the doubled adaptability of principles in Høffding's philosophy: "The idea of a working hypothesis points in two directions: on the one hand…back to the nature of the thinking consciousness, since our consciousness can perform no function, however economical, which is entirely foreign to its own nature; on the other, to the reality to which the phenomena to be understood belong. *A tool must be adapted both to the hand that is to use it and to the object to be worked on.*"[86]

The agency of the worker with principles is thus not restricted to their arbitrary creation or modification, nor is the gradual alteration of a given principle merely due to an evolution toward the truth of Being. In Høffding's vision, both the subjective consciousness and the objective world exert adaptive pressures on principles. Truth is dynamic rather than static not merely because it changes, but because it is changed. That this adaptation is never-ending comes from a final aspect of principles, already alluded to above: principles cannot perfectly cover experience. That is, unlike the situation posited by Machian phenomenology, where one might, by chance, produce the True formulation of a principle of physics, but be unable to either positively identify or prove it, here the very possibility is denied. A rigid principle is, by definition, one that can no longer represent the truth of a world in which continuity and discontinuity, rationality and irrationality stand in dynamic and constant opposition:

> Even after fruitful principles or working hypotheses have been attained, will Being be completely rendered by them? Or will there always remain an irrational relation between the principles which may compose our consciousness and the Being itself from which our experiences are derived? We shall find that under three different forms there is always an irrational remainder....[87]

Flexibility was thus married to utility. Principles must mediate between a world of being and a world of consciousness, evolving toward and adapting to each. Yet neither world can be explained completely by means of a rational principle, and neither can remain static. Bohr's and Høffding's universe was dramatically different from Planck's or Einstein's. In the latter, a deist rationality guaranteed the equivalence of the worlds of thought and of being. In the former, irrationality was a real and irreducible component. Planck and Einstein could believe in the Absolute, Bohr and Høffding only in the absolutely Dynamic. It thus seems unsurprising that the principles that described and explained such universes were different in kind. For Planck, principles had the lasting validity of a holy commandment. For Einstein, who spoke in his early career of principles as heuristics, they became by the war's end fixed truths of Being. For Bohr, as his espoused practice with regard to the correspondence principle attests, principles were both approximate and necessarily changeable. Alteration had to be part of their essence, part of the means of capturing a paradoxical world in action.

7 Crafting the Quantum: Sommerfeld, Bohr, and the Older Quantum Theory

Ich kann nur die Technik der Quanten fördern,
Sie müssen ihre Philosophie machen
—Sommerfeld to Einstein, 1922[1]

Kepler should have experienced today's quantum theory. He would have seen the boldest dreams of his youth realized, not, admittedly, in the macrocosm of the stars, but in the microcosm of the atom. The shell structure of the atom is even more wonderful than the cosmography longed for by Kepler. —Sommerfeld, 1925[2]

Technik and *Atom-Mystik*

A significant portion of chapter 6 sought to elucidate Niels Bohr's efforts in the early 1920s to "develop systematically the principles of quantum theory." This chapter is concerned with Sommerfeld's equivalent statement of purpose, offered, like Bohr's, in the first months of 1922. A letter to Einstein written in January reported the many successes of Sommerfeld's recent work and that of his students, including a young man in his third semester, Werner Heisenberg, who had just completed pioneering work in providing a model for the anomalous Zeeman effect. In spite of this, the situation was not ideal: "Everything works, but remains at the deepest level unclear." That, however, was not Sommerfeld's problem. Laying out the division of labor for the new physics, he wrote: "I can only advance the craft of the quantum, you have to make its philosophy." What it meant to craft the quantum, and more generally, what the physics of problems became in the Weimar period, is one of the questions considered below.

Before proceeding to an answer, however, a brief discussion of terminology is in order. In contemporary German, *Technik* is commonly accorded one of three different English equivalents: "technique," "technology," or "engineering." The phrase above has uniformly been translated as expressing Sommerfeld's preference for "the techniques of the quantum."[3] Yet "technique" in English has a connotation not necessarily intended in German, for it describes only "the mechanical or formal part of an art"; that part "distinct from general effect, expression, sentiment, etc."[4] A late-nineteenth-century text on music put it this way: "A player may be perfect in technique, and yet have neither soul nor intelligence."[5] Given Sommerfeld's defense of *Technik* against those who charged it with possessing "a smaller degree of scientific rigor" than their own disciplines and the importance of technical applications in his own work, one would hardly expect him to share this sentiment.[6] More generally, however, the formalistic implications of the word in English are less prevalent in the German. Grimm's nineteenth-century *Wörterbuch*, for example, offers the following definition: "the artistic or craft activity and the sum of experiences, rules, principles, and know-how according to which, through practice, an art or craft is pursued." Central to this understanding, as Norton Wise has argued in analyzing Helmholtz's use of the term, was a sense of the importance of an aesthetic sensibility for the *Techniker*.

Aesthetics was essential for Sommerfeld's work on the quantum theory of spectral lines. The same man who would speak of an anti-philosophical, "nuts and bolts" approach to the quantum would also, in the same period, wax lyrical about the harmonious "number mysteries" that a study of spectral lines allowed one to glimpse.[7] The preface to the first edition of *Atombau* would, one suspects, make any hard-headed technician blush:

What we are nowadays hearing of the language of spectra is a true "music of the spheres" within the atom, chords of integral relationships, an order and harmony that becomes ever more perfect in spite of the manifold variety. The theory of spectral lines will bear the name of Bohr for all time. But yet another name will be permanently associated with it, that of Planck. All integral laws of spectral lines and of atomic theory spring originally from the quantum theory. It is the mysterious organon on which nature plays her music of the spectra, and according to the rhythm of which she regulates the structure of the atoms and nuclei.[8]

On the other hand, Sommerfeld reacted with palpable fury when the *Süddeutsche Monatshefte* contacted him—one suspects after reading utterances like the one above—to write an article on astrology. "Doesn't it strike one as a monstrous anachronism," he raged, "that in the twentieth century a respected periodical sees itself compelled to solicit a discussion about astrology? That wide circles of the educated or half-educated public are attracted more by astrology than by astronomy? That in Munich probably more people get their living from astrology than are active in astronomy?" In spite of having "no illusions" about his ability to hold back the growing tide of irrationalism that threatened to wash away the remnants of a reasoning European culture, Sommerfeld pledged to "throw myself decisively against it."[9]

Yet it was not only the editors of newspapers and the "half-educated public" that perceived Sommerfeld as espousing an irrational or at least arational approach to the physical world. His colleague Wilhelm Wien, a professor of experimental physics at Munich, snidely referred to Sommerfeld's work not as *Atomistik*, but *Atom-Mystik*, a phraseology that Wien's students apparently adopted.[10] This private (or at least intra-faculty) jibe became public when Wien delivered a rectoral address, at Munich in 1926, on the "Past, Present and Future of Physics." Among the lecture's main topics was the modern atomic theory, and the developments that followed the introduction of the Bohr model. It did not mention Sommerfeld—almost certainly in the audience at the time—by name, a reproach (even an insult) in itself.[11] Even more telling, Wien closed with a discussion of the hope offered by Schrödinger's new wave theory of removing talk of mysticism from quantum-theoretical research:

Now Schrödinger is trying to ascribe the whole numbers of the quantum theory to similar characteristic vibrations [*Eigenschwingungen*], the actual physical meaning of which admittedly still remains dark. If that were really to succeed, then the special role which whole numbers play in the quantum theory would also here be eliminated, also here number mysticism would be supplanted by the cool logic of physical thought; probably not to the joy of everyone. Because mysticism exercises a greater attraction on many minds than the cold and clinical mode of thought of physical contemplation. I am far from wanting to attack mysticism as such. There are many areas of spiritual life from which mysticism cannot be shut out, but it does not belong in physics. A physics in which mysticism rules or only participates leaves the ground from which it draws its strength and ceases to deserve the name.[12]

If Wien took the talk of mysticism in Sommerfeld's work at face value, and perceived in it a disturbing and widespread tendency toward the denial of the "cool logic of physical thought," Max Born took a more cynical (if thereby more supportive) position. On the occasion of Sommerfeld's sixtieth birthday, Born remarked:

He occasionally speaks with gentle coquetry of number mysticism in spectral laws; but he means by that nothing philosophically dark, but only the statement that so far one has still not come directly behind these laws. The word should be once again a lure for spurring on young brains to restless research; because he who is in Sommerfeld's school, for him is mysticism only there to be vanquished.[13]

How are we to judge the meaning of Sommerfeld's talk of *Zahlenmystik*: as a fine aesthetic sense of harmony in physics, as mere pandering to forces of irrationalism, or as a wily way to win students to one's school? The answer, I suggest here, is by understanding it as a constitutive element of the *Technik* of the quantum. To see this requires a close examination of the practices and methodologies of Sommerfeld's work in the 1920s. It was on these practices and methodologies that Pauli, a former member of the Sommerfeld School, commented in 1945 on the occasion of receiving a Nobel Prize for the discovery of the exclusion principle:

At that time there were two approaches to the difficult problems connected with the quantum of action. One was an effort to bring abstract order to the new ideas by looking for a key to translate classical mechanics and electrodynamics into quantum language which would form a logical generalization of these. This was the direction which was taken by Bohr's Correspondence

Principle. Sommerfeld, however, preferred, in view of the difficulties which blocked the use of the concepts of kinematical models, a direct interpretation, as independent of models as possible, of the laws of spectra in terms of integral numbers, following, as Kepler once did in his investigation of the planetary system, an inner feeling for harmony. Both methods, which did not appear to me irreconcilable, influenced me.[14]

In this account, Sommerfeld's emphasis on Keplerian harmony is neither window dressing nor a flight from science and reason. It is, rather, a direct response to the problems of the quantum theory. Bohr's use of the correspondence principle, in contrast, since it relied on the translation of ideas between the classical and quantum realms, becomes an indirect method of dealing with similar difficulties.

Pauli's words, of course were written two decades after the events he was describing, but there are good reasons for seeing in them a clue to Sommerfeld's own understandings of the methods he would employ. Consider, for example, the only mention of number mysteries in the text of the first article in which he would use that term.[15] There the turn to a study of numerical regularities is portrayed as precisely a result of the failure of model-based analysis;

The musical beauty of our number table is thereby not derogated at all by the fact that it for the time being represents a number mystery. In fact, I see at the moment no way toward a model-based [*modellmässig*] explanation, neither of the doublet-triplet phenomena, nor of their magnetic interaction. In the same sense all spectroscopic laws were, until a few years ago a number mystery.[16]

Why Sommerfeld should have felt, around 1919, inspired to abandon a *modellmässig* approach, and how his contemporaries reacted to such a move, are the subjects of this chapter's second section. The third section explores and seeks to explicate his reaction to Bohr's use of the correspondence principle, while the final section examines the variegated problems taken up by members of the Munich school from the end of the war to the mid 1920s. There, as we shall see, *Technik* would take on a double meaning, as students aided Sommerfeld in his quantum craftsmanship and continued to explore the technological problems that had been so central to the school before 1918.[17]

Unsettled Questions of Atomic Physics

In the early weeks of 1920, Sommerfeld began receiving letters congratulating him on the publication of the first edition of *Atombau*. Theoreticians and experimentalists alike lauded the achievements of the text. David Hilbert spoke of reading the "masterly" volume "with daily increasing joy." Hans Beggerow cited the book's "clear and simple" language, which proved its character as a "true classic." Pieter Zeeman claimed that it read like a "thrilling novel."[18] Yet not all of Sommerfeld's correspondents were entirely positive. Although Max Born insisted that little should be changed, and that the book was "marvelous as it is," he also offered some rather pointed criticisms. Sommerfeld, he suggested, had indulged in an excess of "local patriotism," overemphasizing the achievements of those connected to his own school at the expense of those—including Bohr—whose methods differed from those regularly deployed in Munich. In addition, Born criticized the representation of the state of the field that the book appeared to put forward: "You represent some matters in such a way that the lay reader must believe that everything is in order; but that is certainly often not the case. E.g. the molecular model of H_2, etc., as well as the whole theory of Röntgen-spectra." Alfred Landé, Born reported, had recently informed him that the precise opposite was the case: that the theory of x-ray spectra was in disarray. "Would it not be good to emphasize the doubts a little more?"[19]

Sommerfeld took the criticisms to heart. Later editions contained much more detailed and sympathetic portrayals of Bohr's successful utilization of his "analogy principle." The second edition, completed in September 1920, also made more of an effort to bring to light areas of the theory that remained either incomplete or positively confusing. In advance of the publication of the text, Sommerfeld penned a two-page essay for the *Physikalische Zeitschrift* on "Unsettled Questions of Atomic Physics," noting that such questions outnumbered those that had, thus far, been fully and satisfactorily answered.[20] The two areas that Sommerfeld emphasized were both ones in which he had toiled since his first papers on the Bohr-Sommerfeld quantization conditions: the relationship between hydrogenic and non-hydrogenic spectra, on the one hand, and the question of the "topology of the atomic core"—explored through data on Röntgen spectra—on the other.[21] His failure, at least in his eyes, to produce a model-based understanding of these two subjects, one that could provide a dynamical foundation for the empirically based regularities that Sommerfeld *could* derive, would lead by 1922 to his abandonment of the *modellmässig* approach for all aspects of the atomic theory, excepting only, perhaps, the model of the hydrogen atom with which he had begun.

Fine Structure, the Zeeman Effect, and the Birth of a Number Mystery
Sommerfeld began his discussion by laying out the basic theory used to understand atomic spectra. Referring the reader to the diagram reproduced here as figure 7.1, he reminded them of the fact that emission series of spectral lines were produced when an electron moved from a higher to a lower energy level, giving off electromagnetic radiation of a frequency determined by the expression $\Delta E = h\nu$. Each line was thus, according to the "principle of combination," made up of two terms, corresponding to the energy of the initial and final states. The differences between energy levels may be organized into various series, denoted by the letters *s, p, d, b* (sharp, principal, diffuse, Bergmann). Each energy level was described by a quantum number, *m*, made up of the sum of the fixed "azimuthal" quantum number *n* and the variable "radial"

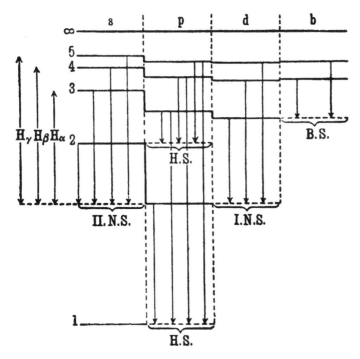

Figure 7.1
Energy-level diagram for sodium. Source: Sommerfeld, "Schwebende Fragen der Atomphysik." *Physikalische Zeitschrift* 21 (1920): 619–20.

quantum number n' (i.e., $m = n + n'$). In Sommerfeld's formulation, $n = 1, 2, 3, 4$ for the *s, p, d, b* series, respectively.[22]

Not all transitions from a higher to a lower energy level are allowed. A "selection principle" offered, in Sommerfeld's words, "an essential supplement to the principle of combination, through which its boundlessness is contained and its practical value raised."[23] The only jumps possible were those for which $\Delta n = \pm 1$ *or* 0. In other words, an electron could move from its current level to a lower, neighboring level (or a level beneath it in the same series), but to no others. The origin of the differing heights of energy levels with the same value of m was explained by the action of electrons closer to the nucleus than the transitioning electron. Their negative charge "shielded" part of the nucleus's positive charge, resulting in a diminution in the attraction experienced by the outer electron from the normal Coulombic force; a diminution that depended for its exact value on the spatial organization and motions of the inner electrons. Since a hydrogen atom only possesses a single electron, one would assume that no such alteration to the Coulomb force could occur, and, in fact, corresponding energy levels for all four series were the same, i.e., $ms = mp = md = mb$. Only when one took relativistic effects into account did a step-like structure for hydrogen's energy levels emerge, and with it came the element's characteristic spectroscopic fine structure. Where, in the absence of a relativistic mass correction, all transitions from the third energy level to the second overlapped, now three distinct transitions were possible, producing a triplet: $3d \to 2p$, $3s \to 2p$, $3p \to 2s$. As Sommerfeld phrased it, hydrogen's fine structure corresponded to "non-hydrogenic gross-structure."[24] The link was a neat one, but it left an obvious question unanswered, for no analogue in the hydrogen spectrum existed that corresponded with the observed fine structure of non-hydrogenic elements. The p and d levels in the diagram had to be imagined as yet further divided, and "for the model-based meaning of these subordinate levels no satisfactory explanation has yet been given."[25] Sommerfeld held out hope that some of Bohr's recent work might aid the situation, but his closing remarks on the subject served merely to indicate the field's deep lack of certain knowledge. "That we have here to do with completely new matters," he wrote, "is shown, in particular, by the anomalous Zeeman effect, which is bound up with the existence of such more subtle divisions of energy levels."[26]

If model-based understanding of non-hydrogenic fine structure was, indeed, inextricably tied to a concomitant understanding of the splitting of lines in a magnetic field, then Sommerfeld was well aware of the problems ahead. In 1916, flushed with the success of his extensions of Bohr's model, he had turned his hand to the question that had first occupied his mind after reading of Bohr's results in 1913: the Zeeman effect. The result was strikingly disappointing. In common with Debye, who had taken up the problem independently, Sommerfeld showed that a quantum calculation of the size of line splitting led to a result that "does not coincide with the result of the classical theory, but is very close." The splitting induced by the magnetic field in Sommerfeld's treatment was equal to a whole number times the classical value obtained through Lorentz's electron theory.[27] The relativistic Zeeman Effect, to Sommerfeld's considerable surprise, was no different from the non-relativistic, meaning that the existence of a relativistic fine structure seemed to have no effect on the results.[28] The calculation could say nothing about the line multiplicities of the anomalous Zeeman Effect.

To deal with that problem, Sommerfeld moved toward a *gesetzmässig* approach. In the middle of August 1919, he wrote to Carl Runge, reporting "half-empirical" results that relied on "Runge's rule" for the anomalous Zeeman effect. In 1907, Runge had determined that the separations between "anomalous" lines were integral fractions of the separations predicted by the original Lorentz theory, i.e., $\Delta v = a \cdot s/r$, where a is the Lorentz separation and both s and r are integers. Sommerfeld's starting point was the suggestion that the size of the line splitting be governed by the combination principle. Then $\Delta v = \Delta v_1 - \Delta v_2$, where the terms on the right-hand side correspond to the splittings of the first and second terms respectively. If $\Delta v_1/a = s_1/r_1$ and $\Delta v_2/a = s_2/r_2$, then

$$\frac{\Delta v}{a} = \frac{s_1 r_2 - s_2 r_1}{r_1 r_2} = \frac{s}{r}.$$

Sommerfeld termed the result his "magneto-optic splitting rule," arguing that it demonstrated that the "Runge number" (r) of each term combination was made up of two components: the Runge numbers of the first and second terms.[29]

The first public discussion of this new spectroscopic rule came the following month, when Sommerfeld lectured in the southern Swedish town of Lund.[30] A version of that paper was published early the next year in *Die Naturwissenschaften* under the title "A Number Mystery in the Theory of the Zeeman Effect."[31] The "mystery" was less the rule itself than the implications Sommerfeld drew from it. The article reproduced a table that Sommerfeld had included in his letter to Runge, laying out the empirically determined Runge components. (See figure 7.2.) All single lines displayed values corresponding to the normal Zeeman effect ($r_1 = 1$) and all s terms in doublets and triplets were, Sommerfeld claimed, "single without exception, according to general spectroscopic experience."[32] These observations provided the data for the first row and column of the table. Runge numbers for lines corresponding to a combination of an s term and a p term, he noted next, were $2 = 2 \times 1$ for triplets, and $3 = 3 \times 1$ for doublets, numbers that accordingly filled out the p column of the table. According to data supplied recently by Paschen, it was further reported, Runge numbers for lines related to the combination of a p and a d term were $6 = 2 \times 3$ for a doublet, and $15 = 3 \times 5$ for a triplet. "With these numbers," he wrote, "we fill in the third column of our table and see a wonderful number harmony complete itself thereby." This number harmony could then, itself, provide the basis for predictions of future, experimentally determined, Runge numbers. "No one will doubt," Sommerfeld claimed, with perhaps excessive optimism, "that we must complete the fourth column with the numbers 4 and 7," corresponding to a pattern of increasing ordinal numbers in the second row and increasing odd numbers in the third.[33] The measurable Runge number for doublets made up from b and d terms should then be 12, for triplets, 35. Of the reason for these numbers, and the origin of the doublets and triplets, Sommerfeld could say nothing.

	s	p	d	b
Einfachlinien	1	1	1	1
Triplettlinien	1	2	3	(4)
Dublettlinien	1	3	5	(7)

Figure 7.2
Sommerfeld's number table for Runge components. Source: Sommerfeld, "Ein Zahlenmysterium in der Theorie des Zeeman-Effektes," *Die Naturwissenschaften* 8 (1920): 61–4.

"The actual cause for the doublets and triplets," he confessed to Runge, "and therefore also the cause of the anomalous Zeeman effect is still unclear to me. Only this much is certain, that in all whole-numbered relationships, quanta are involved [*stecken*]."[34] The paper on number mysteries promised that a "more fundamental" discussion of his results would follow in the *Annalen*, yet that article also eschewed an extensive model-based analysis. "General Spectroscopic Laws, in Particular a Magneto-optic Splitting Rule," published in 1920, began with a description of the hopes of the teens—all too soon to be dashed—that the hydrogen spectra would provide the key to lines in the spectra of all elements. After hydrogenic spectra had been explained, Sommerfeld noted somewhat wistfully, it had seemed only a short step until one could treat non-hydrogenic spectra along the same lines, "for example, tracing the doublet of the *D* lines to path differences in the atomic model." But nature had not proved so accommodating, and for the explanation of the line structure of series other than the Balmer series hydrogen, Sommerfeld wrote, "provides no clue."[35] He continued:

> Thus it is, that at the moment we are at a loss with the *modellmässigen* meaning of the line multiplicities of the non-hydrogenic elements, in spite of repeated efforts from various sides. All the more valuable are all the lawful regularities [*Gesetzmässigkeiten*] that present themselves empirically for the line multiplicities, above all when they are of such a radical and simple kind as those here at hand.[36]

The intention of the article in general was to further explicate several such known laws, and to introduce several new ones. In particular, the magneto-optic splitting rule "unveils a harmony of whole-numbered relationships of a purity that will even surprise those accustomed to the modern quantum theory."[37] There was more than harmony at stake here, however, for the new empirical laws were of both a theoretical and a practical importance as well, corresponding, Sommerfeld seemed to suggest, to the needs both of experimental and theoretical physicists. "The practical significance of multiple lines for the classification of spectroscopic series is well known," he wrote. "The theoretical significance of the line structures for the model-based investigation of the atom will doubtless be no less. We may promise for our magneto-optic splitting rule, apart from a theoretical also a practical importance for the meaning and ordering of the complex Zeeman effect."[38] It could serve, he opined later in the paper, as a "pointer [*Fingerzeig*] toward the correct interpretation of the observations."[39]

Sommerfeld's discussion served to locate the search for model-based explanations—as opposed to those based on empirical regularities—at some point in the future. It was not that models were to be removed from physics altogether, but that, since they had proved less useful than the direct search for regularities in empirical data, their use should be postponed. Spectral laws were concrete, atomic models conjectural. The five sections of the paper then laid out five distinct rules, none of which relied on the Bohr model. The first, Sommerfeld designated as "the law of permanence of multiplicities." It had long been known that elements in the same column of the periodic table displayed similar patterns of line splitting. Sommerfeld wanted to add to this rule the "until now little noticed fact" that the patterns of *p*-term and *d*-term splitting were identical for any given element: If the *p* term was a doublet, for example, then so, too, was the *d* term. The second rule, an extension of the already-enunciated "selection principle," sought to provide a rationale to explain why not all transitions between *d* lines and *p* lines were observed. Transitions denoted by broken lines in figure 7.3 were allowed by the selection principle (since here $\Delta n = 1$), but did not occur in reality. Sommerfeld's suggestion was that another quantum number was involved, one that obeyed its own selection rule. If the azimuthal quantum number was assigned to the angular momentum of the entire atom—its "external rotation"—then "the distinguishing characteristic of the various *d* and *p* terms must be, rather, an inner quantum number, perhaps corresponding to a hidden rotation. Of its geometric significance we are quite as ignorant as we are of those differences in the orbits which underlie the multiplicity of the series terms."[40] Assigning, for the inner quantum number, the values written to the left of each line—beginning with the highest values of *n* for that line—the selection principle for the outer quantum number holds here, too. The number can change by ± 1 or 0, but by no more. Re-emphasizing the fact that no geometrical or model-based representation of the inner quantum number was being offered with this rule, Sommerfeld emphasized again its "rather formal"

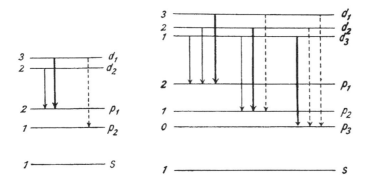

Figure 7.3
Transition diagrams for doublets (left) and triplets (right). Broken lines represent transitions possible in terms of the selection rule for *n*, since here $\Delta n = 1$. Such transitions, however, are not observed; hence Sommerfeld's suggestions that an additional selection rule, corresponding to an "inner quantum number," was in effect. Source: Sommerfeld, "Allgemeine spektroskopische Gesetze, insbesondere ein magnetooptischer Zerlegungssatz," *Annalen der Physik* 63 (1920): 221–63

character:

> I would not like to refrain from pointing out that, in transferring our selection principle to the "inner" quantum number and by the choice of the latter, we have proceeded rather formally. The physical fact is the exclusion of certain term combinations; that the cause of this is to be sought in quantum conditions appears certain to me. But the assumption that these quantum conditions have the same form as with the external quantum number is somewhat arbitrary.[41]

The basic form of rule three had been put forward by Sommerfeld and Walter Kossel in 1919, and was termed by them the "spectroscopic displacement law."[42] It held, in essence, that the spectrum for an element that had been ionized (i.e., had lost an electron) displayed the same splitting pattern as the un-ionized elements in the column preceding it in the periodic table; further, in numerical terms, it was best compared with the element immediately preceding it in the table. As Sommerfeld noted, the rule had a simple model-based explanation. Since one could take it as "empirically guaranteed" that the line character of an element depended solely on the number of its external (valence) electrons, then the removal of an electron should clearly produce line patterns like those of elements with one less electron in the outermost ring. Yet, as Sommerfeld continued, the "theoretical origin" of this rule "touches only on the most general lineaments of the model of the atom, on the incrementally increasing number of external electrons. Of the specific interpretation of the series terms…and of their allocation to the quantum numbers 1, 2, 3, 4 it is completely independent."[43] If *modellmässig*, that is, the rule drew barely, if at all, on the dynamical specificities of the Bohr model. So, too, with the fourth spectroscopic law, which extended a rule first put forward by Johannes Rydberg. The Swedish physicist had noticed that elements with an odd number of valence electrons characteristically produced doublet systems, those with even valence, triplets. Suggesting that the same logic should apply to ionized atoms, Sommerfeld outlined his "exchange law," which held that if an un-ionized element displayed a doublet in its arc spectrum, then its singly ionized form would display a spark-spectrum triplet, and vice versa.[44]

The fifth and final rule was the magneto-optic splitting law, which Sommerfeld introduced in terms similar to those used in his paper in *Die Naturwissenschaften*. Two new columns had been added to the number table: four further bracketed terms "developed only through analogy, not by observation." The table, he wrote, "is perhaps the most perfect example of those number harmonies which the new theory of spectra has bestowed upon us. It represents, for the time being, as I have noted elsewhere, a "number mystery." In fact, our table is of an essentially empirical origin and theoretically just as little understood as the origin of the line multiplicities more generally. Only so much appears to be certain: that the integral harmony of our Runge numbers has its final cause in the action [*Walt*] of hidden quantum numbers and quantum relations."[45]

In noting the similar language used to describe the lack of understanding of the geometrical significance of the inner quantum number and of the theoretical origin of Sommerfeld's number table, Forman has suggested that what ties the paper on "General Spectroscopic Laws" together is "a state of ignorance, conceived as a transitive relation connecting the complex structure with the anomalous Zeeman effect." "We are as ignorant," he argues, "of the significance of the inner quantum number as we are of the cause of the complexity of the spectral terms, and that is precisely how little we understand the cause of the Runge denominators of the terms."[46] Certainly, Sommerfeld made no bones about the extent of his ignorance, but a more charitable (and also more accurate) reading might note that there are not two but five parts to the paper, and all five hold in common their avoidance of dynamical, model-based explanations in favor of empirically based rules. In other words, ignorance about causes was traded for a functionalist understanding of regularities within phenomena. Finding no way to proceed using the theoretical tools he had been so instrumental in developing, Sommerfeld gave up the search for *modellmässig* foundations in order to develop a praxis—or craft—involving "half-empirical" *Gesetzmässigkeiten*.

The shift away from his earlier approach to the quantum theory of spectral lines was a dramatic one. Gone was the realist emphasis on dynamical model building. However, in its place was not, as one might expect, a Planckian emphasis on trans-historical, transcendental, purely theoretical principles. Indeed, in many ways the opposite was the case, since experimental data was not the end but the starting point of Sommerfeld's analysis. Yet an important resource for the formulation of what might be termed his aesthetic phenomenology is not hard to find. In 1913, while reworking Voigt's coupling theory of the Zeeman effect, Sommerfeld laid particular emphasis—as, indeed, did Voigt himself—on the phenomenological character of the theory. The particular values of the coupling coefficients of the mutually bound electrons were all drawn from empirical data, and no attempt to explain the cause of the coupling on electromagnetic grounds was offered. (See figure 7.4.) As Voigt put it, his investigation "restricts itself to the phenomenological level and is therefore satisfied with offering the necessary *preparatory work* for the later formulation of a model for the extremely complicated processes via an explanation of the quantitative relations which underlie the properties of the different degrees of freedom. In any case, the proof of the fundamental significance of the *cyclic coupling of the degrees of freedom* may be regarded as the *beginning* of the construction of a model for the explanation of recent observations."[47] For Sommerfeld, this conclusion was far too weak. The numerical relations between the coupling coefficients—brought to light in a particularly clear manner by his simplification of Voigt's equations—seemed to point to a more concrete result[48]:

The simplicity and symmetry of these equations is highly suggestive with regard to the problem of the construction of an atomic model. Herr Voigt has remarked this already in the context of his general hypothesis of cyclic coupling; it applies to a greater extent to the simplified representation of the specific case of D-line coupling given here. The appearance of three roots of unity as weighting factors seems to point to a ring in which three electrons follow one another

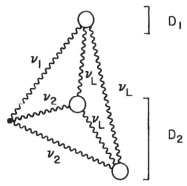

Figure 3. The Voigt-Sommerfeld Model for the Sodium D-Lines.
In a magnetic field the three electrons (circles) are coupled harmonically to a fixed center with frequencies ν_1 and ν_2 and to each other with frequencies ν_L, the Larmor frequency. In zero external field $\nu_L = 0$, the internal coupling disappears, and the electron generates an internal field $\mathbf{H}_i$ oriented along $\mathbf{k}^*$. The core is oriented along the resultant of $\mathbf{H}$ and $\mathbf{H}_i$.

Figure 7.4
Voigt's coupling model for the anomalous Zeeman effect. Source: David C. Cassidy, "Heisenberg's First Core Model of the Atom: The Formation of a Professional Style," *Historical Studies in the Physical Sciences* 10 (1979): 187–224.

equidistantly.... The number 3, corresponding to Runge's rule, plays the determining role in these equations; it is the only parameter of our equations. One may well suspect that with other atomic constructions and correspondingly altered types of Zeeman phenomena, other whole numbers will enter in some fashion that is also designated by Runge's rule. Perhaps the phenomenological study of the Zeeman effect after the fashion of Herr Voigt offers the most secure means for the building up of the structure of the atom; in this respect, the example of the D line's remarkably simple, number-theoretical *Gesetzmässigkeiten* gives us hope.[49]

Where Voigt, in other words, postponed model building, for the most part, to some future date, Sommerfeld saw phenomenology-based equations as the starting point for an immediate discussion of the nature of atomic structure. The more such equations were simplified, he concluded, "the closer we can expect to come to a real understanding of them and therewith to a certain insight into the atomic processes to which they correspond."[50]

Five years later, as the anomalous Zeeman effect seemed to provide many more questions than answers, Voigt's more cautious approach clearly seemed much more appealing. Writing to Zeeman at the beginning of 1920, Sommerfeld lamented Voigt's recent death, noting that the reference to the "musical beauty" of the number table in his *Zahlenmysterium* paper had been directed toward his former colleague. It is surely no coincidence that Sommerfeld's first paper after his complete conversion to a *gesetzmässig* method was his "Quantum mechanical reformulation of Voigt's Theory of the Anomalous Zeeman effect of the Types of the D Lines," a paper that sought to provide "the adequate expression of the facts" in the language of the quantum theory.[51] The first page of the preface to the third edition of *Atombau*, completed a month later, made his position clear:

I attach particular importance to the introduction of the inner quantum numbers and to the systematic arrangement of the anomalous Zeeman effects. The regularities that here obtain throughout are primarily of an empirical nature, but their integral character demands from the outset that they be clothed in the language of quanta. This mode of explanation, just like the regularities themselves, is fully established and is unique.[52]

The only doubts that can arise, he concluded, are those "with respect to the interpretation in terms of the models."[53]

Rings, Cubes, and the *Ellipsenverein*: X-Ray Spectra and Atomic Structure[54]

In 1924, the year in which the fourth edition of *Atombau* appeared, Sommerfeld published an article in the *Annalen* on "The Theory of Multiplets and Their Zeeman Effect." As had become common by then, his introductory remarks included a methodological discussion, outlining his reasons for deferring model-based analysis for some point in the future. This element of his work is now familiar to us, but Sommerfeld's identification of the exemplar of his new approach was more novel:

In the present state of theory, it seems to me to be most secure to put the question of model-based meaning in the background and to first bring the empirical relations to their simplest arithmetical and geometrical form. This rejection [*Verwahren*] has proven itself, e.g. in the theory of Röntgen spectra. While the specific models (circular rings, *Ellipsenverein*, cubes) have proven themselves to be unfruitful, the half-empirical systematics of Röntgen spectra, based on the principle of combination and selection rules, has led to valuable and secure results. We will proceed here, in the question of term structure, in a similar manner.[55]

The idea that the understanding of X-ray data might provide a model for the visible spectrum was one that had been unthinkable only five years previously. Early in 1919, Sommerfeld had written to Bohr of his exasperation with the problem. "I find it genuinely difficult to advance with the Röntgen spectra. There is much too much that is hypothetical in the assumptions about the position and occupation of the rings." Instead, he reported, "I have turned again more toward the visible spectra, about which I have received new material from Paschen."[56] In July he told the experimentalist Manne Siegbahn of recent lectures he had given on the problems of Röntgen radiation, "where certainly so much still lies in darkness."[57]

No doubt part of the vehemence of Sommerfeld's reaction stemmed from the frustration of hopes raised when he had first taken up the question of X-ray spectra in relation to Bohr's model in 1915.[58] A year before, Kossel had put forward a model that seemed to explain both X-ray emission and absorption data simultaneously. The feat was an impressive one, for the asymmetry between the two processes for Röntgen radiation had been a puzzle. For the visible spectrum, Bohr's model provided an easy explanation for emission and absorption as inverse phenomena: in absorbing a quantum of energy, an outer electron "jumped" to a virtual orbit; in emitting energy, it leaped back down again. Yet for X-rays, a given metal, which emitted its K_α line at a frequency $v = v_{K_\alpha}$, did not absorb radiation at the same frequency. There was, rather, a lower limit (an absorption "edge" at a frequency higher than any observed K lines for that metal), below which no characteristic K-ray absorption was possible. Once that limit was reached, however, all K-ray lines seemed to appear simultaneously (that is, not merely K_α, but the higher frequency K_β and even the newly observed K_γ). Kossel interpreted this result as implying that the frequency of the absorption edge, v_K, was a measure of the ionization energy of the K shell. A metal, struck by high-energy cathode rays, according to this model, ejected an electron from the K shell. An electron from the L shell then dropped down to give the K_α line. Röntgen emission spectra, in other words, were only produced in the aftermath of ionization. This explanation implies, of course, that (since $E = hv$) $v_{K_\alpha} = v_K - v_L$ (hv_L being the ionization energy of the L shell), a result that experiments appeared to verify. The identification of two L absorption edges then implied the existence of two L shells.[59]

Sommerfeld combined these ideas with his own relativistic extension of Bohr's model. Assuming the presence of a single electron in an L orbital (orbits designated, beginning with that closest to the nucleus, K, L, M, N,...), he envisioned it revolving about both the nucleus and any electrons in lower orbits. This resulted in an effective reduction of the nuclear charge, Ze—and hence the Coulomb force—which Sommerfeld represented by the term $(Z - l)e$. Since he had already explained the relativistic origin of a fine-structure doublet for hydrogen (of width Δv_H), he now had a near-equivalent explanation for the existence of the observed L doublet and could apply his previous results to determine a value for its size, Δv. To first approximation,

$$\Delta v = (Z - l)^4 \cdot \Delta v_H.$$

Including higher-order terms and comparing the more precise expression with experimental data for Δv for elements ranging from lead ($Z = 46$) to uranium ($Z = 92$),[60] Sommerfeld calculated the value of the screening constant, $l = 3.5$. The result was both remarkable and perplexing. On the one hand, since l appeared to be the same for all elements across a significant part of the periodic table, this would suggest that they shared the same electron structure for their innermost regions—a gratifying result for anyone interested in *Atombau*. On the other hand, Sommerfeld's reasoning would suggest, as he himself admitted, that the value of l should be integral: regardless of their arrangement, n electrons inside the relevant orbit should reduce the effective nuclear charge by *ne*, not by a fractional amount. "How to explain this in terms of the model," Sommerfeld wrote, "remains open."[61]

That was not the only problem to face this first attempt at tackling X-ray spectra. If the problem of the "nuclear charge defect," as Sommerfeld would term it, seemed to depend for its answer on a reconsideration of the dynamical specificities of Bohr's model, a difficulty with the "combination principle" appeared to strike at the most fundamental assumption of Bohr's analysis. Since the frequency of any spectral line was to be obtained from the "combination" of two orbits, of energy E_1 and E_2 ($\nu = (E_2 - E_1)/h$), this could also be understood as a transitive property. Consider a transition from the *N* to the *L* orbital (the line L_α in Sommerfeld's notation) and another from *M* to *L*. Since $(L - N) - (L - M) = M - N$, one should be able to calculate the frequency of the line corresponding to a transition from an *N* to an *M* orbital solely on the basis of data on the other two lines.[62] Empirically, however, the difference on the left-hand side was calculated to be 133.6 for uranium, while the lowest experimentally determined value for a transition from any *N* to any *M* orbital was measured to be 233.5. Similarly, since $(L - O) - (L - M) = M - O$, the two terms on the left-hand side should provide a measure for the frequency corresponding to a transition from an *O* to an *M* orbital. Instead, the data from the *L* series gave a difference equal to 350.7 for uranium, while the highest experimental value was 274.2. Sommerfeld wrote: "We must, therefore, declare that the combination principle for Röntgen radiation fails at precisely the place where it could be most precisely verified."[63]

A paper read to the Bavarian Academy in early November supplied, as its title suggested, a number of extensions to the original quantum theory of spectral lines, but had little concrete to offer concerning the two defects discussed above.[64] Completing a detailed calculation of the energy levels of quantized elliptical orbits near the nucleus, Sommerfeld nonetheless had to conclude that his original suggestion for the reason behind the non-integral value of the screening constant—the effect of "external" electrons—didn't enter into the picture. "The external electron ring," he wrote, "therefore produces no trace" of such a value.[65] Without actually performing the calculation, Sommerfeld suggested that the problem may have lain in the assumption of co-planar rings. Tilting the orbits in relation to one another—"stepping from the planimetry to the stereometry of the atomic core"—held out a hope of explaining the screening values: "important natural constants that are characteristic for the constitution of all elements of the natural system."[66] Sommerfeld's calculation offered no insight into the cause of deviations from the combination principle either. Again, all that could be supplied was a suggestion for future work. Attempting to determine the effect of orbits aligned so that they intersected those of external electrons might serve the purpose, but Sommerfeld was more than a little vague on details. "One can say perhaps," he wrote with more optimism than certainty, "that at least in a qualitative respect, the way has been pointed toward the explanation of the apparent deviations from the combination principle."[67]

There the matter rested until the final months of the war, when Sommerfeld once again took up the question of "Atomic Structure and Röntgen Spectra." The paper opened with a description of the continuing effort to explain the origin of the non-integral value of l. Accepting that it was apparently impossible to derive the right value while assuming that a single electron orbited around a bunch of its fellows who screened it from the full nuclear charge, Sommerfeld noted attempts by both Bohr and Moseley to incorporate the possibility of multiply occupied rings, evenly spaced in a circular orbit. Of course, such an explanation could not work for Sommerfeld's elliptical orbits: rotating electrons equally spaced around an ellipse under a Coulomb force will not be stable. If, however, one imagined n electrons on n identical ellipses, with each ellipse spaced at an angle of $360/n$, then—assuming that each particle begins its motion at the same position—the electrons will trace out paths such that, at any moment, each stands at the corner of a regular n-sided polygon. This was the famous *Ellipsenverein*. Since a circle may always be drawn that touches all the corners of a regular polygon, the electronic motion may be envisioned in terms of a single circular orbit, expanding and contracting—indeed, pulsing—around the nucleus. A footnote to the description of this rather elaborate structure sought to deflect the obvious criticisms. "For my feeling," wrote Sommerfeld, "the artful interlocking of the n electronic paths in our 'Ellipsenverein' is nothing unnatural; I see much more a sign therein for the high harmony of motion that must rule within the atom."[68] (See figure 7.5.)

The question of the spatial orientation of electrons within orbits having been solved, Sommerfeld could turn toward what had become the central questions that many looked to X-ray spectra to answer. How was an atom structured? How many electrons were to be found in how many rings around the nucleus, and how was this structure related to the organization of the periodic table of elements? One might suspect, for example, that each ring corresponded to a row of the table and thus that the lowest, the K ring, could hold two electrons, the L and M rings eight each, the N ring 18, and so forth. While Sommerfeld had pursued other problems for the years between 1916 and 1918, several researchers had sought to determine the occupancy number of each ring by calculating its dynamical stability within the overall planetary model of the atom. None had been entirely successful.[69] Debye, undertaking one of the first analyses, arrived at the unconvincing result that, in a transition from the L to the K ring, the latter went from containing two to containing three electrons, while the former dropped from one to zero. Sommerfeld offered his own calculations as "essentially a theoretical basis" for results obtained by Jan Kroo, who had come up with the decidedly more appealing result: $K = 3$, $L = 9$. Kroo had taken the simple tack of writing down an expression for the energy difference between an atom in which the K ring contains $p - 1$ electrons and the L ring q electrons, and an atom where a transition has occurred, and the K ring has p electrons, the L ring $q - 1$. Determining the frequency corresponding to this energy (which, of course, must be the frequency of the

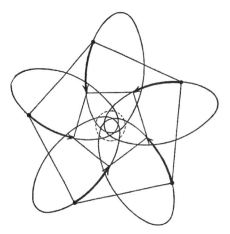

Figure 7.5
Sommerfeld's *Ellipsenverein*. "These ellipses are traversed by the q electrons in such a way, the same for each, that all the q electrons pass through the corresponding aphelia and perihelia at the same moment, respectively. If the electrons are joined up by a sequence of straight lines, then the latter will at every moment constitute a regular polygon (of q sides) which alternately contracts and expands. It is clear that in this pulsating polygon the repulsions exerted on one electron by all the remaining electrons must by symmetry give a resultant which passes through the nucleus...." Source: Sommerfeld, *Atomic Structure and Spectral Lines* (1923).

K_α line), Kroo fitted the data to his theoretical expression to obtain the numbers above. Sommerfeld claimed that he would "found mathematically the theory of the multiply occupied electron rings" but his actual achievement was rather more modest. What he could demonstrate was that one could essentially ignore the contribution of external rings when calculating the value of K_α, an assumption implicit in Kroo's analysis.[70] However, he ruefully acknowledged, "in further pursuit of the problem a series of difficulties emerge, which will possibly compel the foundation of the theory to be reformulated."[71] Of leading importance among those difficulties were, once again, the nuclear charge and combination defects, for which the *Ellipsenverein* could offer little cure.

Sommerfeld's report on the paper to Bohr was less than enthusiastic. "I find many difficulties here, particularly with the combination principle. The aim of determining the number of electrons in each ring appears to me still to lie in the far distance."[72] The sentiment stayed with him, as we have seen, into the new year. Although a diagram of the *Ellipsenverein* would be included in the first three editions of *Atombau*, so too would challenges of it, the elaborations of critiques gaining in length with the years.[73] Thus it was that atomic structure and X-ray data should be at the heart of the second of Sommerfeld's "unsettled questions of atomic physics." By 1920, when the paper was published, a series of results had appeared challenging the stability of "pancake" models with electrons arranged in co-planar rings.[74] It seemed necessary to consider atomic structure as a three-dimensional problem. Although he had, at first, appeared more resigned to than convinced by the arrangement proposed by Alfred Landé, with electrons in the second shell at each of the eight corners of a cube, by early September 1919 Sommerfeld had come to regard the die-shaped shell as the "salvation" of the theory of Röntgen spectra.[75] The religious fervor did not last long. The open question that he took up the next year was whether spectral data confirmed the details of the model. The simple answer was no, but an explanation was at hand. One need not conclude that the cubical picture was false, only that "the cube did not remain a regular cube."[76] If the L shell had the form of a die, then a single electron orbiting in the M shell would deform its shape, causing the Coulomb force acting upon it to change. There was room to hope that calculations (difficult though they might be) that dealt with orbits around this shifting three-dimensional structure might better accord with the results of experiment.

Sommerfeld was not the one to undertake such calculations. His own approach to Röntgen spectra began to take on the shape of his work on the anomalous Zeeman effect, being completed at the same time. Just as 1920 marked the point at which he began to elaborate a series of "half-empirical" laws for understanding the visible spectrum, the same year saw his turn toward an increasing use of selection principles in his analysis of X-ray data. Bohr's response to Sommerfeld's results, soon to be published as "Remarks on the Fine-Structure of Röntgen Spectra" was enthusiastic and captures Bohr's clear realization of the methodological distinction Sommerfeld was making between empirical regularities and model-based accounts:

> I was extremely interested in what you reported in your letter. The *Gesetzmässigkeiten* in the Röntgen spectrum that you mention are wonderful, and one cannot doubt that you have grasped the kernel of the problem. How one can represent it all in terms of a model [*modellmässig*] in its details and how one can overcome the difficulties associated with the limited space in the core of the atom is another story, about which I also feel—the more I try to think about it—that one still comprehends so little that scarcely any ground for skepticism may be found therein.[77]

The most important parts of the paper dealt with the application of the selection principle to the fine structure of the various series within X-ray spectra. Recent dissertation work by one of Siegbahn's students, W. Stenström, had revealed the existence of 1, 2, and 3 absorption edges in the K, L, and M series respectively, a fact that Sommerfeld immediately took as evidence that the splitting of the orbitals was governed by the same rules as in the hydrogen spectrum. If the quantum sum for the M level, for example, was accorded the value $n + n' = 3$ (in a manner analogous to the hydrogen triplet), then three energetically distinct orbits are possible: a circle and two ellipses. Similarly, the N level would be, in fact, a quartet, as depicted in figure 7.6.[78] Applying

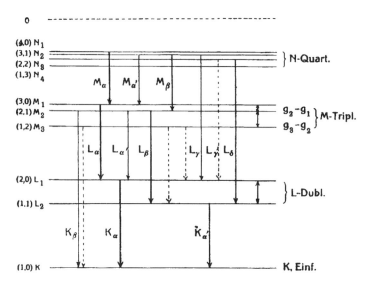

Figure 7.6
Splitting levels, based on absorption-edge and emission data. Source: Sommerfeld, "Bemerkungen zur Feinstruktur der Röntgenspektren I," *Zeitschrift für Physik* 1 (1920): 135–46.

the same selection principle deployed in his paper on spectroscopic laws, certain orbits—those for which the rule $n = \pm 1, 0$ was not obeyed—could now be disallowed, *a priori*. Coupling this principle with an "intensity rule," derived by Bohr's assistant H. A. Kramers, which held that more intense lines were to be found in transitions where $\Delta n = +1$ than when $\Delta n = -1$ or 0, Sommerfeld could immediately explain a result which he had been emphasizing for some time. From his earliest papers on the topic, he had argued that the size of the L doublet was not given by the difference between the lines L_α and L_β—the two strongest lines in the L series—but by the difference between L_β and the weaker line $L_{\alpha'}$. One could now see why: pairs of strongest lines held neither their starting nor their end state in common.[79] Applying the same logic to the M series, it is clear that the difference between measured values for the second-strongest line M_β and the weaker $M_{\alpha'}$ should give the size of the gap between two of the lines of the M triplet ($M_1 - M_2$). Stenström's work offered an independent measure of this quantity, as the difference between two of the *absorption* edges of the series, which he labeled $g_2 - g_1$. The empirical match between the two quantities was remarkably close. For both uranium and thorium, the difference was less than 1 percent.[80]

Sommerfeld was now using a similar methodology for his work on both the Zeeman effect and X-ray spectra, but modeling had not yet been abandoned entirely for the latter problem. In 1918, he had predicted, on *modellmässig* grounds, the existence of a not-yet-observed partner to the known K_β line. In 1921, precision measurements carried out by another student of Siegbahn's, E. Hjalmar, appeared to confirm the fine structure of the line and Sommerfeld returned to his earlier analysis. The argument was a simple one. The energy of a transition was calculated as the difference between the energy of the atom before and after an electron moves from a higher to a lower orbit (or the reverse, in the case of absorption). For a transition from the M to the K orbital, this should have two components. One will simply be the energy difference between the two levels. The other component will come from the effect of the transition on the two orbitals associated with the L doublet. An extra electron in the K shell adds to the negative charge shielding the intermediary L orbitals from the nucleus and each will thus expand, changing the net energy of the atom. For Sommerfeld, however—and this point would prove crucial—the elliptical and circular orbits of the L doublet were not to be found within the same atom. The point was essential when considering the *Ellipsenverein*, otherwise the elliptical orbits corresponding to $n + n' = 1 + 1$ would cut into the circular orbit, with its evenly spaced electrons. By 1921, Sommerfeld had little faith in his "elliptical complex," but the notion that different kinds of orbits corresponded to different "species" of atoms remained: "One kind of atom, in which the more common L_1-level rules, gives the main line, β; the other kind of atom with the more rare L_2 level gives a neighboring line, β'." The point was reiterated in the article's final line: "One must always keep in view the fact that the processes, which we relate to the emission of β' and β", in either case, even if they occur in the very same element, must take place in different atoms (atoms of the L_1 or L_2 states)."[81]

The removal of this single plank from the overall structure of Sommerfeld's model-based argument caused the entire edifice to fall. An addition at the proof stage of the article read: "The last remark stands in contradiction to the latest results that Mr. Bohr has indicated in his letter to *Nature* of February 17 of this year. How the rest of the preceding remarks agree with Bohr's new viewpoint cannot, at present, be gauged."[82] Bohr's four-page letter, outlining how the correspondence principle could be used to determine the details of atomic structure, had shaken the work of the Sommerfeld school to its foundation.[83] "Your comment that Bohr has battered in like a bomb," he wrote to Landé in March, "is true for Munich as well. I received a copy of his letter to Nature from Bohr. We must thoroughly relearn [*umlernen*]."[84] To Bohr himself he raved that the letter "clearly signifies the greatest advance in atomic structure since 1913."[85] In fact, Bohr's letter was light on details, serving more as a promissory note for a paper to follow, but Sommerfeld took from it one essential claim: all orbitals, of whatever shape, that corresponded to a given element, had to be found within a single atom of that element. How profound was the effect of this realization may be easily discerned within the pages of the third edition of *Atombau*. Although the first and second editions had already sounded a skeptical note about the *Ellipsenverein*, the discussion in the third served to bury not only it, but the very project of atomic model building altogether. After describing the functioning of the interlocking ellipses, Sommerfeld noted that "a number of weighty objections speak against the truth of this picture." A few lines below he would list these objections:[86] first, as J. M. Burgers had noted, the ellipses meant to model the L shell actually intersected the path marking out orbits in the smaller K shell (the broken circle in the diagram)[87]; second, the coplanar arrangement was problematic, since it was unclear how the electrons were to be made to confer distinction on one plane over another; third, Landé's solution—a cubical arrangement—will not work either, since the question of the shape of the atom once one of its electrons is removed in the excited state is unclear. "But the most serious objection to polygonal as well as to polyhedral symmetry," Sommerfeld wrote, was the fourth:

> ...in order not to destroy the symmetry we should have to assume that the elliptical and circular modes of motion occur in *different* atoms, and that, accordingly, *one* part of the atom exemplifies the L_1 level, *another* part the L_2 level. Now in addition to the L_1 and L_2 level there is also an L_3 level. Moreover...there are 5 M levels and not less than 7 N levels....But if we distribute L_1 and L_2 among different atoms we must also do the same with L_3, and with the M and N levels. Hence, we should have to postulate not two, but at least 1.3.5.7 = 105 different species of one and the same atom, corresponding to the possible combinations of the various levels with one another. That is already absurd in itself.[88]

The "intermediary doublet" explanation of the fine structure of the line had depended entirely upon this "absurdity." Now, as Sommerfeld phrased it, the theory "falls to the ground" with the conclusion "that the orbits that give rise to the different levels must actually all occur in the same atom." The credit for his realization of the necessity of this about-face was granted entirely to Bohr. "This conclusion," he acknowledged in a footnote, "is entirely contradictory to the view that the author held formerly, and was maintained in earlier German editions of this book. But it coincides with the views of Bohr expressed in his letter to *Nature*.... According to Bohr, it is an indispensable condition for the stability of the atom that the orbits of the various shells be interlocked, in a manner similar to that depicted for the K and L orbitals in [the figure of the *Ellipsenverein*]."[89] Along with the now-defunct intermediary doublet theory went any hope of explaining either atomic structure or the specificities of hard-won quasi-empirical expressions in a *modellmässig* fashion, yet Sommerfeld refused to regard this loss pessimistically:

> ...there can be no such pronounced symmetry as we assumed in the grouped ellipses or in the cubic arrangement. The problem of the arrangement of the electrons within the atom, regarded from an elementary point of view, becomes hopeless. It seems equally hopeless to explain the defect in the nuclear charge namely s = 3.50 in an elementary and pictorial manner....[Yet] our formula for the L doublet does not hereby lose any of its practical value. It cannot, indeed, be regarded as an equation that has been derived from theory, like the formula for the hydrogen doublet, but it stands as an empirical equation that has been brilliantly confirmed.[90]

A sense of optimism, in fact, characterized the 1922 edition of *Atombau*. Sommerfeld declared in his preface that he let the text out of his hands with a "somewhat easier conscience" than he had for previous editions. Much had seemed "unripe and uncertain" in 1919 and, of course, the fields of theoretical and experimental spectroscopy were still in a state of "violent ferment" three years later. Yet the focus on *Gesetzmässigkeiten* allowed much greater hope that the claims of the present would be upheld in the future. For the case of Röntgen spectra, Sommerfeld noted, *modellmässig* interpretation "has been left out almost entirely. Whatever the further researches of Bohr may reveal to us concerning the shell structure of the atom, I feel certain that nothing will be changed in the laws...here described."[91] The point was true, moreover, for almost all of the subject matter of the book: "[T]he way in which the facts of Röntgen spectra, of term multiplicities, of Zeeman effects, have been put together, half-empirically and half by means of the quantum theory, will presumably remain unaffected by later developments. Bohr's recent far-reaching ideas will, indeed, add much that is new, but will not throw doubts on what appears to be established."[92]

In Praise (and Defense) of *Gesetzmässigkeiten*

Given his outspokenness with regard to the "unique" nature of his mode of explanation, it is perhaps unsurprising that Sommerfeld's method met with mixed responses. Max Born, in early 1923, professed to be both supportive and admiring, even if he felt himself unable to emulate his colleague's achievements. "Unfortunately," he wrote, "I do not have your ability to read such connections out of empirical spectral data, but must feel my way forward slowly on the path toward the gradual cleaning up and clarification of principles."[93] After the publication of the fourth edition of *Atombau* in 1924, both Wolfgang Pauli and Erwin Schrödinger wrote letters of effusive praise, each emphasizing in particular the avoidance of model-based explanations. By that time Pauli had become convinced that modeling was entirely inappropriate for the needs of the quantum theory, and clearly saw in Sommerfeld a solution to contemporary problems:

I found it particularly beautiful in the presentation of the complex structure that you have left all *modellmässig* considerations to one side. The model idea now finds itself in a difficult, fundamental [*prinzipiellen*] crisis, which I believe will end with a further radical sharpening of the opposition between classical and quantum theory.... One now has the impression with all models, that we speak there a language that is not sufficiently adequate for the simplicity and beauty of the quantum world. For that reason I found it so beautiful that your presentation of the complex structure is completely free of all model prejudices.[94]

Schrödinger was less critical of other approaches than the characteristically blunt Pauli, but no less admiring, admitting that it "remained incomprehensible to him" that Sommerfeld should have been able to draw such fundamental laws from "not at all so very rich factual materials" without a "proper [*eigentliches*] model." "I have trouble," wrote Schrödinger, "slowly clarifying to myself the admittedly really complicated building up of these whole-number formulas, and you've incorporated them into the observational material so that they now sit as tightly as a guard's uniform!"[95]

If some found the avoidance of models worthy of high praise, others found Sommerfeld's direct engagement with the empirical data—and especially his arguments in favor of particular empirical rules drawn from this data—far more problematic. Otto Klein described Bohr's reaction to Sommerfeld's *Zahlenmysterium* lecture in Lund as follows:

During Sommerfeld's lecture I was sitting at the side of Bohr. Sommerfeld had a few numbers for these anomalous levels. The first was so and the second was so—they were something like 1, 3, and so. And then Sommerfeld said the next must be 5, or something like that. Then Bohr smiled and said to me, "I don't believe that."[96]

If it can be assumed that Landé was proceeding in a Sommerfeldian manner when he arrived in Frankfurt after working in Munich, Born's reaction must be adjudged far more extreme than Bohr's. Born largely ignored Landé, whose way of relating the intensities of multiplet lines and Zeeman effect lines through whole number ratios he could only describe as "horrible," in its obsessive "guessing about numerical values."[97] According to Robert Friedman, Carl Oseen's distaste for Sommerfeld's methods of research was, in large measure, responsible for Sommerfeld's failure to be awarded the Nobel Prize for physics in 1924, in spite of the enormous number of nominations he had received.[98]

Sommerfeld thus had good reasons for trying to establish his method as both subtle and coherent, neither number crunching nor numerology. And it is as an attempt to pitch the method precisely between these two extremes—emphasizing both the hard work involved in the technique and the knowledge and imagination required in its application—that we should read the following lines from the third edition of *Atombau*, perhaps the single clearest statement of the detailed practices in which Sommerfeld was engaged:

It must not be imagined that the combination of the lines into series and their resolution into two terms is a mere trifle. Rather it demands special experience and ingenuity. First of all, the lines of the various series are all mixed together and must be separated out in accordance with the criteria indicated at the beginning of this section. There are usually only a moderate number of lines of a single series present, as the higher members of the series, on account of their feeble intensity, are less accurate than the more intense lower members. To derive the series limit and hence the constant first term of the series by extrapolation, the analytical expression for the current term, for example in the Ritz form, must be used as a basis. The series limit is then obtained, as well as the indeterminate parameters that occur in the series law...by a graphical or arithmetical process of approximation. It almost always appears that the first member (or members) of the series is not given with sufficient accuracy. From this we must conclude that not only Rydberg's but also Ritz's form represent only an approximation to the strict series law and are true only for the greater values of m.... The task of calculating the series becomes much easier if other series or series limits of the same element are already known. On account of the relationships of "combination"...between the different series, we have always to strike a balance between the calculations of several series.[99]

This, then, provides an understanding of a method that might be described as simultaneously a craft of the quantum and an investigation of number mysteries: A craft, in that, working directly with the data one extrapolates and interpolates, drawing conclusions not from model-based deductions, but from arithmetic and graphical approximations, drawing on special experience to strike a balance between different sets of the always-insufficient information from spectroscopic data. Finding that balance, however, and identifying the "correct" regularity in a sea of possibilities requires art and ingenuity, an inbuilt sense of the aesthetic, of the harmony that must rule within the atom. It was this second aspect—a rejection of the instrumentalist implications of what Forman has termed an *a posteriori* method—that Sommerfeld would emphasize in speaking of the values of mysticism in 1925.[100] "Precisely the most successful researchers in the area of theoretical spectral analysis, Balmer, Rydberg, Ritz, were genuine number mystics. They based their research, either consciously or unconsciously, on the claim that the connections of the wave numbers to spectra would have to be so harmonious, so aesthetically simple as to be compliant with the facts; and success justified their standpoint."[101] Conventionalism and positivism, he argued "sink into a nothingness before the beauty and security of our newest physical conclusions."[102] His was phenomenological practice fused with metaphysical belief, the robustness of empirical laws, gleaned with the aid of a delicate sense of aesthetics. Sommerfeld's was a music of the spheres composed by a craftsman: a beauty and a truth to be calculated and constructed, analyzed and approximated.

From the Old World of Waves to the New World of Quanta

In discussing Sommerfeld's postwar work on quantum spectroscopy, I have focused on detailing what his methodology was, or at least what it was meant to be. Just as important, however, is the question of what it was not. Throughout his papers and his texts, Sommerfeld sought time and again to distinguish his approach from that of Bohr. Even in 1919, Sommerfeld had been troubled by the implications of what he termed Bohr's "analogy principle" (labeled the "correspondence principle" soon thereafter). And in general, one can characterize the early 1920s for Sommerfeld as a period in which he drew away from the use of particular forms of analogical thinking in the atomic theory. Eschewing both mechanical models as analogies and the specific classical/quantum analogy that the correspondence principle denoted, in 1922 Sommerfeld turned instead to the "direct" empirical method we have discussed. The aim, above all, was to avoid the mixing of classical and quantum concepts, precisely the crime of which he would, as we shall see, later accuse Bohr.

In his examination of the role of "formal analogies" in the development of the quantum theory, Olivier Darrigol has presented the disagreement between Sommerfeld and Bohr over the correspondence principle as a central tension. Bohr's use of that principle provides Darrigol's prime example of the fruitfulness of "concepts that are formal in the sense that they do not provide a visual picture of the sort one is accustomed to in the explanations with which natural philosophy deals."[103] Moving from classical electromagnetic theory to the realm of the quantum, Bohr's bridges were less illustrative or visualizable resemblances, and more "entire pieces of logical and mathematical structures."[104] In this way, Bohr constructed the correspondence principle as a *rational generalization* of classical radiation theory:

> …although the process of radiation cannot be described on the basis of the ordinary theory of electrodynamics, according to which the nature of the radiation emitted by an atom is directly related to the harmonic components occurring in the motion of the system, there is found, nevertheless, to exist a far-reaching *correspondence* between the various types of possible transitions between the stationary states on the one hand and the various harmonic components of the motion on the other hand. This correspondence is of such a nature, that the present theory of spectra is in a certain sense to be regarded as a rational generalization of the ordinary theory of radiation.[105]

Central to the utilization of such an analogy was the flexibility emphasized both by Bohr and his followers, as well as Darrigol himself, who writes of the "crucial ambiguity" that underpinned the connection between the antithetical classical and quantum worlds. Bohr contrasted his deployment of a flexible "rational generalization" to "the tendency of considering the quantum theory…as a set of formal rules," which Darrigol rightly construes as a swipe at the Sommerfeld school.[106] Darrigol takes this characterization, however, as a legitimate representation of Sommerfeld's own understanding of his approach to the quantum theory, and uses it as a means of explaining "Sommerfeld's reticence" in acknowledging what is described as the "superiority of Bohr's philosophy of correspondence."[107] "Sommerfeld and many other quantum theorists of lesser importance were out of sympathy with Bohr's strategy of rational guessing," Darrigol writes in his introduction. "Sommerfeld's concept of rationality demanded a sound complete mathematical framework or, as long as nothing better was available, a set of clear mathematical models."[108] Elsewhere he argues that "Sommerfeld was unlikely to accept Bohr's characterization of the correspondence principle as a 'principle of the quantum theory.' Such an evaluation conflicted with his struggle for a mathematically closed theory of atoms."[109]

The reader of the previous sections of this chapter will no doubt find such a characterization of Sommerfeld peculiar.[110] This depiction of a logic-bound hyper-rationalist does not seem consistent with the writings of a self-proclaimed mystic. And

Sommerfeld's stated antipathy to models—an antipathy that profoundly shaped his methodology in the early 1920s—seems utterly at odds with the claim that his purpose was to find a set of clear models, mathematical or not. Darrigol appears to have taken Bohr's understanding of Sommerfeld's aims as Sommerfeld's own, and to have read what Bohr saw as the advantages of his own approach as precisely those elements that Sommerfeld opposed. In other words, the debate has been taken as defined by Bohr, without contemplating the possibility that Sommerfeld's objections might spring from another source. To locate that source, one needs to offer a symmetrical analysis of Sommerfeld's and Bohr's positions and to focus equally upon Sommerfeld's understanding of the relationship between classical electrodynamics and the quantum theory. We need, that is, to see Sommerfeld as productive, rather than merely reactive. Just as Bohr made use of two principles—one to deal with the mechanical properties of a quantized atom (orbits, for example), and one to deal with its electromagnetic properties (emitted and absorbed radiation)—so, too, did Sommerfeld make use of two widely disparate sets of techniques. It is true that the methods of Bohr-Sommerfeld quantization were essential when one sought to obtain the *frequencies* of spectral lines, but an entirely different method was used to understand other properties of the radiation given off by an excited atom, for example, its polarization.[111] Bohr and Sommerfeld—both once proponents of an all-encompassing electromagnetic view of nature—thus had distinct solutions to a common problem; that of the relation between the electromagnetic and the quantum theories. While Bohr sought an analogy and hence some form of connection between the two, however, Sommerfeld strove—as he had during the Solvay conference—to keep them apart. It would be precisely what he saw as Bohr's gifted intermingling of conceptually disparate worlds to which Sommerfeld would object, but the point of contention was not solely one of mathematical closure or completeness. If he found Bohr's analysis incoherent, this did not mean that Sommerfeld assumed that the only way forward was a fully logical or rational one. To the contrary, he repeatedly emphasized not only the incompleteness of his own solution, but the conceptual benefits that could be derived from such an incompleteness. Mathematical clarity would always be important to Sommerfeld, but there was more at stake here than neatness. The difference between his open-ended solution and that of Bohr was that Sommerfeld's kept classical electromagnetic theory and the world of the quantum largely distinct, while Bohr's principle of correspondence appeared magically—and unjustifiably—to bring them together.

Sommerfeld's efforts to maintain a separation and an equality between the electromagnetic and the quantum theories are tracked here in temporally consecutive sources: the ever-longer discussions in the three editions of *Atombau und Spektrallinien* published between 1919 and 1922, and a paper on "the Foundations of the Quantum Theory and of Bohr's Atomic Model," written in 1924.[112] The book that would become an indispensable guide through the quantum maze for all researchers in the field,

Atombau was originally conceived of as a text for a general readership. It grew, Sommerfeld claimed, from a lecture delivered in Munich in the winter semester of 1916–17 to an audience of chemists and medical colleagues.[113] A few months before its publication he referred to it in a letter to Einstein as a "popular book about atomic models."[114] The first edition was structured with this audience firmly in mind. Early chapters were kept as simple and as free from mathematical apparatus as possible. Later, the discussion grew steadily more complex, but even here the most mathematically abstract and difficult derivations were reserved for an appendix—a structure maintained in subsequent editions.[115]

The first two chapters were largely preparatory. Chapter 1 introduced the reader to the components of an atom, and the means of its study: ions and electrons, cathode and canal rays, α and β particles, X- and γ-rays, and radioactivity. Chapter 2 explained the system of the elements built from such atoms, and sketched the simplest examples of atomic models.[116] Chapter 3 introduced research much closer to Sommerfeld's heart—X-rays—and began to explain the intricacies of spectral analysis, to begin the process of what Sommerfeld referred to as learning "to understand the language of spectra."[117] But chapter 4 was *Atombau*'s centerpiece. In that chapter (the longest one), readers learned the details both of Bohr's atomic model and of Sommerfeld's extensions and re-interpretations of it. The chapter serves as a fulcrum; the culmination of the more descriptive chapters before it, it provided the necessary tools and theoretical background for the next, Sommerfeld's fine-structure theory. Of the last, most recently worked out and speculative chapter, Sommerfeld wrote:

The most difficult and at the same time most interesting question is treated in the sixth chapter, the relationship between wave theory and quantum theory. Even two years ago it appeared futile to want to build a bridge from the old world of waves to the new world of quanta. Now that bridge is made through the idea of regarding the momentum, and the angular momentum of radiation as well as its energy, and of comparing this energy with the corresponding magnitudes in the atom. This thought was first made known to me in several conversations with my collaborator Rubinowicz, but was simultaneously also developed and published by Bohr; it constitutes, as I endeavored to explicate in the sixth chapter, the first and essential step toward a reconciliation of wave theory and quantum theory.[118]

In spite of initially twinning his and Rubinowicz's "bridge" with that of Bohr, Sommerfeld soon made clear that he felt the two were not equivalent. Of his own conception, he noted that it was, at least for the moment, incomplete but that it was both capable of further development and "suited to the sense of the quantum theory." Bohr's "analogy principle," on the other hand, appeared "in spite of its brilliant efficacy for the time being to stand as something foreign to the quantum theory and to meet our causal requirements in an unsatisfactory manner." The battle lines drawn, a fuller critique awaited the reader in the book's final chapter.

Chapter 6, titled "Wave theory and Quantum theory," began by quoting Heinrich Hertz on the "certainty" of the then-current wave theory of light. That certainty had not been shaken since then, claimed the once-strident proponent of the electromagnetic worldview. To the contrary, it had only been amplified through studies of X- and γ-rays, and the kilometer-long waves of wireless telegraphy. Yet the quantum theory provided very little information to do with the qualities and properties of the electromagnetic radiation that propagated out from an atom after a transition. One of Bohr's fundamental assumptions ($h\nu = W_a - W_e$) provided the frequency, but nothing about the intensity, polarization or coherence of the wave. The situation, Sommerfeld claimed, was like that of all purely energetic treatments: the energy equation only provided one equation to determine the course of the phenomena. For systems with more than one degree of freedom, one had to supplement the energetic conception with a deeper "dynamical" treatment.

That dynamical treatment would be an electromagnetic one. Sommerfeld's starting point was abstract, almost phenomenological. He began by positing the existence of a "practically monochromatic" spherical wave that propagated out from the atom.[119] To be able to fully describe such a wave, Sommerfeld noted, one needed values for a certain number of observable "determining factors." He proceeded to list seven such factors, which would be sufficient for the task: the wave and coherence lengths, the amplitudes of the electromagnetic field in any two perpendicular planes through an arbitrary direction of emission, the phase difference between their partial vibrations, and two angle measurements to define a directional axis. A complete theory of the emission and propagation of light from an atom would need to provide information on all seven such determining factors. This would suggest that the next step was an electromagnetic theory of emission, but how was one to reconcile this with the postulates of the quantum theory? As Sommerfeld explained, in the ordinary view of the wave theory, one assumed a constant coupling between electrons and the ether. A moving electron always produced radiation. "According to this view," he wrote, "we consider an electron active at the origin of every spherical wave, which generates in unison with the rhythm of its own motion the electromagnetic spherical wave."[120] In the quantum theory, on the other hand, electrons moving in stationary orbits produced no radiation and hence were uncoupled from the ether unless actually involved in a transition. This meant that one had to disengage from one's thoughts the image of a moving electron producing a wave. "We must speak not of an electron," Sommerfeld declared, "but of a solution of Maxwell's equations, which is determined by conditions of coupling in the process of emission between the atom and the ether. The more abstract mode of expression, to which we are forced, is inevitable if we wish to follow out logically the view of the quantum theory."[121]

The effect of this "abstract mode of expression" was to divorce explanations of the quantum and electromagnetic worlds. By the third edition of *Atombau*, the split would be laid out explicitly from the outset: the classical theory dealt with the *propagation* of light, the quantum theory with its *production*. The electron within the atom—the electron of the quantum theory—had fundamentally different properties, was almost a different entity to the electron of electromagnetic theory. Even at the point at which it was responsible for the production of electromagnetic radiation, one was to forget that it was an electron. What produced radiation was a solution of Maxwell's equations, which then propagated out from the atom in a spherical wave. Only certain conservation laws—energy, momentum, angular momentum—connected the quantized atom and the electromagnetic ether, but these functioned in some ways more as markers of difference than equivalence. In the expression $h\nu = W_a - W_e$, for example, the "equals" sign served to distinguish two utterly different modes of explanation. Within an atom, where energy levels were defined by the various W's, there was absolutely nothing that corresponded to the frequency, ν. That frequency only had meaning for the spherical wave that moved out from the atom as a result of a transition:

The duration of vibration of the radiation has nothing to do with the revolution of the electron in its stationary paths. Even during the transition there is nothing in the atom that occurs in rhythm with the vibration number ν. The ether demands its $h\nu$, the atom furnishes it by giving up an amount of energy $W_a - W_e$. The duration of vibration follows if these two quantities are equated; at the same time, the polarization follows if we equate the two corresponding moments of [i.e. angular] momentum. It has, indeed, been suggested that the transition from the stationary initial orbit to the stationary final orbit takes place along a spiral, which is traversed with the frequency ν. This too specialized picture seems to us unfruitful. *It is not the atom that vibrates, but the ether.* ...The atom gives the ether a certain amount of energy and moment of momentum. The ether does with this, what its nature compels it to do, namely, it converts these amounts into vibrations of a definite state of polarization.[122]

One could bridge the divide between the wave and the quantum theory, in other words, by insisting that energy, momentum and angular momentum be conserved between them, but this did not bring the two sides any closer together. Sommerfeld's mode of explanation further accentuated the divide by leaving part of the problem undetermined. Without additional information, conservation laws could provide only five of the seven determining factors for the process at hand. "The problem of building a bridge from the quantum theory of light to the wave theory," wrote Sommerfeld, "is thus in fact supported and pointed in the right direction, but not yet completed, through our principle; we could say perhaps: 5/7 of this problem is solved, 2/7 remains open."[123] Rather than a flaw, however, this half-finished explanation was painted as an advantage. "A theory," remarked Sommerfeld aphoristically, "should of course determine the observable process, but it should also not over-determine it." Clearly it was important that the coupling conditions described several significant elements of the vibrations of the ether. But it was perhaps equally important that both atom and ether were allowed the freedom necessary to behave in their characteristic fashions.

Both were contained within a closed system—what was lost by one was gained by the other. But under-determination supplied connections without exacting unification. If one can speak, indeed, of the lands of the quantum and the wave, Sommerfeld's partial solution described a common world with two separate ecologies.

It is possible that Sommerfeld might have found his 5/7 of a solution less satisfactory were it not for the fact that so much could be determined through it. Following arguments first laid out by Adalbert Rubinowicz in 1918, Sommerfeld was led very rapidly to a selection principle governing allowed transitions within the atom, and a polarization rule for the light given off.[124] As long as momentum was conserved and the radiation given off was emitted in a spherical wave, Sommerfeld argued, the momentum transferred between atom and ether was zero. Calculations beginning with Maxwell's equations and making use of a series of formulas drawn from vector analysis, however, led to a non-zero value for the angular momentum, N, of the radiation of a spherical wave, in terms of its frequency ν and total energy W:

$$N = \frac{W}{2\pi\nu} \frac{2ab\sin\gamma}{a^2 + b^2}. \tag{1}$$

Here a and b denote amplitudes of vibration and γ denotes a difference in phase. Sommerfeld then equated the total energy of the wave with the energy given off in a transition ($W = h\nu$) to give

$$N = \frac{h}{2\pi} \frac{2ab\sin\gamma}{a^2 + b^2}. \tag{2}$$

This quantity of angular momentum, Sommerfeld argued, must be equal to the change in the angular momentum of an atom due to the transition. For the hydrogen atom, this momentum, p, was given in terms of the azimuthal quantum number, n, by the expression $2\pi p = nh$. Thus Δp, the change in angular momentum, is given by

$$\Delta p = \frac{h}{2\pi} \Delta n. \tag{3}$$

By equating (2) and (3), Sommerfeld produced an expression for the range of values of the change in the azimuthal quantum number:

$$\Delta n = \frac{2ab\sin\gamma}{a^2 + b^2}. \tag{4}$$

Since the right-hand side of equation 4 must be less than or equal to 1 (with equality holding only when $a = b$ and $\gamma = \pm\pi/2$), and since n can only be integer, it follows that Δn is limited to the following three values:

$$\Delta n = \begin{cases} +1 & a = b \quad \text{and} \quad \gamma = +\pi/2 \\ 0 & a = 0 \quad \text{or} \quad b = 0 \quad \text{or} \quad \gamma = 0 \quad \text{or} \quad \pi. \\ -1 & a = b \quad \text{and} \quad \gamma = -\pi/2 \end{cases}$$

As impressed by his mode of derivation as by its outcome, Sommerfeld wrote: "In this way by a remarkably rigorous process of deduction, reminiscent of the incontrovertible logic of numerical calculation, we have arrived from the principle of the conservation of angular momentum at a principle of selection and a rule of polarization." The first stated that "the azimuthal quantum number can at the most alter by one unit at a time in changes of configuration of the atom." The second demanded that "if the azimuthal quantum number changes by + or –1, the light is circularly polarized; if the number remains constant, the light is linearly polarized."[125] The derivation, and Rubinowicz's work in general, would remain a central element of quantum studies in the Munich School. Sommerfeld, Werner Heisenberg would later recall, "[emphasized] this point about the conservation in the spherical waves very strongly. So as soon as one had to go away from it, then one would always meet Sommerfeld's criticism."[126]

Sommerfeld did not herald Bohr's general "analogy principle" as a similarly logical and incontrovertible mode of explanation. Indeed, while both Bohr and Sommerfeld began with the same problem, they moved in opposite directions in their solutions. The problem was one intrinsic to the quantum theory. By disconnecting the orbital frequency of an electron from its spectral frequency, one was forced to abandon the normal classical means of expressing the intensities and polarizations of emitted radiation in terms of the decomposed Fourier components of the motion. For the classical case, one can write

$$\xi = \sum_\tau C_\tau(n)\cos(2\pi(\tau\omega_n t + c_\tau)) \tag{5}$$

for the electron's displacement, ξ, in an orbit corresponding in size to the nth stationary state. Here ω_n is the orbital frequency of the electron, and hence $\tau\omega_n$ represent the fundamental frequency and the overtones of the optical spectra, the intensities of which are related to the Fourier coefficients by $I \propto |C_\tau(n)|^2$. The polarizations are given by the orientation of the complex vector $C_\tau(n)$.

For the quantum case, on the other hand, the frequency emitted in a transition from the nth to the $n - \tau$th stationary state is given by

$$\nu_{n,n-\tau} = \frac{E_n - E_{n-\tau}}{h}. \tag{6}$$

This frequency is related to the orbital frequency only insofar as the latter aids in defining the energy of the stationary states. No other aspect of the mechanical motion can be ascribed to the emission, and hence there is no information in the above equation that relates to the polarization or intensity of the emitted radiation. Sommerfeld's solution had been to affirm this fundamental distinction between the quantum and classical motions, and to gain information about the polarization only through the use of various conservation principles. Bohr, on the other hand, in a paper "On the

Quantum Theory of Line Spectra," noted first that in the limit of high quantum numbers, where τ is small in comparison to *n*, the energy difference between the two states in (2) is equal to $\tau\omega h$.[127] Hence the classical and the quantum results coincide for *frequencies* in the slow vibration limit. Bohr now extrapolated from this result in two directions, first to move from frequencies to intensities, and second to move from the limit of high quantum numbers to lower ones. "In order to obtain the necessary connection...to the ordinary theory of radiation in the limit of slow vibrations," he wrote, "we must further claim that a relation, as that just proved for the frequencies will, in the limit of large *n*, hold also for the intensities of the different lines in the spectrum." Following Einstein,[128] Bohr further related these intensities to the *probabilities* of transitions between given stationary states and then added:

> Although, of course, we cannot without a detailed theory of the mechanism of transition obtain an exact calculation of the latter probabilities, unless *n* is large, we may expect that also for small values of *n* the amplitude of the harmonic vibrations corresponding to a given value of τ will in some way give a measure for the probability of a transition between two states for which *n'* − *n"* is equal to τ.[129]

The "correspondence" of the "correspondence principle" (so named in 1920) was thus between a quantum transition and a *single* component of the classical spectrum, decomposed into a Fourier series.[130] With this, Bohr could provide information about the intensities, the transition probabilities, and the polarization of the radiation emitted during a transition. And in cases where the relevant Fourier coefficient could be shown to be zero (e.g. for the harmonic oscillator for all cases where τ was greater than 1), Bohr's analysis suggested the existence of a selection principle, governing the possibility of certain transitions. Sommerfeld could hardly argue with the potential of such a heuristic, and he acknowledged that Bohr had been able, through its use, to arrive at results as good as, sometimes better than, his own. Yet he insisted that the principle was introduced as something foreign to the quantum theory. Justified only in the limit of infinitely great quantum numbers, Bohr's extrapolation from these to finite, even small values made him uneasy. "The validity for this is in no way self-evident and can only be proven through concurrence with experience," wrote Sommerfeld, although he was quick to note that such a concurrence did, indeed, appear to support Bohr's argumentation.[131]

In general, Sommerfeld made a point of not only emphasizing the difference between his own and Bohr's approaches, but of depicting them as fundamentally antithetical. And it was clear that he saw Bohr's as one that had very shaky foundations. In the lines below, the reference to Bohr's principle as a "magic wand" is no compliment. An empirically based mysticism was one thing, a wobbly magic quite another:

> Our object in the above discussion was the reverse of Bohr's: by relinquishing completeness in our inquiry into the wave-theoretical determining elements of the process of emission, we wanted

to align the wave theory and the quantum theory with each other, according to the self-evident maxims of the conservation of energy and momentum and to prove the conceptual compatibility of both standpoints. By contrast, Bohr has found a magic wand in his analogy principle (he himself denotes it as a "formal" principle), which without clearing up the conceptual difficulties allows him to make the results of the classical wave theory directly useful for the quantum theory.[132]

Again, one can see that for Sommerfeld incompleteness was a virtue, since it served almost to reify the conceptual distinctions between the wave and quantum theories while nonetheless emphasizing certain overarching connections. The analogy principle, on the other hand, blended results that were, strictly speaking, applicable only in one realm with those of the other. Sommerfeld's frustrations with Bohr's justification of the correspondence principle continued, even as he came to accept its heuristic utility. In November 1920, he wrote to Bohr telling him of the changes made to the second edition of *Atombau*, which was completed in September, and would appear the next year:

In the appendixes to my book you can see that I have gone to some pains to treat [*würdigen*] your correspondence principle better than in the 1. edition.... Nonetheless, I must admit that your principle, the origin of which is foreign to the quantum theory, is even so still distressing to me, however much I recognize that through it a most important connection between quantumth. and classical electrodynamics is revealed.[133]

Even the appendixes themselves remained neutral on the question of the validity of Bohr's principle. Bohr, Sommerfeld noted, used the correspondence principle as a alternative to his own analysis, and through it was able to determine not only the polarizations of the light emitted after a transition, but also its intensity. In the light of such an achievement, Sommerfeld concluded that "the question of whether Bohr's method is just as satisfactory as our own more incomplete method must recede."[134] It was a question, in any case, he conceded, that depended subjectively on the standpoint one was to adopt in the first place. Yet even in his espousal of the theory, Sommerfeld fails to convince the reader that he is convinced himself. His explanation of the functioning of the principle in producing the probabilities of quantum events—as opposed to its practical efficacy—reads as a series of contradictions:

Although we are convinced that the quantum theory is right in regarding the events that lead to the emission of different spectral lines as independent phenomena, and although we know that the classical calculation is incorrect in treating these events as conditioned mechanically by the motion in the orbit, we yet repose trust in the classical theory to the extent that we derive from it the conditions of intensity of the spectral lines. The classical theory is in error in regarding these conditions of intensity as determined by mechanics; in reality it furnishes the quantum theory with the missing statistics of the individual processes, as it were, without wanting to do so and without giving grounds for it in its foundation. This interlocking of the quantum theory becomes intelligible to some extent of course only from the side of great quantum numbers. The classical theory here hits on the correct vibration numbers. We believe therefore, that for great quantum numbers, too, it will yield the correct conditions of intensity, actually then, the true statistics of the individual phenomenon. Consequently we can understand that we may enlist the aid of the classical theory to get at least approximate results for the statistics in the case of finite or small quantum numbers, too.[135]

By 1922, when the third edition was published, the discussion of the wave theory in relation to the quantum theory had ceased to be either the most troubling or the most exciting area of current research for Sommerfeld. The structure of the book changed dramatically, as what had once been the concluding chapter now came to operate as an *intermezzo* between the older and most secure material (culminating with Bohr's theory and a discussion of the hydrogen spectrum in chapter 4), and later chapters (6–8) which took up Sommerfeld's new empirical method. The rules of selection and polarization laid out in what was now chapter 5 offered a basis for chapters 6 and 7, which examined the data of series and band spectra in detail. Sommerfeld's theory of fine structure—situated in the first two editions before the chapter on the wave and quantum theories—was "now placed at the conclusion of the book to crown the whole."[136] Sommerfeld left open the question of whether his theory of the spherical wave would "stand the test of time."[137] "It is possible," he wrote in the book's forward, "that we are even now passing through a critical period in the history of the wave theory. Yet in this as in other scientific revolutions we shall certainly take much of the older view over into the new one."[138]

The text of the third edition, while retaining a note of suspicious agnosticism in the appendixes and continuing to refer to Bohr's "magic wand," offered an enthusiastic appraisal of the achievements gained through use of the correspondence principle. In common with previous editions, Sommerfeld expressed his satisfaction that he and Bohr should arrive at such similar results from two completely different starting points. Now he added a list of areas in which Bohr's method was in fact superior to his own: it not only provided information as regards the intensity of waves, but even offered "sharper and more definite results" for polarizations; it allowed the extension of Maxwellian electrodynamics even into the quantum realm; and it had proved its "complete superiority...in the matter of atomic models."[139] Yet even here, Sommerfeld's words speak perhaps more to his generosity than his conviction. The words of an author whose book began by eschewing models should be taken with a grain of salt when they appear complimentary toward another's use of them.[140] And certainly one can see why a proponent of the electromagnetic worldview should find a great advantage in the postulation that "Maxwell's theory be generally valid for long waves (Hertzian vibrations of wireless telegraphy), and that it does not throw overboard the many useful results, which the classical theory gives for optical waves and Röntgen rays, but makes fundamental use of them."[141] Yet Sommerfeld had long ago given up the hope that the generalizing dream of electromagnetic theory could be maintained, and for all his enthusiasm he continued to regard what he saw as Bohr's fusion of incommensurable categories as deeply flawed. "The magic of the correspondence principle," Sommerfeld wrote toward the end of 1924, "has proved itself generally through the selection rules of the quantum numbers, in the series and band spectra. The principle has become the guide for all new discoveries of Bohr and his students." Yet Sommerfeld could not regard the principle as truly satisfying "on account of its mixing of quantum-theoretical and classical viewpoints." He would continue, he acknowledged, to regard it as a particularly important *consequence* of a future, complete quantum theory, but could not see it as such a theory's foundation.[142]

With sections finalized as early as August 1923 and a preface completed more than a year later, the fourth edition of *Atombau* offers a glimpse into the rapidity with which Sommerfeld's views on atomic theory could fundamentally change. Among the most striking alterations to the new edition, prompted by news of the "Compton effect," was the effective abandonment of his once-beloved spherical-wave theory. At the beginning of December 1922, Arthur Holly Compton presented a paper to the American Physical Society on "A Quantum Theory of the Scattering of X-rays by Light Elements."[143] It is this paper that is now credited with explicitly framing, for the first time, the "wave-particle duality," with its evidence for the particulate nature of light. Sommerfeld, on a lecture tour of the United States for the winter semester of 1921–22, was among the first Europeans to hear of the result and quickly penned a letter to Bohr.[144] He called Compton's work "the most interesting thing I have experienced scientifically in America," and added (with astonishing praise for a leading X-ray theorist): "after it the wave theory of Röntgen rays will become invalid." Noting that he was not yet completely sure as to whether Compton was correct, Sommerfeld nevertheless predicted that "eventually we may expect a completely fundamental and new lesson."[145] By the time the fourth edition of *Atombau* appeared, any doubts concerning the validity or significance of Compton's results had been assuaged:

The most important theoretical question is, at the moment as before, the darkest: the question as to the nature of light. Whereas, earlier, I had sought to maintain the wave theory for pure propagation phenomena for as long as possible, I have been pushed ever more to the ground of the extreme quantum theory of light. I have assumed the Compton effect as one of the fundamental facts of experience in the first chapter. It is probably the most important discovery that could be made in the present state of physics.[146]

Although its importance could not be gainsaid, Compton's result left Sommerfeld in something of a quandary. The spherical-wave theory, after all, was Sommerfeld's alternative to Bohr's correspondence principle. As this alternative was devalued, Bohr's principle entered the main text of *Atombau* in an extended form for the first time, serving as the introduction to a chapter on polarization and intensity rules for spectral lines. As such, it now preceded a much-abridged discussion of the spherical wave in the chapter's second section. On the other hand, that discussion began with Sommerfeld hedging his bets. "The correspondence principle has not grown out of the soil of the quantum theory," he wrote. "A formal procedure is needed to transplant the results of the classical theory into this soil."[147] Indeed, Sommerfeld continued to insist that his spherical wave was more closely fitted to a quantum-theoretical standpoint. In spite of the fact that the Compton effect appeared to exclude the mediating capacities of the *Kugelwelle*, it retained a place—18 pages, whereas Bohr's principle was granted 10—in the new edition.

If Sommerfeld's overall attitude toward the correspondence principle had not substantially changed between the third and fourth editions of his classic text, his position with regard to some of Bohr's other theoretical claims had altered dramatically. As we saw earlier, Bohr's letter to *Nature* in March 1921 caused Sommerfeld to retract a number of claims regarding electronic structure in atoms. The body of the fourth edition devoted a considerable amount of space to detailing the results that Bohr had hinted at in his letter. Together with a close analysis of the anomalous Zeeman effect and complex structure, the inclusion of Bohr's theory of the periodic system constituted the two principle novelties of the 1924 edition. Bohr's theory of atomic constitution and of "penetrating orbits" is discussed in the next chapter. Here, it is more important to note that Sommerfeld's original enthusiasm had largely evaporated by the time he penned his preface in October 1924.

As Helge Kragh has pointed out, Sommerfeld, like most German theorists, had assumed that Bohr's theory of the periodic system had been deduced from detailed and difficult mathematical calculations, which the Dane had chosen not to publish in his short note to *Nature*.[148] For such theoreticians, Bohr's work promised not merely new (even path-breaking) results, but new *techniques*. Thus, discussing the prodigious problems then facing models of the neutral helium atom in the 1922 edition of *Atombau*, Sommerfeld suggested that "new *methods* will have to be thought out to overcome [their] extraordinary mathematical difficulties." He hoped, however, that such new methods "are already available in Bohr's newest ideas about atomic structure."[149] Further on in the same chapter, he alluded to "the later views of Bohr, *which were accompanied by calculations*, and which he communicated to us at the conclusion of his letter to *Nature*."[150]

By the end of 1922, however, the shine had gone off Bohr's new theory as the promised techniques failed to materialize. Writing from Munich, Gregor Wentzel, who had completed his dissertation with Sommerfeld in June 1921 and who would be thanked most profusely of all those who aided Sommerfeld in producing the fourth edition of *Atombau*, penned an article on "Advances in Atomic and Spectral Theory" for the newly established journal, *Ergebnisse der Exakten Naturwissenschaften*.[151] Wentzel framed his discussion by describing two main approaches to the investigation of atomic structure and spectra: the "inductive, empirical" and the "mathematical, deductive" methods.[152] In the first, one began with empirical materials and sought to build a model for the atom. Examples included Kossell's shell model, Landé's cube, and Sommerfeld's *Ellipsenverein*. Using the purely deductive method, on the other hand, one began with a model and derived results from it. The example Wentzel gave was Bohr's model of the hydrogen atom, where one used perturbative techniques—after the fashion of astronomical analysis—to derive detailed trajectories for electrons moving in planetary orbits. In transitioning from this (essentially classical) procedure to the quantum case, one then applied the correspondence principle to provide the probabilities of transitions between such orbits. The problem for this method, of course, lay in moving beyond the two-body problem of the hydrogen atom to the multi-body problem of higher elements. In making this shift, according to Wentzel, Bohr's methods lost their purely deductive character:

> In their application to the hydrogen atom and its behaviors in electric and magnetic fields, Bohr's methods have, as is well known, proven very fruitful. But, according to Bohr's own comments [*Mitteilung*], they also, apart from this, provide a suitable foundation for the understanding of the remaining atoms. Unfortunately, Bohr has not indicated how this assertion is to be understood in detail; the calculations so far apply only to hydrogen. From Bohr's Copenhagen and Göttingen lectures (October 1921 and June 1922) we know, however, that his investigations of the construction of complex atoms do not quite possess any longer the purely deductive character of his studies of hydrogen.[153]

Wentzel's essay was written while Sommerfeld was in the United States. Sommerfeld's own critiques of Bohr's methods did not appear in print for more than eighteen months. When they did, in the preface to *Atombau* (1924), his distaste for what he clearly saw as Bohr's idiosyncratic use of the correspondence principle, "intuitive" arguments, and vague invocation of symmetries came through clearly. Empirical data, appearing after the galley proofs for the relevant chapters had been completed in August 1923, had already led to significant alterations in Bohr's model. Exceptions to Bohr's system were, in any case, Sommerfeld added, to be expected, since it "was certainly not mathematically grounded but was, rather, intuitively grasped."[154]

As in the third edition, indeed to an even greater extent, Sommerfeld felt justified in vaunting his less *modellmässig*, more purely quantum-empirical approach, extending it to include the details of all spectra. "Throughout," he wrote, "since I have restricted myself, in essence, to the quantum-theoretical ordering of the facts and have placed in the background, in a fashion similar to that done earlier for X-ray spectra, all atom-mechanical speculations, I hope to achieve a situation where the representation will not age too quickly."[155]

In view of this attitude toward modeling, it should not be surprising that Sommerfeld maintained his criticisms of the correspondence principle. Here, at least, he and Wentzel would appear to be in striking disagreement. Where, for Wentzel, Bohr's principle—when coupled with the astronomers' perturbative methods and a concrete model—provided the ideal example of a "deductive" research method, for Sommerfeld it did nothing of the sort. The use of the correspondence principle, to the contrary, was but one example of a broader problem that Sommerfeld identified in Bohr's approach to physics. And in that problem schema one can see how the two halves of Sommerfeld's own approach to the quantum theory—his empirically based *Atom-Mystik* and his determinedly incomplete bridging of the quantum and electromagnetic theories—fitted together. "The difficulties that emerge ever more clearly in atomic physics," he wrote in a 1924 essay on "The Foundations of Quantum Theory and of Bohr's Atomic Model," "appear to me to arise less from an exaggerated application of the quantum theory and much more from a perhaps exaggerated belief in the reality of concepts of models [*Modellvorstellungen*]." The next sentence made clear to whom he ascribed such beliefs. "Obviously the hydrogen model functions correctly in all cases (except perhaps in strong magnetic fields) and surely Bohr's explanation of the chemical systematics of the periodic system is generally reliable. But the phenomena are much simpler than we would expect according to the models."[156] Landé's analysis of the term splitting—which had proceeded along Sommerfeldian quasi-phenomenological and numeric lines—had provided certain laws, but for all its success it had made clear that such laws could not proceed from a simple model. It would appear, Sommerfeld concluded, that "the model of the atom would then be more a means of calculation [*Rechenschema*] than a representative of reality [*Zustandsrealität*]. Admittedly this would be very regrettable for the visualizability of our model conceptions. But it would be nonetheless reconcilable [*erträglich*] with our demand for the unique mathematical determination of theory."[157]

As well as such specific analyses of the state of the field, this paper on the foundations of the quantum theory expressed a broader and more overarching intent. A central theme contrasted what Sommerfeld put forward as the ideal scientific method with an alternative methodology that made extensive use of analogies. It was a contrast that Sommerfeld denoted as that between deductivism and inductivism. One hears the echoes of Wentzel's paper, but the dichotomy had appeared in a similar context several years earlier, during Sommerfeld's visit to Denmark in 1919. In his "Speech of Appreciation," Bohr described Sommerfeld's early years as a mathematician, and then his turn toward mathematical physics, a field in which his talents compared very favorably with the often "dilettantish" efforts of those with considerably less mathematical training and talent. "But more and more you are attracted to the physical side of the problems...you are now absorbed with all your energy with the treatment of a question where one could perhaps expect least from the mathematical deductive method of treatment and could hope most of the physical inductive direction of thought: the question of the construction of the atoms and molecules of the elements."[158] Five years later, Sommerfeld appeared to have accepted the distinction but inverted its valence. "Every fundamental physical theory," he wrote, "must at the end of the day proceed deductively." Newtonian mechanics, with its axiomatic basis, provided the first exemplar of the true scientific methodology. In contrast, electromagnetic theory only achieved such a status after the dynamic analogies introduced by Maxwell in its early development had been replaced by Hertz's system of partial differential equations. Immediately forestalling the objection that the quantum theory was not yet "ripe" for such a purely deductive representation, Sommerfeld acknowledged that "That is very possible, [yet] in spite of this we may sketch for ourselves that ideal form of the quantum theory that is contrary to the one that presumably [*mutmasslich*] has developed."[159]

The rhetorical ploy was a neat one. To the critic who noted (justifiably) that Sommerfeld's methods—particularly that which proceeded *a posteriori*—might hardly be called deductive, Sommerfeld could reply that he at least had a deductive quantum theory in mind as a final goal. More significantly, the point here is clearly less about deduction *per se* and more about the production of a *Rechenschema* independent of any model. Indeed, working from a model, in Sommerfeld's (admittedly idiosyncratic) scheme, counts as inductive rather than deductive, since deductions proceed directly, without connection to an external referent. The invocation of Hertz is telling. Just as the hero of Sommerfeld's youth had abandoned Maxwellian models, claiming famously that "Maxwell's theory is Maxwell's equations," Sommerfeld's aim was to strip models out of quantum theory. His description of the functioning of his phase-integral method certainly sounded almost syllogistic as he laid it out: "By defining certain areas of this phase space in the canonical variables p and q, we come to the so-called phase integral. By asking when these integrals can be carried out independent of one another, we are led to multiply periodic systems and to the method of separation. The arithmetic character enters from the fact that the areas defined are whole number multiples of h."[160] Perhaps more important, then, than whether Sommerfeld's approach might be called deductive is the fact that Bohr's methodology was "inductive," and hence the exact opposite of the true scientific method that Sommerfeld had defined. In contrast to the phase-integral method, Sommerfeld wrote, "Bohr seeks in his correspondence principle to tightly connect the quantum theory to the classical radiation theory. He proceeds, as far as is possible, *inductively and physically*, by allocating at each step a period of motion to each quantum number." Sommerfeld accorded the same criticism to Bohr's use of Schwarzschild's action-angle variables in combination with the adiabatic principle: "To each quantum number is allocated a certain periodicity and finds in this its visualizable kinematic counterpart [*Gegenbild*]." Both principles, that is, functioned via analogy with classical physics, by the stepwise pairing of quantum and classical concepts. In contrast, according to Sommerfeld, his method, together with the sharpening offered by Schwarzschild and Epstein's method of separation, "led by means of the radial quantum condition *immediately* to the series formula...while by Bohr's system the quantization emerges somewhat *indirectly.*"[161]

Here, then, are all Sommerfeld's critiques melded into one: Bohr's method was indirect, inductive, and founded on a problematic mixing of incommensurable worlds. Rather than seeing Sommerfeld's lack of enthusiasm for the correspondence principle as the result of an excessive literalism and rigidity of thought, we are led by a symmetrical consideration of his parallel attempts to solve similar quantum problems to a much deeper explanation. Sommerfeld's objection was to the use of analogical thinking, "formal" or not, in the consideration of the quantum. Distinct worlds were, for him, to be treated in distinctly different fashions. One can begin now also to see Pauli's description of the two main currents of thinking about the quantum in the 1920s as a particularly perspicacious one. While Bohr had sought "a key to translate classical mechanics and electrodynamics into quantum language which would form a logical generalization of these," Sommerfeld had looked for "a direct interpretation, as independent of models as possible." Whatever the positions of Bohr and Sommerfeld themselves, for students like Pauli, the methods "did not appear to me irreconcilable," and the new quantum mechanics that would begin to emerge after 1924 would exhibit facets of both. It is to the students of the Sommerfeld School that we turn in the next section.

The Sommerfeld School, 1919–1926

My discussion of Sommerfeld's *Atombau* has been at once a study of research and of pedagogy in the Munich School. The book was both based upon, and became the basis of, lectures throughout the late 1920s. Many of Sommerfeld's students (both past and present) were contributors to the text, as the book's footnotes and acknowledgements testify. Heisenberg described it as "the common work of the whole institute," noting that students would write certain parts (he estimated about 20 percent of the whole) and then Sommerfeld would revise them.[162] Their work was clearly significant enough as research that Sommerfeld mentioned it explicitly in reports to his faculty colleagues regarding students' dissertations.

This section is nonetheless different in motivation to those before. Until now my aim has been to uncover and recover the logic (or logics) of Sommerfeld's broad scientific methodology in the first half of the 1920s, and to demonstrate that the Munich school offered a viable—indeed, powerful, flexible, and comparably encompassing—alternative to Bohr's search for general heuristic principles. What has been less clear has been the object to which Sommerfeld's methodology was applied. For an approach that was earlier characterized as the "physics of problems," there has been altogether too little discussion of the range of concrete questions and puzzles with which the Munich school was engaged after the war. I aim to rectify this failing here, with a discussion of the problems of Weimar physics. As will become clear, while the day-to-day functioning of the Sommerfeld School after 1918 tended, in general, to perpetuate patterns laid out before the war, at the more detailed level of particular problems—especially those of the quantum theory—teaching and the work of students mirrored Sommerfeld's own shift away from earlier approaches.

In chapters 2 and 3, I discussed Sommerfeld's teaching methods for the first dozen years of the Sommerfeld School, emphasizing four elements characteristic for the pedagogical practice of the physics of problems: the eclecticism of the problems solved, their often practical bent, their shared nature, and the prevalence of problems that were formulated within the context of a programmatic commitment to the electromagnetic view of nature. As we have seen, this last aspect continued in a modified form in the postwar period, as Sommerfeld sought—as he had at the Solvay conference—to develop an understanding of the atomic world in terms of both the electromagnetic and quantum theories. Separate but equal, both modes of explanation were necessary for his study of atomic structure and spectral lines. Indeed, describing his fine-structure theory, Sommerfeld would write of the "confluence of the three main currents of modern research in theoretical physics, namely, the theory of electrons, the theory of quanta, and the theory of relativity." In 1917, one of Sommerfeld's students, Karl Glitscher, showed that the new spectroscopic theory allowed a decisive determination of the relative merits of the "rigid" (Abraham) and "deformable" (Einstein-Lorentz) electron theories.

It would be wrong, however, to focus exclusively on atomic theory. Twenty-five students completed dissertations between 1919 and the winter semester of 1926–27. While quantum spectroscopy was the focus of a bare majority of dissertation topics (13), many older subjects recurred. Werner Heisenberg, for example, received his degree not for the work in which he introduced half quantum numbers, but for a study "On Stability and Turbulence in Streams of Fluids." The embeddedness of that piece in the work of the Sommerfeld school may be seen from the fact that, of the five different names listed in the footnotes of the first page, three were those of Sommerfeld students or collaborators (Fritz Noether, Ludwig Hopf, and Otto Blumenthal) and a fourth was that of Sommerfeld himself.[163] The problem of hydrodynamic stability was taken up by a second Sommerfeld student in 1926, as Karl Schlayer met the challenge of extending von Karman's two-dimensional analysis of his so-called vortex street to deal fully with the three-dimensional case.[164]

The former emphasis on non-quantum electromagnetic problems continued as well, albeit to a lesser extent. Four of the dissertations dealt with some aspect of optics, for example; two of these exploring the operation of the ultra-microscope, a device developed in 1903 to make submicroscopic colloids visible for the first time. Two other students continued the Sommerfeld School's leading work on the theory of wireless telegraphy. Hans Pfrang took up a problem that Sommerfeld had explored, but not completed, during the war, namely that of wireless communication between airplanes and earthbound stations. To the first approximation apparently used in war work, one could treat the earth as an infinitely good conductor. Pfrang's dissertation, like the work of Sommerfeld and, after him, Hermann Weyl, provided a more exact analysis by taking the Earth's finite material properties into account. For the most significant result (a fully general reciprocity theorem), Sommerfeld made a strong claim of practical utility. "It cannot be missed," he wrote in his report to the faculty, "that this theorem will also be valued and applied in the practice of wireless telegraphy."[165] Ernst Guillemin's topic was similarly lauded for its practicality. After noting that Guillemin came from a "particularly good German-American" family he had met during his travels in the United States, Sommerfeld described the project:

> The work at hand connects up with an old tract of mine on arc oscillations in the Proceedings of the Bavarian Academy and, like that one, arises from problems that have been treated experimentally by Professor Zenneck. It deals with a task essential for the practice of radio telegraphy; transforming the frequency of an AC machine through an intermediary iron core. Due to the similarity of its characteristics, an iron core works in a similar fashion to an arc, but it has great advantages over it in practice.[166]

Further dissertation topics included a theoretical investigation of the functioning of the eardrum as a pressure-measuring device (a "manometer"); an attempt to represent the mathematical operations of general relativity in a more geometrically visualizable form, after the fashion of Minkowski's and Sommerfeld's geometrical reworking of special relativity; a statistical-mechanical treatment of the theory of concentrated solutions; and a largely mathematical study of crystal structures by a former student of Max Born's. This last drew heavily enough on work originally published by Paul Ewald that Sommerfeld requested his former student to add his own comments to the report to the faculty on the thesis.

One can see, then, close ties binding the Sommerfeld School before and after the war. Similar areas of research were explored, similar sets of problems recurred, and the work of former students became the starting point for newer dissertations. The range of problems remained broad, Sommerfeld would often emphasize their practical import, and a significant portion took up the problems of classical Maxwellian electrodynamics and the electron theory. Yet the focus on the quantum theory that had already begun to make itself apparent during the war years became thoroughly entrenched during the first half of the Weimar republic. And it was most clearly in the context of quantum problems that the shared nature of work in the Sommerfeld school in this period became clear. Writing of Karl Bechert's dissertation on the nickel spectrum, for example, Sommerfeld began by noting that the topic "ran along the lines of investigations particularly valued in my institute on the disentanglement of complicated spectra."[167] And *Atombau*, in general, it has already been noted, was seen as the collective work of the institute for theoretical physics. Yet, as before, it is in the social functioning of the school, as much as the details of particular problems, that one sees the ties that bound students to each other and to their *Doktor-Vater*. Word of new developments from the other great center of quantum research, for example, was dissected in a common space, worked over until the institute as a whole had understood and digested it. Heisenberg described the process as follows:

> Quite frequently, if somebody brought the news of this kind, say news from Copenhagen about quantum theory, then people would stand together at the blackboard. Pauli would be asked what he thought about the latest news; he would try to analyze it at the blackboard, but other people would interrupt him and so on. And so we tried to get a common opinion on the recent development.
>
> We heard about these developments either by new periodicals which came in, or by letters which Sommerfeld had gotten. If he had gotten a letter from Einstein or Bohr or others of this group, then he would usually hand the letter to Pauli and say, "Well, Pauli, here look, Einstein writes this—what do you think about it?" So this was the eternal life in the Sommerfeld Institute.[168]

This golden age of the Sommerfeld School may well have seemed eternal to a man who had not yet turned 20 when he began making contributions to the most cutting-edge research of contemporary physics. But life in Munich was, of course neither everlasting nor unchanging. Indeed, at the time Heisenberg arrived at the institute in 1920, Sommerfeld was shifting toward the *a posteriori* method discussed above. It was a shift that marked a sea change in his own work as well as that of his students. The first task assigned the precocious son of Munich's professor of Byzantine languages and literature was the disentanglement of levels in the spectra of the anomalous Zeeman effect. Sommerfeld merely passed over the experimental data, and asked his student to draw the positions of the initial and final states of an emitting electron from the frequencies of the lines. It was only later (in 1922) that Sommerfeld assigned Heisenberg the task of providing a model for the phenomena that Landé in particular had described in phenomenological terms, and Heisenberg's memory of the logic of that task makes clear the primacy of the empirical laws over any form of model-based deduction. "I tried to do my best," Heisenberg later related, "using this material of Landé and of Sommerfeld, to work *back* to a model."[169] That is to say, one did not begin from a model and then derive results to be compared with experience. One began with the data, derived empirical laws from them, and then sought a model that might produce (or at least reflect) such empirical regularities.[170]

Even such a limited faith in (or at least search for) models may have seemed impossible in the first years after the war. In early 1920 Sommerfeld described Josef Krönert's dissertation on the anomalous Zeeman effect as providing an insight into an area of study that at the time was "truly dark." Of the polarization rules Krönert was able to derive empirically, Sommerfeld noted that "naturally it is theoretically for the time being incomprehensible, just as are all other phenomena in the anomalous Zeeman effect."[171] Writing of Gregor Wentzel's study of the systematics of the X-ray spectrum in mid 1921, Sommerfeld claimed similarly that "all the results of this work are essentially of an empirical nature. The time is not yet ripe for a model-based derivation of them. The security and the soundness of Wentzel's work lies in this more empirical than calculational character of the results."[172] Sommerfeld greeted Pauli's dissertation two and a half weeks later with all the praise due a prodigy, but it is noteworthy that no similar affirmation of the "soundness" of his results was forthcoming. The latter part of the paper dealt with a model of molecular ionized hydrogen (H_2^+) and attempted, as Pauli put it, to calculate stable quantum orbits "according to *general theoretical principles*."[173] After pointing out the general considerations that arose from the work, and which "could be of considerable consequence," Sommerfeld's description of the project ended with the rather more laconic hope "that the results of the work...can, sooner or later, be tested experimentally."[174] Certainty and security, in other words, were to be found, at least for the present, not in model-based deductions, but empirically grounded rules. Indeed, the logic of an aesthetically grounded *a posteriori* method would dominate the research of both Sommerfeld and his school until the mid 1920s, when the rise of the methods of wave and matrix mechanics offered vastly more powerful means to solve the problems of the quantum.

Conclusion

I was always satisfied if I could explain a certain complex of facts mathematically, without troubling myself too much with the fact that there were other things that didn't fit. Einstein, who always looks at the whole picture [*auf's Ganze blickt*], makes life more difficult for himself.
—A. Sommerfeld to J. Sommerfeld, December 24, 1928[1]

By the middle of 1925, Wolfgang Pauli was in a better mood than he had been in for some time. Only two months earlier, his exasperation with the state of theoretical physics had boiled over into a fit of pique. At the moment, his discipline, he acknowledged to Ralph Kronig in May, was developing apace; "nevertheless, it is much too difficult for me and I wish that I were a film-comedian or something similar and had never heard anything of physics!"[2] A letter written to Hendrik Kramers at the end of July revealed both his changed perspective and the personal reasons for his earlier dismay: "I feel at the moment a little less lonely as about half a year ago, when I found myself (spiritually as well as spatially) pretty much alone between the Scylla of the number mysticism of the Munich school and the Charybdis of the reactionary Copenhagen putsch propagated by you with the excesses of a zealot!"[3]

The timing of Pauli's last comment is significant. Six months before the letter to Kramers he had, in fact, produced the work for which he remains most famous and for which he received the Nobel Prize in 1945: the exclusion principle. Worked through at the end of 1924 and received by the *Zeitschrift für Physik* on January 16 of the new year, "On the Connection of the Closing of Electron Groups in the Atom to the Complex-Structure of Spectra" laid out Pauli's new quantum rule in simple, declarative terms: "There can never be two or more equivalent electrons in an atom for which, in strong fields, the values of all quantum numbers...coincide. If an electron is to be found in an atom for which these quantum numbers (in an external field) possess determinate values, then this state is 'occupied.'"[4]

As we saw in chapter 7, Pauli's sense that there were two main elements running through his work in the 1920s would remain with him. Shorn of its characteristic sarcasm, his sense of the significance of both the Sommerfeld and Bohr schools to his work on the exclusion principle reappeared in his Nobel speech.[5] Yet, although a considerable amount of space within a growing secondary literature on the history of quantum mechanics has been devoted to the importance to Pauli of the "Copenhagen putsch," almost nothing has been written on the corresponding significance of the Sommerfeld School's "number mysticism."[6] The point is even more striking for the absence of any mention of Sommerfeld in the development of quantum mechanics more generally. Although acknowledged as one of the most prolific and important contributors to the older quantum theory, not a single paper by Munich's professor of theoretical physics is to be found in B. L. van der Waerden's *Sources of Quantum Mechanics*.[7] Daniel Serwer's 1977 paper on Pauli and Heisenberg's reactions to the "unmechanical force" [*unmechanischer Zwang*] introduced by Bohr in the early 1920s excluded from the outset Sommerfeld's role in the further development of his former students' thinking in the crucial period 1923–1925. Insisting that his particular interest was in the *differences* between the two young men, Serwer avoided any extensive discussion of the ongoing impact of the similarities in their training:

> Sommerfeld was important for the basic education he offered Pauli and Heisenberg in the details of atomic spectra and in the use of the quantum-theoretical tools available to "save the phenomena." Since I assume Pauli and Heisenberg to have this basic education by 1923, I shall not have much to say about Sommerfeld. After they left Munich, Pauli and Heisenberg became increasingly concerned with whether the quantum mechanical postulates were adequate.[8]

Bohr's work, Serwer continued without further elaboration on what would appear to be the crucial point, "was more directly relevant to this concern than Sommerfeld's during 1923 to 1925."[9] Most recently, in a detailed study of the work leading up to the paper by Heisenberg that would launch the new mechanics, Michel Janssen and Anthony Duncan have argued that the problems central to the Sommerfeld School in the early 1920s were not only not useful to Heisenberg, but were actively obfuscatory: "…many of the other preoccupations of the old quantum theory, such as the detailed understanding of spectral lines, the Zeeman and Stark effects, and the extension of the Bohr-Sommerfeld model to multi-electron atoms (in particular helium), mostly added to the overall confusion and did little to stimulate the shift to the new mode of thinking exemplified by the *Umdeutung* paper."[10]

The aim of this concluding chapter is to delineate the ongoing historical and historiographical payoffs for the history of physics (and the history of science more generally) of the detailed study of Sommerfeld's "practice of theory." It proceeds in two parts. The first section takes the case of Pauli's path to the exclusion principle as a means of drawing out the importance of Sommerfeld's work—and that of his school—to the origins of the exclusion principle and to the birth of quantum mechanics. Accepting Janssen and Duncan's claim that the problems of spectroscopy were less significant to Heisenberg's *Umdeutung* than the more central dispersion problem, it is nonetheless argued that at the level of *method*, rather than of content alone, Sommerfeld's phenomenological turn in the 1920s was one element crucial to the development of the *discourse* of the new physics: its exclusive focus on "observable" quantities. The second section examines the reception within Sommerfeld's Munich school of Heisenberg, Born, and Pascual Jordan's matrix mechanics on the one hand, and Schrödinger's wave mechanics on the other. Schrödinger's approach, while acknowledged in many quarters to be less rigorous in conceptual terms, was lauded by Sommerfeld for its simple and powerful problem-solving capacities. Based as it was on familiar mathematical techniques, wave mechanics was almost tailor made for a physics of problems. In spite of its power and successes, however, Sommerfeld did not join many of his contemporaries (nor most subsequent analysts) in celebrating a new quantum-mechanical revolution. Indeed, as we shall see, his insistence that the new approaches constituted improvements but not ruptures in physical thought offers a deep insight into the connection between practice and participatory understandings of the nature of scientific change.

Exclusion

Magic, Harmony, and the Exclusion Principle

Pauli's path to the exclusion principle proceeded by way of two topics dear to Sommerfeld's heart: the anomalous Zeeman effect and the problems of *Atombau* and Röntgen spectroscopy.[11] Indeed, as the title of his paper suggests, the union of the solution to these apparently distinct problems provided the key to his success. He took up the anomalous Zeeman effect in late 1922, together with Bohr, who had agreed to write a short paper on the topic for a special issue of the *Annalen der Physik*, due in January 1923, celebrating the 70th birthday of the experimentalist Walther Kayser. The fact that Bohr did not send the article off until March led Pauli to quip that his mentor had effectively made Kayser younger by several months. Pauli himself was still stumped in June, writing to Sommerfeld with evident exasperation that it "would not, but would not come out."[12]

The problem with which Pauli was grappling was his attempt to explain the *Gesetzmässigkeiten* determined previously by Alfred Landé, like Pauli himself a former Sommerfeld student. In the early 1920s, Landé had incorporated and redefined the "inner" quantum number, j, to obtain a value for the size of line splittings in a weak magnetic field. Classically, such a field, when applied to a system like Bohr's planetary model of the atom, should cause all electrons to precess together at a given frequency around the field axis. The induced separations between given lines should then be proportional to this precessional (Larmor) frequency (i.e., $\Delta E = m\omega h$, with m integral). In the quantum case, however, Landé determined (empirically, after the fashion of Sommerfeld's *a posteriori* method) that the energy separation was given by the classical value multiplied by another term, the so-called *g*-factor (i.e., $\Delta E = g \cdot m\omega h$).[13] In 1921, when Landé first introduced this term, he could offer no physical interpretation of its meaning, other than that it was related to *j*, now understood as the total angular momentum of the atom.

A *modellmässig* explication of the *g*-factor had to await Heisenberg's publication, in 1922, of his *Rumpf* model of the atom. Infamously introducing half-integral quantum numbers, Heisenberg suggested that the observed splitting could be explained by the existence of a "coupling" between the angular momentum of a single, outermost electron (governed by *k*, the azimuthal quantum number) and the sum of the angular momenta of the remaining electrons (given by *r*), which formed the atomic core [*Rumpf*]. The sum of these two net momenta then gave, according to Landé, the total angular momentum of the atomic system, *j*. Doublets, in Heisenberg's peculiar vision, arose from a "sharing" of angular momenta: in effect, the outer electron gave a half quantum's worth of momentum to the core. This *ad hoc* move gave the right number of lines, but did so at the expense of any customary physical understanding.[14] In 1923, leaving most of the model's difficulties to one side, Landé deployed a conception similar to Heisenberg's to offer an expression for g[15]:

$$g = \frac{3}{2} + \frac{1}{2} \cdot \frac{R^2 - K^2}{J^2 - \frac{1}{4}}.$$

Pauli wrestled for months to provide, as he put it in a letter to Landé, "a satisfactory *modellmässige* meaning for such astoundingly simple *Gesetzmässigkeiten*."[16] His efforts were not crowned with success. When he published on the topic in April, he did so, he wrote to Sommerfeld, "with a tear in the corner of my eye," deeply unimpressed with what he had achieved.[17] Instead of *explaining* a peculiarity of the *Rumpf* model—the fact that the magnetism of the core appeared to be twice that which one would calculate from classical electron theory, while the magnetic moment of the external electron conformed to the classical result—Pauli proceeded "purely phenomenologically."[18] The odd doubling of the core's magnetic moment was, in effect, simply taken as an empirical datum.[19]

Heisenberg lauded the result, but Pauli had clearly had enough. For a year he withdrew from work on atomic physics, returning to the field only after accepting a brief to complete an encyclopedia article on the principles of the quantum theory in mid 1924. By the time he did so, his attitude toward the importance of modeling in quantum theory had undergone a radical transformation. As we saw in chapter 7, Pauli's highest praise for the fourth edition of *Atombau* (completed in October 1924) was reserved for its avoidance of *modellmässig* explanation for the problems of complex structure. A severe blow to a model-based method, however, had already been struck by Born and Heisenberg in early 1923. Taking up the problem of the helium atom, the two men demonstrated that, for the excited atom, detailed calculations using perturbation theory gave results irreconcilable with experiment. The failure of the Bohr-Sommerfeld model could no longer be attributed to insufficiently precise analysis. Pauli's negative reaction to the "reactionary Copenhagen putsch" and its faith in the powers of the correspondence principle had reached its zenith with the joint publication of a treatment of the dispersion problem using "virtual" oscillators and violations of the energy principle by Bohr, Kramers, and the Englishman John Slater. "I would always much rather say," he wrote to Kronig in 1925 with regard to the so-called BKS theory, "that I have so far no complete picture of the phenomena, than even temporarily to put up with a hideousness of this kind which hurts my physical sensibility."[20] Sommerfeld's number-mystical method no doubt appeared more appealing to Pauli as the reputation of its alternative seemed ever more tarnished.

In the midst of writing his encyclopedia article, Pauli hit upon a novel idea: a means of testing a fundamental assumption of the *Rumpf* model. If an alkali core had a non-zero magnetic moment (as it must when sharing some of the magnetism of the external electron), then g should depend on Z, the atomic number. Landé, tapped for data on the topic, soon returned the results: no dependence on Z, therefore no magnetic moment for a closed shell. The "sharing" model was dead. The cause of the anomalous Zeeman effect was not to be found in a peculiar coupling of the *Rumpf* with an external electron, but in some peculiar property of the electron alone. "The doublet structure of the alkali spectrum, as well as the violation of Larmor's theorem comes about through a peculiar, classically non-describable kind of *Zweideutigkeit* [ambiguity, doubled signification] of the quantum-theoretical characteristics of the light-electron."[21]

In a stroke, Pauli had solved (or, rather, dissolved) the problem of the *modellmässig* understanding of j and the cause of the apparent doubling of the magnetism of the atomic core. The *Rumpf* ceased to play any role in the production of spectral lines and hence the notion of a "total" angular momentum ceased to have any great significance. Since the violation of Larmor's theorem—the fact that g was not equal to unity—was the result of a "classically non-describable" characteristic of the outer electron, no model-based understanding should (or could) be sought. The *Zweideutigkeit* was a quantum property with no classical counterpart.

The "exclusion principle" proper was predicated on the notion that electronic behavior (and only that of electrons) was governed by four quantum numbers. The problem that it solved, however, arose not from the complexities of the Zeeman effect, but rather from the attempt to explain the processes of *Atombau* and the reason for the periodic structure of the table of elements. That question had been in Bohr's mind since his first quantum paper in 1913. It was not until his "second atomic theory," however, that he could develop what appeared to be a complete solution.[22] The theory was first revealed to the world in a letter to *Nature* in 1921, the same letter that, according to Sommerfeld, "battered in like a bomb," causing seismic shifts in his and his school's understanding of the processes of atom-building.[23] Two novel elements—together with the all-powerful correspondence principle—formed the essence of Bohr's second theory: the notion of "penetrating orbits" and the "construction principle" [*Aufbauprinzip*]. The first made explicit and detailed use of the Bohr model and argued that outer electrons in elliptical orbits about the atomic nucleus must penetrate the orbits of internal electrons, causing deviations in the Coulombic force acting upon them. Where, for Sommerfeld, the fact that the *Ellipsenverein* for the L shell intersected the circular K shell was an argument against the harmonious interlocking of electronic orbits, for Bohr the coupling between penetrating and inner electrons was essential to the theory.

Bohr's arguments proceeded by way of both analogy and disanalogy with the Bohr-Sommerfeld model of the Hydrogen atom. Characterizing each orbit by two quantum numbers, n and k ($n = 1, 2, 3, \ldots, \infty$; $k = 1, 2, 3, \ldots, n$), Bohr took up the question of the shape and size of each n_k orbit for elements with $Z > 1$. As was known from Sommerfeld's work on hydrogenic atoms in 1915, when $n = k$, orbits are circular; elliptical with increasing eccentricity as $n - k$ increases. The area enclosed by each orbit increases with n, so that any 4_k orbit, for example, is larger than any 3_k orbit. It is precisely this aspect that would change in Bohr's second theory. (See figure C.1.)

The notion that electrons "filled" lower orbitals in a fashion that reproduced the structure of the periodic table (two electrons in the lowest orbital, eight in the next, and so on) was one common to almost all models of atomic structure after Bohr's

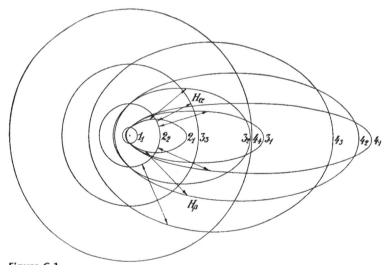

Figure C.1
Hydrogen orbits. Source: Niels Bohr, "The Structure of the Atom (Nobel Lecture, delivered December 11, 1922)," *Nature* (1923): 10.

first model in 1913. This implied, of course, that outer electrons in elements of higher atomic number than helium were "shielded" from the full nuclear force experienced by inner electrons. According to Bohr's analysis in the 1920s, however, some orbits for $n \geq 3$ penetrated lower orbitals. "Even though a 3_2 orbit will not penetrate into the innermost configuration of 1_1 orbits," he wrote in 1921 in a paper titled "The Structure of the Atom and the Physical and Chemical Properties of the Elements," "it will penetrate to distances from the nucleus which are considerably less than the radii of the circular 2_2 orbits."[24] In doing so, a 3_2 electron would feel the same force upon it—for this part of its orbit—as an electron in a lower orbital.[25] The effect would be to radically change the size and shape of its trajectory around the nucleus, shifting some elliptical orbits to lower overall energy states and causing the electron to trace out the shape of a rosary, rather than a single, repeated ellipse. Given that, as figure C.2 indicates, 4_1 and 4_2 orbitals are—contrary to the model of the Hydrogen atom—energetically more stable than 3_3 orbitals, one would expect these to be filled first. And so Bohr argued, at least for the first and second elements of the third period of the periodic table. After that, Bohr claimed that a "simple calculation" demonstrated that the increasing nuclear charge caused the higher 4_1 and 4_2 orbitals to increase in size in relation to a 3_3 orbit.[26] Electrons added to the structure of elements following Calcium ($Z = 20$) thus ceased filling the 4_1 orbits and began filling the 3_3 instead. Since the *Aufbauprinzip* held that the electronic structure of elements with $Z = N + 1$ was identical to that for element $Z = N$ except for the placement of the $N + 1$th electron, one could now explain the puzzling chemical

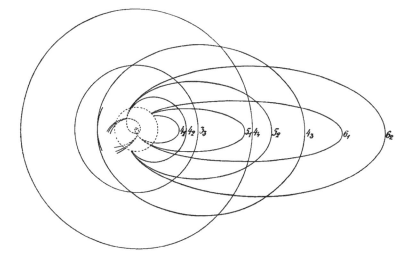

Figure C.2
Orbits of multi-electronic atoms. Source: Bohr, "Structure of the Atom," 12.

properties of the transition metals—and especially the "Lanthanides" and "Actinides"—in the table of elements. The elements from lanthanum ($Z = 57$) to lutetium ($Z = 71$) are, in spite of their increasing atomic numbers, chemically near-identical. In Bohr's theory, this followed from the fact that each such element possessed an identical number of *valence* electrons in a 6_1 orbital, even as each added more electrons to a previously empty (because previously higher energy) 4_4 orbit. Less convincing were Bohr's explanations of the *periodicity* of the periodic table. Why, for example, were there eight (and only eight) electrons in the second shell? For this, Bohr could only invoke somewhat vague symmetry arguments, which few outside Copenhagen seemed to find believable.[27]

Sommerfeld, originally enthusiastic about Bohr's seemingly miraculous calculations, soon became disillusioned as he and others in Germany realized that Bohr's results did not follow from detailed mathematical analysis at all.[28] Instead, even after being pressed for his method, Bohr could only invoke the power of the correspondence principle and his own intuition. The fourth edition of *Atombau* and a number of subsequent papers, as we have already seen, contained several explicit rebukes concerning Bohr's approach to the problems of atomic structure and reaffirmed Sommerfeld's distrust of *modellmässig* accounts. Much more promising seemed to be a new direction proposed by an English researcher, E. C. Stoner, in the *Philosophical Magazine* in October 1924. "According to Stoner," Sommerfeld wrote in his preface, dated the same month, "the shells in the interior of the atom are to be further subdivided, the number of electrons in the subgroups of the atomic shells are to be differentiated among themselves and are determined through the formal rules of the inner quantum

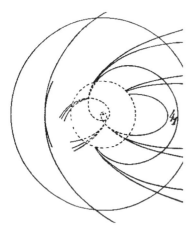

Figure C.3
Changed orbital shapes as electrons penetrate lower shells (section of figure C.2). Source: Bohr, "Structure of the Atom," 12.

	1_1	$2_1 2_2$	$3_1 3_2 3_3$	$4_1 4_2 4_3 4_4$	$5_1 5_2 5_3 5_4 5_5$	$6_1 6_2 6_3 6_4 6_5 6_6$	$7_1 7_2$
1 H	1						
2 He	2						
3 Li	2	1					
4 Be	2	2					
5 B	2	2(1)					
– –	–	– –					
10 Ne	2	4 4					
11 Na	2	4 4	1				
12 Mg	2	4 4	2				
13 Al	2	4 4	2 1				
– –	–	– –	– –				
18 A	2	4 4	4 4				
19 K	2	4 4	4 4	1			
20 Ca	2	4 4	4 4	2			
21 Sc	2	4 4	4 4 1	(2)			
22 Ti	2	4 4	4 4 2	(2)			
– –	–	– –	– – –	–			
29 Cu	2	4 4	6 6 6	1			
30 Zn	2	4 4	6 6 6	2			
31 Ga	2	4 4	6 6 6	2 1			
– –	–	– –	– – –	– –			
36 Kr	2	4 4	6 6 6	4 4			
37 Rb	2	4 4	6 6 6	4 4	1		
38 Sr	2	4 4	6 6 6	4 4	2		
39 Y	2	4 4	6 6 6	4 4 1	(2)		
40 Zr	2	4 4	6 6 6	4 4 2	(2)		
– –	–	– –	– – –	– – –	–		
47 Ag	2	4 4	6 6 6	6 6 6	1		
48 Cd	2	4 4	6 6 6	6 6 6	2		
49 In	2	4 4	6 6 6	6 6 6	2 1		
– –	–	– –	– – –	– – –	– –		
54 X	2	4 4	6 6 6	6 6 6	4 4		
55 Cs	2	4 4	6 6 6	6 6 6	4 4	1	
56 Ba	2	4 4	6 6 6	6 6 6	4 4	2	
57 La	2	4 4	6 6 6	6 6 6	4 4 1	(2)	
58 Ce	2	4 4	6 6 6	6 6 6 1	4 4 1	(2)	
59 Pr	2	4 4	6 6 6	6 6 6 2	4 4 1	(2)	
– –	–	– –	– – –	– – –	– – –	–	
71 Cp	2	4 4	6 6 6	8 8 8 8	4 4 1	(2)	
72 –	2	4 4	6 6 6	8 8 8 8	4 4 2	(2)	
– –	–	– –	– – –	– – – –	– – –	–	
79 Au	2	4 4	6 6 6	8 8 8 8	6 6 6	1	
80 Hg	2	4 4	6 6 6	8 8 8 8	6 6 6	2	
81 Tl	2	4 4	6 6 6	8 8 8 8	6 6 6	2 1	
– –	–	– –	– – –	– – – –	– – –	– –	
86 Em	2	4 4	6 6 6	8 8 8 8	6 6 6	4 4	
87 –	2	4 4	6 6 6	8 8 8 8	6 6 6	4 4	1
88 Ra	2	4 4	6 6 6	8 8 8 8	6 6 6	4 4	2
89 Ac	2	4 4	6 6 6	8 8 8 8	6 6 6	4 4 1	(2)
90 Th	2	4 4	6 6 6	8 8 8 8	6 6 6	4 4 2	(2)
– –	–	– –	– – –	– – – –	– – –	– – –	–
118 ?	2	4 4	6 6 6	8 8 8 8	8 8 8 8	6 6 6	4 4

Figure C.4
Table depicting electronic structure for elements of the periodic table. Source: Bohr, "Structure of the Atom," 12.

numbers, [and] the construction of atoms is to be joined even more tightly than before to the structure of Röntgen spectra."[29] Stoner's scheme, he wrote soon afterwards, "possesses more of an arithmetical than a geometric-mechanical character, makes no assumptions about the symmetry of orbital arrangements and uses not only a part, but the entirety of Röntgen-spectroscopic facts."[30] This last point was essential. Further subdividing Bohr's n_k orbitals, Stoner used an extra quantum number, j, and spoke of

n_{kj} orbits.[31] Each such n_{kj} orbit was populated by $2j$ electrons. Thus, whereas in Bohr's conception there were two $n = 2$ orbits—an elliptical 2_1 orbit and a circular 2_2 orbit—each containing four electrons, in Stoner's version there were three orbits. The 2_{11} and 2_{21} ($j = 1$) subgroups each contained two electrons, while the 2_{22} ($j = 2$) contained four. The periodicity of Mendeleev's table could be maintained with no mention of the dynamical character of electronic orbits. The new scheme was, to Sommerfeld's satisfaction, formal, arithmetic, and purely quantum in character: it contained neither argument nor conceptual representation in terms of classical models.

Having passed over Stoner's paper when he first received the relevant copy of the *Philosophical Magazine*, Pauli returned to it after perusing Sommerfeld's preface. His "generalization" of Stoner's ideas must have followed almost immediately. In the notation used in his exclusion principle paper (discussed below), four quantum numbers were necessary. Instead of a single quantum number k, he used two: k_1 and k_2 ($k_1 = 1, 2, 3, \ldots, n$; $k_2 = k_1, k_1 - 1$). In place of the once-crucial j he used m_1, following Sommerfeld in defining this as the (quantized) component of the momentum parallel to an externally imposed field. j was then defined as the maximum value of m_1, given by $j = k_2 - \frac{1}{2}$. Clearly, the maximum number of values m_1 can take is equal to $2k_2$.[32] Stoner's $2j$ identical electrons in each of his n_{kj} orbitals become, in Pauli's formulation, $2k_2$ electrons, each distinguished by the possession of a different value of m_1. The exclusion principle then simply amounts to the fact that there are no such things as equivalent electrons within a given atom. The rule abandons, as Pauli would emphasize to Bohr, any talk of orbits and provides instead a formal quantum rule connecting the number of terms into which a single spectral line could split with the periodic structure of the table of elements.[33] "We cannot give a more precise foundation for this rule," he emphasized in public, "nevertheless, it seems to present itself very naturally."[34]

Reporting his result to Sommerfeld, Pauli portrayed it as a victory of quantum *Gesetzmässigkeiten* over model-based analysis. "Should my generalization of Stoner's ideas also, in the future, stand up to experience in more complicated cases, this would then simultaneously signify that you were completely right as regards the problem of the closing of electronic groups within the atom 'to place greater hope in the magic [*Zauberkraft*] of quanta than in correspondence- or stability-considerations.'" I do not, in fact, believe, he continued, "that the correspondence principle has anything to do with this problem."[35] Sommerfeld, studying the published version for one of his lectures, declared it "very beautiful and doubtless correct."[36] Bohr was equally impressed, but resisted Pauli's conclusion that his result signified the death of the correspondence principle. Even he, however, noted the elements of Sommerfeldian number-mysticism—the magic of quanta—in Pauli's paper. "I am also not entirely sure whether you do not step over a dangerous line if you—as you intone your old 'Carthage must be destroyed'—pronounce the final death sentence of a correspondence-based explanation of group closure." Pauli's own "lovely number-harmonies" [*smukke Talharmonier*], he suggested, were not entirely free from "poor, classical conceptions of space," a claim intimated by the fact that there was one quantum number for each spatio-temporal dimension required to describe an electronic trajectory.[37]

Zweideutigkeit about *Zweideutigkeit*

The discussion thus far has centered on the years after Pauli completed his dissertation with Sommerfeld in 1921. It should be clear that it is impossible to ignore the ongoing intersections between their work. Although Pauli took up the problem of the anomalous Zeeman effect with Bohr's encouragement, his approach to its many puzzles—indeed, in many ways the very framing of the problem itself—was molded by the work of Sommerfeld and his present or former pupils, including Heisenberg and Landé. Pauli's method, in developing his exclusion principle, was similarly Sommerfeldian both in its skepticism about the power of a *korrespondenzmässig* approach and in its focus on formal quantum rules gleaned from empirical data, with little analysis of their foundation. With regard to this last point, it is worth emphasizing that Pauli (and Heisenberg) were both at Munich during the precise period in which Sommerfeld abandoned model-based analysis in favor of his harmonious lawful regularities. Both young men participated in preparing the volumes of *Atombau* that recorded the shift in Sommerfeld's thinking. Indeed, Pauli would note in 1924 that the fourth edition of the text was the first upon which he had not worked.[38]

Sommerfeld's work thus offered, as Pauli himself would suggest in 1945, a methodological model for Pauli's development and postulation of the exclusion principle. It also, as we shall now see, offered a more direct resource for the content of the 1925 paper. To see how, one must look more closely at the precise meaning of the term *Zweideutigkeit* as it was used there. The word has a customary meaning in German and can be translated simply as "ambiguity." In Copenhagen in the early 1920s, however, the term had acquired a technical meaning—"two-valuedness"—and signified the two possible positions that the core could take up under the action of Bohr's *unmechanischer Zwang*.[39] Pauli used the term to mean something similar to this in a letter to Landé in September 1923. Rewriting Landé's expression for g as

$$g = \frac{3}{2} + \frac{1}{2}\frac{r(r-1)-k(k-1)}{j(j-1)},$$

Pauli then pointed out that "each momentum will act not through a single number, but through a *pair* of numbers. The momenta appear in a certain sense to be *zweideutig*.... One sees further that this *Zweideutigkeit* also includes k."[40] *Zweideutigkeit* in this sense refers to the two values that a *single* quantum number can take. In understanding what Pauli meant by the term in 1925, van der Waerden has used essentially the same notion.[41] Acknowledging that Pauli's statements about the *Zweideutigkeit* in his exclusion principle paper are "rather vague" and mysterious, he nonetheless argues that what Pauli meant was that *J* (defined as the maximum value of the total angular momentum of the atom in the direction of the field, *M*) can take on two possible values (L + $\frac{1}{2}$, L − $\frac{1}{2}$) for the two terms of an alkali doublet. One number, two values. But this is not at all how Pauli explained the meaning of the *Zweideutigkeit* in a letter to Bohr in December 1924. There he wrote:

> The doublet structure of the Alkalis is, in essence, a property of the valence electron alone. This classically non-describable *Zweideutigkeit* expresses itself first through the fact that a different quantum number is responsible for the size of the relativistic correction as for the central force between the valence electron and the core and second, through the fact that a different quantum number is responsible for the magnetic moment than for the moment of momentum.[42]

Pauli's point here was not that one quantum number could take two values, but rather that two quantum numbers were required to explain one set of phenomena: "a different quantum number is responsible for the size of the relativistic correction as for the central force." *This* is why it was classically non-describable. The issue here had little to do with spatial orientation. It concerned, rather, a more basic dynamical issue. As Pauli's notation in 1925 would suggest, k_1 and k_2 are *both* azimuthal quantum numbers, each taking *part* of the properties of the single—classically describable—*k*. One part gave the size of the relativistic correction, the other the size of the central force.

Pauli offered no explanation of how this might be possible in *modellmässig* terms. Instead, the reader finds a positivistic claim: k_1 and k_2 are *measures* without classical meanings. In seeking the origins of this conception, one finds, again, Pauli's Scylla and Charybdis. Around the end of 1922, independently of one another, Dirk Coster, who had completed his dissertation in Leiden, and Gregor Wentzel (working with Sommerfeld) hit upon a means of ordering the increasingly numerous Röntgen lines then being found. Both men introduced a third quantum number (added to *n* and *k*) to explain the existence of "extra" lines (three L lines rather than two, five M lines rather than three, and so on). In a paper received at the *Zeitschrift für Physik* in November 1922, Bohr and Coster offered the $n(k_1, k_2)$ labeling. Explaining what were essentially *two* new (or newly defined) numbers, the two men rewrote Sommerfeld's fine-structure formula to include two new parameters, γ and δ. The energy required to remove an electron from an n_k orbit was then

$$W = Rh\frac{(N-\gamma)^2}{n^2} + Rh\frac{(N-\delta)^4}{n^4}\left(\frac{2\pi e^2}{hc}\right)^2\left(\frac{n}{k}-\frac{3}{4}\right)+\cdots.$$

The term involving the fourth power denoted the relativistic correction. γ designated the "total screening number" and was assumed dependent only on *n* and k_1; δ provided a measure of the relativistic correction and depended, conversely, on *n* and k_2. In an allowed transition between levels, k_1 changes by ±1 unit, k_2 by ±1 or 0. Clearly, one recovers the form of Sommerfeld's original expression when $k_1 = k_2$. Bohr and Coster designated levels for which this equality held "normal"; "abnormal" levels, then—those without a corollary in Sommerfeld's analysis of Hydrogen's fine-structure—were those for which k_1 and k_2 possessed different values. L_{11} and L_{22}, by this logic, are normal levels, and L_{21}, located between them in energetic terms, is abnormal. One can see immediately that the difference between the L_{21} and L_{11} terms will be dependent only upon the difference in their screening terms, while the difference between L_{22} and L_{21} can be expressed as the difference in their relativistic corrections. The result was that the theory reproduced the alternating appearance of "screening" and "relativistic" doublets between successive terms in X-ray spectra.

Explaining the peculiarity of having two different values of k in the same expression (one meant to denote a single n_k orbit) was a different matter. Here Bohr and Coster invoked Bohr's second theory, and in particular the suggestion—tied to the notion of penetrating orbits—that adjacent electronic subgroups were coupled together:

With regard to the existence of the other levels, having different values of k_1 and k_2, it may first of all be said that the values of k_1 and k_2 used to classify these "abnormal" levels simply are intended to express that each of these levels, purely formally, is quantitatively closely related to two normal levels. Thus, each of these levels appears in the atom together with a normal level for which k_1 has the same value, and with which it also has the screening constant in common. On the other hand, it is related to a normal level having the same value of k_2, in that both have the same relativity term, provided the difference in the effective nuclear charge occurring in this term be ignored. This state of affairs leads one to seek the origin of these abnormal levels in a close interplay between the electronic motions in two neighboring subgroups of the excited atom, involving an essential change in the harmonious interplay of the electronic motions within *both* subgroups when an electron is removed from *one* of these subgroups. Such a conception is suggested by the theory, inasmuch as it is assumed that the process of forming an electron group depends essentially just upon the interplay of electronic motions within different subgroups.[43]

Sommerfeld and Wentzel's version was strikingly different. They termed their new quantum number m, the "ground quantum number." In the third edition of *Atombau*, Sommerfeld noted explicitly the intellectual connection to the "inner quantum number," itself introduced as a formal concept (i.e., without a model-based meaning) in the study of the Zeeman effect: "For each level we here define a third quantum number m, which, linking up with the visible spectra, we should best like to call "inner quantum number" but which, owing to the fact that it is not actually quite analogous to the inner quantum number defined earlier, we prefer to indicate by the more non-committal name "ground quantum number.""[44]

No understanding of this quantum number in terms of the model was offered. In fact, m played no role, when first introduced, in providing the *size* of doublet separations. It explained only the number of levels and the selection principles behind allowed transitions. This would change in the fourth edition of *Atombau*, where Sommerfeld abandoned m altogether and accepted Bohr and Coster's notation instead. The reason for his about-face was connected, if not directly, to the "doublet riddle," which, as Paul Forman has noted, was "fairly launched" by the first half of 1924.[45] The riddle concerned an oddity with regard to the two distinct explanations on offer in the early 1920s for the origin of doublets in Röntgen spectra on the one hand and complex optical spectra on the other. One of the earliest successes of Sommerfeld's extension of Bohr's model during the war years had been his transference of the analysis of Hydrogen's fine structure to X-ray spectra in multi-electron atoms. The widths of certain so-called fine-structure doublets in X-ray spectra appeared to be governed by a formula of precisely the same kind, but of vastly greater magnitude, than that found in the Hydrogen spectrum.[46] For the case of complex structure, in contrast, one did not invoke relativistic explanations at all. These doublets were explained by the coupling between a valence electron and a magnetic *Rumpf*.

In 1923 Landé proposed the existence of a formal analogy between optical and Röntgen spectra, and in 1924 he undertook a calculation to demonstrate that the "relativistic" effect could, in fact, be explained by transferring the logic of the *Rumpf* model into the X-ray region. Sommerfeld was deeply unimpressed. Not only was the X-ray fine-structure calculation among the most prized achievements of his theoretical work to date; it also—as he had emphasized in *Atombau*—provided one of the strongest arguments in favor of Einstein's special theory. In the midst of increasingly vitriolic arguments with anti-Semitic critics of Einstein's work, Sommerfeld had little desire to fight the battle on two fronts.[47] In any case, Landé's efforts did not produce the results for which he had hoped. Calculations on magnetic grounds could not produce the dependence on the fourth power of Z needed to provide a theoretical justification for the size of Röntgen doublets. A month after raising the possibility of explaining X-ray spectra in magnetic terms, Landé was forced to admit that this was impossible. Sommerfeld was delighted, repeatedly referencing Landé's failure in a number of publications over the next several months.[48] The riddle remained, however, for the analogies between phenomena in each region seemed obvious. Intensities in the X-ray region, for example, appeared to follow a formula derived for complex structure. Yet any attempt to reverse Landé's original intuition by reducing "magnetic" doublets to relativistic ones ran the risk of destroying the theoretical basis of work in the optical realm altogether.

Among Sommerfeld's responses to this paradoxical situation was his retraction of his earlier notation for the classification of X-ray data. A note in *Atombau* (1924) explained that closer examination of Röntgen spectra necessitated the abandonment of m, the "ground" quantum. In its place, one offered a divided azimuthal quantum number: "k splits apart into two quantum numbers k_1 and k_2.... With regard to the n_k electronic orbits k_1 plays the role of the azimuthal quantum number k; on the other hand, with respect to the quantitative representation of the doublets the number k_2

plays this role. Under these circumstances, which are not yet fully explained, the Bohr-Coster labeling seems more correct than our previous one...."[49]

But accepting Bohr's notation and the theory that went with it led to its own oddities. To which k is one referring when one speaks of n_k orbits: k_1, or k_2? The question arises not because of the doublet riddle itself, but because of another problem connected to relativistic explanations for doublet formation. According to Sommerfeld, the theory concerning relativistic doublets in the *Röntgengebiet* argues that one should speak of n_{k_1} orbits. On the other hand, determinations of intensity relationships during the building up of shells suggests that n_k orbits are dependent on k_2, and hence that one should really speak of n_{k_2} orbits. The situation appeared deeply ambiguous:

> The remarkable ambiguity [*Zweideutigkeit*] upon which we strike here clearly depends on the still completely unsolved question of why it is that the L-shell is divided into three and not two, the M-shell into five and not three etc. We have assumed in the theory of the periodic system that the complete eight-fold occupied L-shell contains only two types of n_k orbits, namely 2_2 and 2_1 and we know, on the other hand, that they can be ionized into three different kinds, namely those corresponding to the final L_{22}, L_{21}, L_{11}, of which clearly each must correspond to a different inner configuration of electrons. The relativistic representation of terms takes L_{21} and L_{11} for the elliptical orbital type 2_1, but the theory of the periodic system ascribes L_{22} and L_{21} to the orbital type 2_2. The meaning of the L_{21} level is therefore contradictory. This is true of other shells, namely for all terms for which k_1 and k_2 are different. We are forced to conclude from this that our model-based conception [*Modellvorstellung*] with regard to types of orbits is too narrow and that the real circumstances are not yet satisfied.[50]

Pauli's solution, it now becomes clear, was to take Sommerfeld's *Zweideutigkeit* as an empirical datum; to render it an integral element of an electron's quantum characteristics. His notation is Bohr and Coster's, but the meaning and significance of his four quantum numbers derives from Sommerfeld, as does a shared skepticism concerning *modellmässig* explanation. Pauli, of course, would also offer a crucial addition, for his 1925 paper solved the doublet "riddle" by taking it at face value without attempting to understand it in terms of a model. A notation intended for X-ray spectra and *Atombau* was assumed valid for all spectral lines, including complex structure. Via the deceptively simple step of effectively replacing Stoner's j with Bohr and Coster's k_2 Pauli had unified the problem-sites that Sommerfeld had designated, in 1920, the two "unsettled" areas of atomic physics. The price of settlement was the conversion of an "ambiguity" into a reality, a "doubled meaning" for a previously singular term, now with no counterpart in classical mechanics.[51]

Observability and Quantum Mechanics

What may be the most-often-cited part of the letter from Pauli to Bohr reporting his work in late 1924 has little directly to do with the exclusion principle. It concerns, rather, Pauli's statement of a future program for a new quantum physics:

The relativistic doublet formula appears to me now to show beyond any doubt that not only the dynamical conception of force within the classical theory, but also the kinematical conception of motion will have to undergo thorough modifications. (For that reason I have also avoided the designation "orbit" throughout my work). Since this conception of motion also lies at the heart of the correspondence principle, its clarification requires above all the efforts of theoreticians. I believe that the values of energy and momentum for stationary states are much more real than "orbits."

The (not yet achieved) goal must be to deduce these and all other physically real, observable properties of the stationary states from (whole) quantum numbers and quantum-theoretical laws. But we may not bind atoms in the shackles of our prejudices (to which, in my opinion, the assumption of the existence of electronic orbits in the sense of customary kinematics, also belongs), to the contrary we must, conversely, adapt our conceptions to experience.[52]

Only a few months later, Heisenberg, working in Copenhagen with Bohr when Pauli's letter arrived, put forward his own program, now quantum-*mechanical*. It too emphasized the importance of a methodological focus on observable properties in future quantum-theoretical analyses. Heisenberg's aim in that now famous paper, he wrote, was "to establish a theoretical quantum mechanics, analogous to classical mechanics, but in which only relationships between observable quantities occur."[53] The about face from an earlier position of committed realism—one need only think of the *Rumpf* model—was striking and contemporary scholarship suggests that Heisenberg's apparent positivism here be regarded as more tactical than heartfelt; a means, above all, of avoiding accusations of naiveté from an increasingly strident Pauli. That said, there is no denying that the focus on observables soon became a central discursive tenet of the new matrix mechanics, and of the so-called Copenhagen interpretation of quantum theory more generally.

Late in his life, Heisenberg offered an account of the intellectual origins of this positivistic element of his *Umdeutung* paper. Recalling a conversation with Einstein in 1926, Heisenberg recorded his astonishment at discovering that Einstein did not agree with his focus on the construction of a quantum theory based solely on observable quantities. Einstein "thought that every theory in fact contains unobservable quantities," Heisenberg remembered. When the younger scholar pointed out that he had "merely been applying the type of philosophy that he [Einstein] had made the basis of his special theory of relativity," Einstein simply retorted: "Perhaps I did use such philosophy earlier…but it is nonsense all the same."[54] An Einsteinian origin of the observability criterion would appear to be confirmed by the fact that, in June 1925, while Heisenberg was with them in Göttingen, Born and Jordan invoked the operationalist definition of simultaneity that was at the heart of the 1905 special-relativity paper as a model for their analysis of the quantum behavior of aperiodic systems. "Only such terms enter into the true natural laws," they wrote, "that are in principle observable and determinable."[55] Further, while Pauli's published work and his private critiques were of obvious importance to Heisenberg, Pauli himself drew heavily on Einstein's operationalism in the first years after the end of the war, critiquing Hermann Weyl's gravitational theory in 1919 for its use of fundamentally unobservable quantities.[56]

However, as we have seen, there was more than one kind of positivism available to Pauli in the 1920s. Einsteinian operationalism, rooted in the philosophy of Ernst Mach, clearly provided one intellectual resource. Yet Sommerfeld's emphasis on number mysteries and quantum-theoretical *Gesetzmässigkeiten*—which drew on Woldemar Voigt's phenomenological approach to spectroscopy—just as obviously provided another. With this in mind, one can also understand an oft-neglected aspect of *Heisenberg's* quantum mechanics paper: its title. The full title of the now-classic paper, received by the *Zeitschrift für Physik* at the end of July 1925, was "Über quantentheoretische Umdeutung kinematischer und mechanischer Beziehungen." By that time a number of researchers had begun to speak of the need for a fundamental *Umdeutung* (re-interpretation) of classical mechanics. However, a "quantum-theoretical reinterpretation" had been offered several years earlier by Sommerfeld in the paper in which he reworked Voigt's theory of D-line splitting: "Quantentheoretische Umdeutung der Voigtschen Theorie des anomalen Zeemaneffektes vom D-Linien Typus." The connection between this and Heisenberg's paper was far more than merely a titular similarity. Sommerfeld's paper, written while Heisenberg was a student in Munich, may well contain the first printed mention of the future Nobel laureate's name. Received at the *Zeitschrift* on December 12, 1921, Sommerfeld's paper concluded by "heartily thank[ing] Herr stud. W. Heisenberg for his successful collaboration on the entire problem of the anomalous Zeeman effect."[57] Five days later, the journal received Heisenberg's first publication, "The Quantum Theory of Line Structure and the Anomalous Zeeman Effect."

Sommerfeld's *Umdeutung*, as we saw in chapter 7, had been an explicit attempt to convert Voigt's *phenomenological* explanation of D-line splitting from the classical realm to the quantum. What Lorentz's theory provided for the normal Zeeman effect, Sommerfeld argued, Voigt's theory provided for the anomalous Zeeman effect, "namely the adequate expression of the facts in the language of the theory of oscillations." Sommerfeld's aim was to show "how these equations could be translated from a oscillations-theoretical to a quantum-theoretical language."[58] Heisenberg, less than a week later, began his own paper by reviewing the theoretical and experimental work on the anomalous Zeeman effect. Referencing Paschen and Back's most recent work, he then noted that "a further extension of the empirical materials can thus be effected if one translates Voigt's theory of the D-line splitting or the simplification of this by Sommerfeld into the language of the quantum theory." Heisenberg thus understood Sommerfeld's paper as doing two things: it began with a phenomenological classical account and then "translated" this into quantum-theoretical terms. In formal terms, this is precisely what Heisenberg's *Umdeutung* would also do in 1925. Convinced by Pauli and others of the need to abandon—at least publicly—his realist, *modellmässig* commitments, Heisenberg then referenced two different strands of phenomenological thinking in modern physics—one rooted in spectroscopy, the other in relativity. The result was his titular mirroring of the first paper to which he, as a *Wunderkind*, had contributed.

The End of the Older Quantum Theory and the Reception of Quantum Mechanics

As the name would suggest, "the older quantum theory" is a *post hoc* ascription, a somewhat artificial designation for the period between the introduction of Bohr's atomic model in 1913 and the invention of the two new forms of quantum mechanics in 1925 and 1926. Heisenberg, Born, and Jordan's matrix mechanics and Schrödinger's wave mechanics offered a (largely) self-consistent set of techniques to replace what Jammer perhaps somewhat unjustly called "a lamentable hodgepodge of hypotheses, principles, theorems, and computational recipes."[59] Wave mechanics, in particular, with its far more familiar mathematical methods, was taken up with astonishing rapidity. Hans Bethe, a student of Sommerfeld's in the mid 1920s, remembered the enthusiasm with which members of the Munich school adopted what they saw as an almost miraculous new means of solving previously intractable problems: "In the beginning they were just interested that you could now do all this, and one of the fascinating things was the Stark effect—that one could *do* the Stark effect and *do* the intensities.... I think this first seminar was simply fascinated by the fact that it now worked."[60] In contrast, matrix mechanics offered perhaps a greater conceptual clarity, but was ill-suited to the rapid solution of problems. Sommerfeld wrote in 1928: "We need here only to consider the Schrödinger form of the new theory, because it has the closer relationship to experiment and is more suitable for practical calculational manipulation. But we want to emphasize that the Heisenberg theory factually agrees with that of Schrödinger and that it can claim a high knowledge-theoretical interest."[61] Bethe put the same point more pithily: "Sommerfeld said, 'Well, of course we really believe that Heisenberg knows better about the physics, but we calculate with Schrödinger.'"[62]

If the end points of the old quantum theory are relatively clear, characterizations of intervening moments are less so. Particularly fraught is the question of how to understand the meanings and significance of events just before and after 1925, given that the invention of quantum mechanics has been cast as a scientific revolution. Revolutions cannot proceed entirely smoothly. They are defined as much by those who deny as those who support them. Hence the existence of a historical narrative in which major figures hold out against the most radical suggestions of the new theories. Einstein is endlessly cited refusing to admit that God plays dice, and the story of

Schrödinger driven to his sick bed by arguments with Bohr becomes a narrative set piece. Revolutions also cannot be spontaneous events. They come about when situations reach their most dire, and must function as both a break with and a solution to the problems of the past. Revolutions must also be resolutions.

The period just before the rise of matrix mechanics has thus been labeled a time of crisis. In one sense, the label is less problematic here than for the period around 1900. Few physicists spoke of a crisis in the fin-de-siècle.[63] A great many, as Paul Forman showed, spoke of crisis during the early years of the Weimar republic.[64] And there was a self-identification of the problems that caused such a critical outlook. In early 1923, Born and Heisenberg completed a systematic treatment of the helium spectrum on the basis of perturbation theory, a treatment that revealed basic contradictions with experimental data, and with predictions made on the basis of Bohr's correspondence principle. Born called the result a "catastrophe," while Bohr claimed it as "evidence of the inadequacy of the present basis of the quantum theory."[65] Pauli, who had written to Sommerfeld in 1924 concerning the "fundamental crisis" in which ideas of atomic models found themselves, located another of the principal fault lines of the modern physics in the then-persistent inability to explain the anomalous Zeeman effect. It was there, he noted, that one could see "how deep-seated is the failure of the theoretical principles known till now."[66] The problems with helium and the Zeeman effect thus stood as markers for many of the inability of present techniques to deal with the intricacies of spectral data.[67]

Many, but not all. How does one distinguish between difficult questions and critical paradoxes, between problems within a theory and problems for a theory? After a revolution is said to have occurred, it is easy to go back and identify certain research problems as having been signs of the moment of crisis for the older theory. Whether they were seen as such at the time is another question. Retrospective comments made by participants during interviews for the Archive for the History of Quantum Physics make apparent that not every atomic theorist expected the fundamental failure of the older quantum theory. While questions by Thomas Kuhn, in particular, often attempt to elucidate the point at which the crisis was recognized, several of those interviewed simply refused the applicability of the term. Consider the following exchange between Kuhn and Paul Dirac:

Kuhn: Was there also that sense which again people speak of on the continent that something fundamental now had to come to get around these problems that were just not responding. That there was something fundamentally the matter?
Dirac: I am not sure that that is so. They had the Bohr-Sommerfeld method of quantization and they thought it would have to be extended in some way.... I don't think people suspected that one would need such a complete revolution.... It rather came as a surprise to me when Heisenberg's ideas came out.[68]

In these interviews, Kuhn's questions on the issue of crisis often appear leading. It is never the existence, but only the nature and the timing of the crisis that is a matter of investigation. For example, in a 1963 interview with Werner Heisenberg he said:

By 1923, at least around Copenhagen—I'm not sure now it was then in Munich—it's perfectly clear that what I will call a "crisis" exists and is recognized with respect to the problems in quantum mechanics. It's clear in things that Bohr said, it's clear at least by what Bohr thinks Born said, it's clear for many people in that 1923 Bohr Heft of Naturwiss. and in other places. But it is by no means clear to me when that strong attitude came, how it developed, and more particularly, where it developed.[69]

This perspective is perhaps to be expected from a man whose philosophy of science depended so completely on notions of crisis and revolutions. Also uninterrogated by Kuhn, one may notice, is the question of whether one should consider the older quantum theory as a paradigm at all.[70] But it is clear that many of those interviewed had imbibed the same notions about the period just before the development of quantum mechanics. It was clearly inconceivable to Linus Pauling, for example, that Sommerfeld should not have emphasized the deep-seated problems in which quantum theory had found itself, or rather, in which the quantum theory *must* have found itself, since these problems were then resolved by the quantum mechanical revolution. In an interview with John Heilbron he began by remembering that Sommerfeld had talked a great deal about a particular "anomaly" during a visit to the United States: "But while Sommerfeld was here, he emphasized very strongly the anomaly of the inner quantum number and the outer quantum number—well at least he talked about the inner quantum number and the outer quantum number."[71] That memory, however, was subsequently completely reversed upon closer reflection:

So that they introduce an azimuthal quantum number; in fact they introduce two azimuthal quantum numbers and then in talking about the s, p, d separation you use one and in talking about the fine structure splitting you use the other. And the question is how can you have two quantum numbers that describe the eccentricity of the orbit? It only has one eccentricity. So this was brought out, and my memory—*I have to change my memory*. Sommerfeld just talked along glibly about the inner azimuthal quantum number without giving anybody, any auditor, any impression that there was anything funny about it.[72] [emphasis added]

Sommerfeld and the members of his school did not register a sense of crisis, a sense of the existence of paradoxical and insurmountable anomalies, or a sense of the occurrence of a revolution. In 1929, Sommerfeld, looking back on the events of the past several years, was quite explicit: "The new development does not signify a revolution [*Umsturz*], but a joyful advancement of what was already in existence, with many fundamental clarifications and sharpenings."[73] Students' descriptions of the reception of some of the major new results of the new quantum mechanics confirm the sense that what were seen as conceptual breakthroughs elsewhere were viewed as useful

Conclusion

tools in Munich. Thus, to Kuhn's question as to the reception of Heisenberg's uncertainty principle paper, Hans Bethe replied somewhat laconically that it was "Not very deep."

Kuhn: Was it interesting? Had these problems bothered people?
Bethe: No, these problems had not bothered people; at least it was not evident that they had bothered people. The paper was discussed, so that was it; this was a "fine conclusion."
Kuhn: But not a conclusion to anything that had really upset people previously?
Bethe: No. It really should have been discussed as the thing which now finally solved the paradox, but it wasn't.[74]

Sommerfeld's focus was elsewhere entirely, as Bethe makes clear: "It was all concentrated on solving the problems of the atom; then the molecule, then the solid state and not finding out more about the foundations."[75] A comment by Pauling makes a similar point with regard to the Munich response to the Schrödinger equation: "My memory is that everyone was so excited about the possibilities of solving problems, answering questions, the mechanism provided by the new quantum mechanics, that there was little discussion of those details of interpretation."[76] For the physics of problems, in other words, there were not anomalies, crises and revolutions, but problems and their methods of solution. To cite Kuhn and Bethe once more:

Kuhn: Were there great puzzles?
Bethe: No, there was no great puzzle and I think this is the greatest characteristic of the Sommerfeld group; we were not made aware of great puzzles. We were given the impression that here was a wonderful tool. Now you could do things, now you could solve all these interesting problems like all the complicated atoms and so on and chemistry, but fundamental problems, no.[77]

An immediate response to these descriptions might be to simply argue that Sommerfeld, and hence his school, had merely "missed"—failed to see—paradoxical anomalies that led to a real crisis. Yet the claim is a tricky one. When does a problem that one cannot solve yet become a problem that cannot be solved? Sommerfeld—like Dirac in the interview quoted above—saw the situation in the early 1920s as difficult but not dire. The foundations and the philosophy of the new field may have been shrouded in darkness, but one could still proceed with the appropriate (and appropriately modified) techniques. In contrast,[78] both Bohr and Born had called for a fundamental reworking of classical mechanics long before 1924. As Paul Forman noted, "crisis talk" considerably preceded the identification of the particular crises that could be said to have provoked a quantum revolution.[79]

The point here is not that Kuhn or others were "wrong" in describing this period in terms of crisis. Rather, in seeing crisis as an aspect of the community as a whole, scholars have given up the possibility of using the presence or absence of "crisis talk" as a marker for more fundamental divisions. Explaining why some saw a crisis and a revolution and some did not requires a deep rethinking of our understanding of these terms. Although Kuhn drew much of his terminology from extant materials (often the language of scientists themselves), his use of a particular vocabulary has profoundly shaped our perceptions of the processes of scientific change. It is thus to the ideas of crisis and revolution in his classic text, *The Structure of Scientific Revolutions*, that we now turn.

The most fundamental distinction in *Structure* was that between Normal and Revolutionary science. The former involved puzzle solving *within* a given paradigm; the latter involved a dramatic change of the paradigm itself. The two are temporally related by the fact that a revolutionary change ushers in a new paradigm and hence a new regime of normal science. In Kuhn's words:

> Successive paradigms tell us different things about the population of the universe and about that population's behavior. They differ, that is, about such questions as the existence of subatomic particles, the materiality of light, and the conservation of heat or energy. These are the substantive differences between successive paradigms and they require no further illustration. But paradigms differ in more than substance, for they are directed not only to nature but also back upon the science that produced them. They are the source of the methods, problem-field and standards of solution accepted by any mature scientific community at any given time.... The normal scientific tradition that emerges from a scientific revolution is not only incompatible but often actually incommensurable with that which has gone before.[80]

In spite of the connection between normal science (which preserves and extends paradigms) and revolutionary science (which replaces them), a basic asymmetry remains: a change in normal science does not necessarily lead to a revolutionary change, but a revolution overturns the modes of normal science. The asymmetry is occasionally minimized by the fact that certain puzzles that arise in the day-to-day practice of normal science can present difficulties that become anomalies; enough anomalies and a crisis develops, followed by a revolution. Yet no puzzle or set of puzzles is sufficient to cause a revolution. All they can induce is the search for, or perhaps the creation of, an alternative paradigm. A Revolution then proceeds as the choice among competing paradigms. That choice is made on a *substantive* basis, and then (and only then) has an effect on the practices of normal science. Scientific revolutions, in other words, are revolutions of conceptual foundations, not of puzzle-solving techniques. Most simply: Science sees revolutions of principles, not of problems.

Kuhn's book was among the first to focus historians' attention away from either ideas or individuals, and toward scientific groupings. Yet Kuhn's analysis breaks down at what might be termed the *mesoscopic* level that distinguishes between small groups

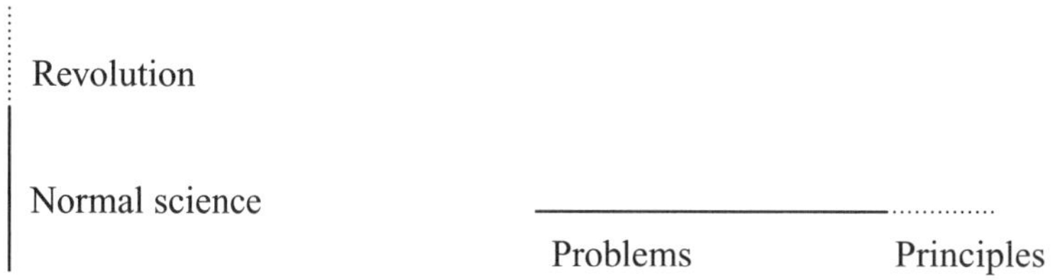

Figure C.5
Principles and problems.

within a community, like those designated in terms of a theoretical physics of principles and of problems. Communities are either homogeneous (defined by a shared allegiance to a common paradigm) or collections of individuals (whose thoughts Kuhn tracks with care and subtlety). The fact, however, that such a substantial part of the quantum physics community—the hugely important Sommerfeld School—should be so completely out of alignment with other (otherwise similar) members on the fundamental question of whether their shared field had undergone a revolution suggests that at the mesoscopic scale one must reconsider the basic distinction between normal and revolutionary science. Rather than imagining a homogeneous community that solves puzzles the vast majority of the time, and then contemplates revolutions at select moments, one can, so to speak, imagine the temporal axis of scientific change rotated until it aligns with that of intra-disciplinary structure. The result is a scientific community made up, in the majority, of those who solve problems (and eschew the pursuit of revolutions) and a much smaller group whose focus on principles and foundations means that the only change that counts is a revolutionary and fundamental one.[81] (See figure C.5.)

What must not be transferred in this rotation of axes is the value judgment that is explicit in Kuhn's book. There is no suggestion here that the physics of problems is, like Kuhn's "normal-science," basically conservative, unimaginative, blinkered, and averse to the calling forth of "new sorts of phenomena."[82] In regard to the utilization of the new methods of quantum mechanics there were probably few places more enthusiastic than Munich. In general, as the preceding pages have surely shown, the Sommerfeld School was one of the centers of innovation in German theoretical physics for more than a quarter of a century. Our loss of memory of these innovations may come down to the fact that, given a choice between a story of similar events told in terms of crises and foundational change and a story told in terms of a succession of puzzles solved—however similar the endpoints—historians have opted for the romance of revolutions.

Notes

Introduction

1. S. Hirzel to Arnold Sommerfeld, 15 February 1909, in *Arnold Sommerfeld: Wissenschaftlicher Briefwechsel*, ed. M. Eckert and K. Märker, volume 1 (2000), 353–4.

2. Max Planck to Arnold Sommerfeld, 24 February 1909, in *Arnold Sommerfeld*, ed. Eckert and Märker, 355.

3. Voigt to Sommerfeld, 25 November 1909. Quoted in Christa Jungnickel and Russell McCormmach, *Intellectual Mastery of Nature: Theoretical Physics from Ohm to Einstein*, volume 2: *The Now-Mighty Theoretical Physics* (1990), 123–4.

4. Arnold Sommerfeld, *Mechanics*, volume I: *Lectures on Theoretical Physics* (1952), ix–x.

5. No book-length study of Sommerfeld's work exists in English. German works and articles in English include the following: John L. Heilbron, "The Kossel-Sommerfeld Theory and the Ring Atom," *Isis* 58 (1967): 450–85; Armin Hermann, "Der Brückenschlag zwischen Mathematik und Technik," *Physikalische Blätter* 23 (1967): 442–9; Armin Hermann, *The Genesis of the Quantum Theory, 1899–1913* (1971); Sigeko Nisio, "The Formation of the Sommerfeld Quantum Theory of 1916," *Japanese Studies in the History of Science* 12 (1973): 39–78; Ulrich Benz, *Arnold Sommerfeld: Lehrer und Forscher an der Schwelle zum Atomzeitalter, 1868–1951*, ed. H. Degen (1975); Michael Eckert and Willibald Pricha, "Boltzmann, Sommerfeld und die Berufungen auf die Lehrstühle für theoretische Physik in München und Wien, 1890–1914," *Mitteilungen der Österreichischen Gesellschaft für Geschichte der Naturwissenschaften* 4 (1984): 101–19; Helge Kragh, "The Fine Structure of Hydrogen and the Gross Structure of the Physics Community, 1916–26," *Historical Studies in the Physical Sciences* 15, no. 2 (1985): 67–125; Michael Eckert, "Propaganda in Science: Sommerfeld and the Spread of the Electron Theory of Metals," *Historical Studies in the Physical and Biological Sciences* 17 (1986): 191–233; idem, *Die Atomphysiker: Eine Geschichte der theoretischen Physik am Beispiel der Sommerfeldschule* (1993); idem, "Mathematics, Experiments, and Theoretical Physics: The Early Days of the Sommerfeld School," *Phys. Persp.* 1 (1999): 238–52; idem and Walther Kaiser, "An der Nahstelle von Theorie und Praxis: Arnold Sommerfeld und der Streit um die Wellenausbreitung in der drahtlosen Telegraphie," In *Chemie-Kultur-Geschichte*, ed. A. Schurmann and B. Weiss (2002).

See also the articles in *Arnold Sommerfeld: Wissenschaftlicher Briefwechsel*, ed. Eckert and Märker (2000, 2004).

6. The term "physics of the principles" was first used by the French mathematical physicist Henri Poincaré. See Jerzy Giedymin, "The Physics of the Principles and its Philosophy: Hamilton, Poincaré and Ramsey," in *Science and Convention: Essays on Henri Poincaré's Philosophy of Science and The Conventionalist Tradition* (1982); Olivier Darrigol, "Henri Poincaré's Criticism of Fin-de-siècle Electrodynamics," *Studies in History and Philosophy of Modern Physics* 26 (1995). No authors of whom I am aware, however, draw a contrast between Sommerfeld's approach to physics and Poincaré's, and Planck's emphasis on principles appears to have an origin independent of Poincaré's. For that reason, no extensive analysis of Poincaré's work is offered here.

7. Kuhn and Uhlenbeck 1962.

8. Cited in Walter Moore, *Schrödinger: Life and Thought* (1992), 269. My thanks to David Kaiser for bringing this quotation to my attention.

9. David C. Cassidy, "Heisenberg's First Core Model of the Atom: The Formation of a Professional Style," *Historical Studies in the Physical Sciences* 10 (1979): 187–224; Jonathan Harwood, *Styles of Scientific Thought: The German Genetics Community, 1930–1933* (1993). See also Joseph Fruton, *Contrasts in Scientific Style: Research Groups in the Chemical and Biochemical Sciences* (1990).

10. On what Warwick has termed the "practice-ladenness of theory" see, in particular, Peter Louis Galison, *Image and Logic: A Material Culture of Microphysics* (1997), Peter Galison and Andrew Warwick, "Introduction: Cultures of Theory," *Studies in History and Philosophy of Modern Physics* 29 (1998), David Kaiser, *Drawing Theories Apart: The Dispersion of Feynman Diagrams in Postwar Physics* (2005), Andrew Warwick, *Masters of Theory: Cambridge and the Rise of Mathematical Physics* (2003). For earlier work on experimental and theoretical practice in the physical sciences, see, e.g., Harry M. Collins, *Changing Order: Replication and Induction in Scientific Practice* (1992); Kathryn Mary Olesko, *Physics as a Calling: Discipline and Practice in the Königsberg Seminar for Physics* (1991), Andrew Pickering, ed., *Science as Practice and Culture* (1992).

11. Max Planck, "Die Stellung der neueren Physik zur mechanischen Naturanschauung," *Physikalische Zeitschrift* 11 (1910). Reproduced in Max Planck, *Physikalische Abhandlungen und Vorträge*, volume III (1958).

12. Albert Einstein to Arnold Sommerfeld, 11 January 1922, in *Arnold Sommerfeld: Wissenschaftlicher Briefwechsel*, ed. Eckert and Märker, volume 2 (2004), 113; Max Born, "Sommerfeld als Begründer einer Schule," *Naturwissenschaften* 16 (1928).

13. The novelty of theoretical physics as a discipline has been challenged by the periodization offered in what is, on most issues, the definitive account: Jungnickel and McCormmach, *Intellectual Mastery of Nature*. Their defense of their dating of the beginning of the field to 1800, however, is weak, and depends on such an elision between theoretical physics and physical theory. If the two are assumed to mean the same thing, one wonders why one should stop in 1800, rather than with Newton's natural philosophy, an earlier contribution to "physical theory."

Jungnickel and McCormmach write: "Although there had been a body of physical theories for centuries, in Ohm's day, the early nineteenth century, theoretical physics was not a specialized field of study, and it was not to emerge as one for a good part of the century, until nearly the time of Einstein. *So, for much of the period covered by our study, our subject, strictly speaking, did not exist.* But of course it did, in a practical sense, right from the start. Major contributions to physical theory were made by Ohm and his contemporaries and by their successors throughout the nineteenth century. In the second half of the century, teaching positions were gradually created for theoretical physics...and these positions laid the foundation for the partial separation of physics into the fields of experimental physics and theoretical physics." (*Intellectual Mastery of Nature*, volume 1, 16) Paul Forman notes this peculiarity of the two-volume history: "[V]olume 1 never does arrive at the announced subject, the history of *theoretical* physics, which scarcely existed as a separate intellectual, let alone institutional, enterprise before 1870." ("Review: Intellectual Mastery of Nature," *Philosophy of Science* 58 (1991): 130) Elizabeth Garber makes a related point, arguing that it is ahistorical to label G. S. Ohm's work "physics": "To label Ohm as a physicist and his mathematical work as physics was to miss the point of what he was actually trying to accomplish....Ohm's experimental results were physically important but his goal was not to produce a physical interpretation of those results in mathematical form." (*Language of Physics: The Calculus and the Development of Theoretical Physics in Europe, 1750–1914* (1999), 159) In general, however, it should be emphasized that Garber's desire to draw a sharp line between mathematical and theoretical physics seems overly dogmatic. In particular, her claim that "no one in Germany used the terms mathematical and theoretical physics interchangeably or simultaneously" (167) is belied by the naming of chairs alone. Note, to take merely one example, that the chair created in 1883 for Woldemar Voigt at Göttingen was in "theoretical (mathematical) physics." See, more generally, Kathryn Mary Olesko, "The Emergence of Theoretical Physics in Germany: Franz Neumann and the Königsberg School of Physics, 1830–1890" (1980); Olesko, *Physics as a Calling*.

14. Lewis Pyenson and Douglas Skopp, "Educating Physicists in Germany circa 1900," *Social Studies of Science* 7 (1977), esp. 356.

15. In some cases the definition has been even narrower, limited to what has been termed "microphysics." Schweber, for example, defines the term for the purposes of an article on theoretical physics in the United States as "refer[ring] principally to quantum theory and its immediate forerunners. Thermodynamicists, relativists, acousticians, fluid dynamicists, and so on do not appear in my story because I have concentrated on theorists who supposed that they were addressing fundamental problems.... For my purposes, "fundamental" refers to physics dealing with the elementary constituents of matter and radiation, their properties, their interactions and dynamics." ("The Empiricist Temper Regnant: Theoretical Physics in the United States, 1920–1950," *Historical Studies in the Physical Sciences* 17 (1986): 58–9) Since the discourse of fundamentality is one way that the physics of principles distinguished itself from Sommerfeld's physics of problems (and vice versa, at times), this distinction would seem particularly problematic for the German case.

16. Max Born, "Arnold Johannes Wilhelm Sommerfeld," *Obituary Notices of the Fellows of the Royal Society* 8 (1952).

17. The historiographically useful distinction between "social" understandings of the term "social construction" (in terms of interests and ideologies) and "constructivist" ones (which focus on socially embedded practices and traditions) is discussed in Alex Soojun-Kim Pang, *Empire and the Sun: Victorian Solar Eclipse Expeditions* (2002), 4–6.

18. Peter Galison, "Einstein's Clocks: The Place of Time," *Critical Inquiry* 26 (2000); Stanley Goldberg, "Max Planck's Philosophy of Nature and His Elaboration of the Special Theory of Relativity," *Historical Studies in the Physical Sciences* 7 (1976); John L. Heilbron, *The Dilemmas of an Upright Man: Max Planck as Spokesman for German Science* (1986); Erwin N. Hiebert, "The Conception of Thermodynamics in the Scientific Thought of Mach and Planck," in *Wissenschaftlicher Bericht Nr. 5/68* (1968); Gerald Holton, "Einstein and the Cultural Roots of Modern Science," *Daedalus* 127 (1998); Gerald Holton, *Thematic Origins of Scientific Thought* (1973); Martin J. Klein, "Max Planck and the Beginnings of the Quantum Theory," *Archive for History of the Exact Sciences* 1 (1962); Suman Seth, "Allgemeine Physik? Max Planck und die Gemeinschaft der theoretischen Physik," in *Der Hochsitz des Wissens: Das Allgemeine als wissenschaftlicher Wert*, ed. M. Hagner and M. Laublichler (2006).

19. David Cahan, *An Institute for an Empire: The Physikalisch-Technische Reichsanstalt, 1871–1918* (1989), 24. Cahan writes: "The Imperial period witnessed an institutional revolution in German physics. Whereas before 1865 Germany had only a group of largely antiquated physical cabinets housed in limited quarters built for other purposes, by 1914 it had an entire set of new physics institutes. Whereas before 1865 research in physics was carried out by only a limited number of physicists, by 1914 it pervaded every institute. Whereas before 1865 only a handful of private laboratories existed, by 1914 every university had its own public laboratory, and many of these were capable of instructing a hundred or more students. Whereas before 1865 attendance at physics lectures rarely reached more than 50 students per class, by 1914 it was not uncommon to find 300 or more students listening to lectures on experimental physics. Whereas before 1865 there was never more than one assistant in physics at any one university, by 1914 many institutes had three or more assistants. And whereas before 1865 cabinets were either without budgets or provided only minimal financial help, by 1914 every institute had a substantial budget, sometimes running to tens of thousands of marks per year." (22–3) See also Cahan, "The Institutional Revolution in German Physics, 1865–1914," *Historical Studies in the Physical Sciences* 15 (1985).

20. Kohlrausch, quoted in Cahan, *An Institute for an Empire*, 23.

21. Arnold Sommerfeld, "Die naturwissenschaftlichen Ergebnissen und die Ziele der modernen technischen Mechanik," *Physikalische Zeitschrift* 4 (1903). The translation here is altered from Arnold Sommerfeld, "The Scientific Results and Aims of Modern Applied Mechanics," *Mathematical Gazette* (1903). The three-person panel ("Referate über den gegenwärtigen Stand der Mechanik"), which also included papers by Karl Schwarzschild ("Über Himmelsmechanik") and O. Fischer ("Physiologische Mechanik"), was organized by Felix Klein.

22. Andrew Warwick, *Masters of Theory: Cambridge and the Rise of Mathematical Physics* (2003).

23. See, e.g., Goldberg, "Planck's Philosophy of Nature"; Heilbron, *The Dilemmas of an Upright Man*; Klein, "Max Planck and the Beginnings of the Quantum Theory"; Allan Needell,

"Irreversibility and the Failure of Classical Dynamics: Max Planck's Work on the Quantum Theory, 1900–1915" (1980).

24. Max Planck, "Allgemeines zur neuren Entwicklung der Wärmetheorie," *Zeitschrift für physikalische Chemie* 8 (1891).

25. Max Planck, "Über die Grundlage der Lösungstheorie, eine Erwiderung," *Annalen der Physik* 10 (1903): 49.

26. Max Planck, "Das Einheit des physikalischen Weltbildes (Lecture, 9 December 1908, University of Leiden)," *Physikalische Zeitschrift* 10 (1909). On the distinction between Planck's philosophy and his physics (a distinction assumed even when connections are sought between the two areas), see Goldberg, "Planck's Philosophy of Nature."

Chapter 1

1. Cohn had held an extraordinary professorship in theoretical physics at Strasburg for 20 years and had published comparatively little in that time. Although he had been ranked equal with Wiechert by the Munich faculty commission, he never received a call. Wiechert declined Munich's offer after Göttingen made him a full professor of geophysics (Jungnickel and McCormmach, *Intellectual Mastery of Nature*, volume 2, 277–8). On Boltzmann and Sommerfeld at Munich more generally, see ibid., volume 2, 149–60, 274–87; Michael Eckert and Willibald Pricha, "Boltzmann, Sommerfeld und die Berufungen auf die Lehrstühle für theoretische Physik in München und Wien, 1890–1914," *Mitteilungen der Österreichischen Gesellschaft für Geschichte der Naturwissenschaften* 4 (1984).

2. Arnold Sommerfeld, "Autobiographische Skizze [Incl. a short addendum by Fritz Bopp]," in *Gesammelte Schriften*, ed. F. Sauter (1968), 677. Sommerfeld's term "Pflanzstätte theoretischer Physik" has also been used by Michael Eckert to describe the Sommerfeld School. See Michael Eckert, *Die Atomphysiker: Eine Geschichte der theoretischen Physik am Beispiel der Sommerfeldschule* (1993), 38–41.

3. In seeking to bridge the gap between histories of pedagogy and research, I make extensive use of Olesko's *Physics as a Calling* and of Warwick's *Masters of Theory*.

4. Arnold Sommerfeld, "Lectures, Theorie der Strahlung" (DM: NL 089 (026), Sommer 1907). The lectures are, nonetheless, difficult sources with which to work. Sommerfeld would return to them year after year, making revisions to the text. Like the examination scripts studied by Andrew Warwick in *Masters of Theory* (1–48), Sommerfeld's lectures thus constitute something close to "real-time records" of theoretical physics "in the making." Precisely because of this, dating lecture entries is a particularly difficult task. Nonetheless, much can be done with internal evidence (handwriting, nib width, discoloration of pages and ink, and so on), and Sommerfeld's lectures on Planck's radiation theory, in particular, seem to have remained largely unrevised. I suspect, given that his opinions of Planck's theory changed rapidly after 1906, that he gave this particular set of lectures only once or twice.

5. Thomas S. Kuhn, *Black-Body Theory and the Quantum Discontinuity, 1894–1912* (1978).

6. Armin Hermann, "Der Brückenschlag zwischen Mathematik und Technik," *Physikalische Blätter* 23 (1967): 449.

7. Peter Debye et al., *Probleme der modernen Physik: Arnold Sommerfeld zum 60. Geburtstage gewidmet von seinen Schülern* (1928).

8. Arnold Sommerfeld, "Untitled, undated paper [presumably written in Königsberg in 1890]" (henceforth cited as Königsberg 1890).

9. For biographical data on Sommerfeld, see Sommerfeld, "Autobiographische Skizze." Most other accounts tend to rely on this for personal information about Sommerfeld's early life. For a short biographical article, see Born, "Arnold Johannes Wilhelm Sommerfeld." For book-length biographies, see Ulrich Benz, *Arnold Sommerfeld: Lehrer und Forscher an der Schwelle zum Atomzeitalter, 1868–1951*, ed. H. Degen (1975); Eckert, *Die Atomphysiker*. See also the articles in Eckert and Märker, eds., *ASWB I*.

10. Kant, *Metaphysische Anfangsgründe der Naturwissenschaft, Vorrede*, A IX.

11. Sheet 1, front side. Kant would come up again in Sommerfeld's dissertation, if somewhat peremptorily. His second thesis was the baldly stated and unelaborated claim that "Kant's remarks about the apriority of space and time are not equally valid." F. Sauter, ed., *Arnold Sommerfeld: Gesammelte Schriften*, volume I (1968), 76.

12. Sommerfeld, "Autobiographische Skizze," 674; Born, "Arnold Johannes Wilhelm Sommerfeld," 275–7.

13. Quoted in Olesko, *Physics as a Calling*, 356. Volkmann's connection to the geothermal station and his involvement with the prize commission is discussed more generally on 354–60.

14. Born, "Arnold Johannes Wilhelm Sommerfeld," 276.

15. Arnold Sommerfeld, "undated, untitled manuscript [presumably written in Königsberg, 1890]" (henceforth Königsberg 1890, #2). The machine is discussed in "Speciellen Teil: Die Maschiene," 35–43. A paper on the topic was published in 1892: "Über eine neue Integriermaschine 1891," *Ber. Physik-oekonom Ges. Königsberg* and *Katalog math. Modelle und Apparate, München (1892)*. The latter is reproduced in Sauter, ed., *GS I*, 77–85.

16. Olesko, *Physics as a Calling*.

17. Arnold Sommerfeld, *Die Willkürlichen Funktionen in der Mathematischen Physik*, reprinted in Sauter, ed., *GS I*, 1–76. The first section dealt with the representation of arbitrary functions in terms of Fourier series and integrals, the second with their representation through double integrals with cylindrical functions, and the third with their representation through series of spherical functions.

18. Eckert and Märker, eds., *ASWB I*, 27. The work on diffraction has recently been translated and republished as Arnold Sommerfeld, *Mathematical Theory of Diffraction* (*Progress in Mathematical Physics*, volume 35, 2004).

19. Michael Eckert, "Mathematics, Experiments, and Theoretical Physics: The Early Days of the Sommerfeld School," *Physics in Perspective* 1 (1999): 240. Sommerfeld was also worried that he was poorly qualified for the position. "I understand nothing about experiments," he wrote to his mother, "and I have told this to Voigt. I am afraid to make a fool of myself." (Eckert, "Mathematics, Experiments, and Theoretical Physics," 242)

20. Arnold Sommerfeld to Cäcilie Sommerfeld, 5 January, 1894. Eckert and Märker, eds., *ASWB I*, 18. "Habt ihr gelesen, dass Hertz gestorben ist? Jammervoll! Der Mann hat vor 5 Jahren seine glänzenden experimentellen Untersuchungen angefangen. Die Hälfte aller Physiker geht augenblicklich in seinen Fusstapfen u. arbeitet über Hertz'sche Schwingungen. Es giebt wenige Entdeckungen, die seinen elektromagnetischen Lichtwellen an die Seite zu stellen sind. Hätte da nicht, wenn es gerade ein Physiker sein sollte, einer von den nichtsutzenden Pape, Volkmann etc. darauf gehen können."

21. Sommerfeld to Lorentz, 12 December 1906, ibid., 257–8.

22. *Maxwell'sche Th. U. Elektronenth* (Maxwell's Theory and Electron Theory) would also be the title of one of Sommerfeld's first lectures at Munich, held in the winter semester of 1906/7 and then again in 1908–09. On electron theory in Göttingen in this period, see Lewis Pyenson, "Physics in the Shadow of Mathematics: The Göttingen Electron-Theory Seminar of 1905," *Archive for History of Exact Sciences* 21 (1979).

23. Arnold Sommerfeld, "Lectures, Wärmeleitung, Diffusion u. Elektrizitätsleitung nebst ihren molekular- und elektronentheoret. Zusammenhangen" (DM: NL (089) 028, Sommer 1908) (pages unnumbered).

24. Ibid.

25. P. P. Ewald, "The Setting for the Discovery of X-Ray Diffraction by Crystals. Speech Given at the First General Assembly of the International Union of Crystallography at Harvard University, 2 August 1948" (DM: NL 089 (026), 1948), 19.

26. This may have had had an additional cause. At the start of the search for Boltzmann's replacement, Röntgen had said "We need no mathematician." Having won the position in spite of a paucity of experience in physics, Sommerfeld may well have felt the need to emphasize how far he had come from his mathematical training. Quotation in Jungnickel and McCormmach, *Intellectual Mastery of Nature*, volume 2, 278.

27. That said, Sommerfeld would also emphasize the importance of experimental analysis in technical mechanics itself. Sommerfeld, "Die naturwissenschaftlichen Ergebnissen und die Ziele der modernen technischen Mechanik."

28. P. P. Ewald, "Erinnerungen an die Anfänge des Münchener Physikalischen Kolloquiums," *Physikalische Blätter* 24 (1968).

29. Eckert, "Mathematics, Experiments, and Theoretical Physics," 244–5.

30. Ibid., 242.

31. Arnold Sommerfeld, *Mechanik* (1943). vii.

32. Biographical data from Sommerfeld, "Autobiographische Skizze," 675–6. On Klein's role in the Aachen offer, see Eckert, *Die Atomphysiker*, 25–6.

33. Arnold Sommerfeld, "Autobiographische Skizze, [Incl. A short addendum by Fritz Bopp]" in Sauter, ed., *GS IV*, 672–82, here 675–6.

34. On the tension between the *allgemeine Abteilung* (general section) and the engineering sections of the college (Architecture, Civil Engineering, and Mechanical Engineering), provoked by the events around Sommerfeld's hire, see Eckert and Märker, eds., *ASWB I*, 127–31; Michael Eckert, "Arnold Sommerfeld: Theoretische Physik und Technik" (1996).

35. Riedler to Kultusminister [Ministerium der geistl. Unterrichts—u. Medizin. Angel.], 16 July 1904. Geheimes Staatsarchiv, Dahlem [Henceforth, GSA] Rep. 76 Vb Sekt. 4, Tit. III, 10, Bd. VIII, Folder titled "Die Lehrer der technische Hochschule in Berlin." Document titled "Besetzung der Physik Professor."

36. Ibid.

37. It was also during his Munich years (1875–1880) that Klein first began to take an interest in technical problems. He and Carl Linde, professor for machine theory, "were members of a small group of friends who got together regularly to discuss problems at the interface of science and technology. Linde's machine laboratory in Munich…later served as a model for those built in Göttingen around the turn of the century." David E. Rowe, "Klein, Hilbert, and the Göttingen Mathematical Tradition," *Osiris* 5 (1989): 190–1, quotation on 191.

38. Grashof was director of the VDI from 1856 to 1890. Fritz K. Ringer, *The Decline of the German Mandarins: The German Academic Community, 1890–1933* (1969).

39. Kees Gispen, *New Profession, Old Order: Engineers and German Society, 1815–1914* (1989), 141.

40. Ibid., 144–5.

41. Their efforts were substantially rewarded only after the accession to the throne of Kaiser Wilhelm II in 1888. In 1892, the German emperor passed legislation awarding professors at technische Hochschulen the same rank as university professors, and granting them the right to wear the same academic garb. In 1898, on the tenth anniversary of his accession to the throne, Wilhelm called the Rectors of the three Prussian technical institutes—Adolf Slaby (Charlottenburg), Wilhelm Launhardt (Hanover), and Otto Intze (Aachen)—to sit in the parliamentary upper house, a right formerly granted only to university rectors. Finally, and most importantly, in 1899 technische Hochschulen were given the right to grant doctoral degrees, creating the title "Dr.-Ing" and establishing the younger schools as complete tertiary institutions. Charles E. McClelland, *State, Society, and University in Germany, 1700–1914* (1980), 306; James C. Albisetti, *Secondary School Reform in Imperial Germany* (1983), 269–70. Albisetti points to the significance of Slaby's close relationship to the Kaiser in the battle over the *Promotionsrecht* (the right to grant doctorates). Slaby often lectured to the royal entourage on recent scientific discoveries, "unfortunately"

(in the words of Count Robert von Zedlitz-Trüzschler) "pass[ing] all bounds in the matter of flattery and servility."

42. Ibid.

43. Alois Riedler, *Zur Frage der Ingenieur-Erziehung* (1895); Riedler, "Die Ziele der technischen Hochschule," *Zeitschrift des Vereines deutscher Ingenieure* 40 (1896); Riedler, *Unsere Hochschulen und die Anforderungen des 20 Jh* (1898); Riedler, *Die technsiche Hochschule und die Wissenschaftliche Forschung, Rektoratsrede* (1899).

44. Riedler, *Zur Frage der Ingenieur-Erziehung*, 22.

45. Sommerfeld, "Die naturwissenschaftlichen Ergebnissen und die Ziele der modernen technischen Mechanik." in Sauter, ed., *GS I*, 446–55, quotation on 454.

46. Felix Klein, quoted in Gert Schubring, "Pure and Applied Mathematics in Divergent Institutional Settings in Germany: The Role and Impact of Felix Klein," *History of Modern Mathematics* I (1989): 187. This interpretation of Klein stands in contrast with that in Karl-Heinz Manegold's *Universität, Technische Hochschule, und Industrie: Ein Beitrag zur Emanzipation der Technik im 19. Jahrhundert unter besonderer Berücksichtigung der Bestrebungen Felix Kleins* (1970). It agrees, however, with most other major accounts. These, however, do not tend to treat Sommerfeld in any great detail, and scholars of the latter have tended to accept Manegold's rather saccharine depiction of Klein's aims and motives.

47. Felix Klein, "Über die Gründung eines physikalisch-technischen Universitätsinstituts in Göttingen," *Zeitschrift des Vereines deutscher Ingenieure* 40 (1896): 76. For more on this issue, see Lewis Pyenson, *Neohumanism and the Persistence of Pure Mathematics in Wilhelmian Germany* (1983).

48. For Riedler's reply, see Riedler, "Die Ziele der technischen Hochschule."

49. Arnold Sommerfeld, "Zur Theorie der Eisenbahnbremsen (1902)," "Das Pendeln parallel geschalteter Wechselstrommaschinen (1904)," and "Zur hydrodynamischen Theorie der Schmiermittelreibung (1904)," all in *Arnold Sommerfeld: Gesammelte Schriften*, ed. F. Sauter (1968).

50. Sommerfeld, "Zur hydrodynamischen Theorie der Schmiermittelreibung (1904)," 511.

51. Sommerfeld, "Autobiographische Skizze," 677.

52. Pyenson, *Neohumanism and the Persistence of Pure Mathematics in Wilhelmian Germany*.

53. See Manegold, *Universität, Technische Hochschule, und Industrie*, 15–55.

54. Renate Tobies, "On the Contribution of Mathematical Societies to promoting Applications of Mathematics in Germany," *History of Modern Mathematics* I (1989): 227.

55. The distinction between Klein's position and that of a newer breed of "entrepreneurial" engineers was profound. Even at the most basic level, like their understandings of the dichotomy between theory and praxis, Riedler and Klein found no common ground. To Riedler, praxis was what was carried out in the industrial world, theory was what was taught in the universities. As far as teaching at the technische Hochschulen was concerned, Riedler believed there should be

less of the latter and a great deal more of the former. Klein, on the other hand, seemed to have seen no difference in the distinctions between pure and applied and between theory and praxis. In the second volume of the textbook on Gyroscopes that Sommerfeld and Klein co-authored, the dichotomies are used indistinguishably: discussing the easy manner in which one can cross from elliptical integrals to the so-called elliptical functions, they write: "We note therefore a remarkable working together of theory and praxis, a *pre-established harmony* so to speak, between pure and applied mathematics...." Felix Klein and Arnold Sommerfeld, *Über die Theorie des Kreisels*, volume 1: *Einführung in Kinematik und Kinetik des Kreisels (1897)*; volume 2: *Durchführung der Theorie im Falle des Schweren symmetrischen Kreisels (1898)*; volume 3: *Die störende Einflüsse, Astronomische und geophysikalische Anwendungen (1903)*; volume 4: *Die technische Anwendungen der Kreiseltheorie* (1910) (Stuttgart, 1897–1910), 393. By equating the two sets of antonyms, Klein effectively removed industry from the discussion altogether. Theory and praxis were both to be found within the university system. In fact, eliding the connection between technology and praxis for which Riedler was arguing, Klein put the distinction between theory and praxis down to one of mental capacity, rather than of sites of production. At a 1912 meeting of the Göttingen Organization for the support of applied physics and mathematics, Klein claimed that the opposition between theory and praxis corresponds "to the different sides of human talent, which are only seldom found in one personality. Only the truly great have united both." Listing such greats, Klein did not look to engineering, but to Archimedes, Newton, and Gauss. See "Auszug aus dem Protokoll der Versammlung der Göttinger Vereinigung zur Förderung der angewandten Physik und Mathematik von 30. November 1912," in Tobies, "On the Contribution of Mathematical Societies to promoting Applications of Mathematics in Germany," 244–8, here 245.

56. Klein and Sommerfeld, *Über die Theorie des Kreisels*, volume 1: *Einführung in Kinematik und Kinetik des Kreisels (1897)*; volume 2: *Durchführung der Theorie im Falle des Schweren symmetrischen Kreisels (1898)*; volume 3: *Die störende Einflüsse, Astronomische und geophysikalische Anwendungen (1903)*; volume 4: *Die technische Anwendungen der Kreiseltheorie* (1910). I make use of a single-volume reprint of all four volumes: Felix Klein and Arnold Sommerfeld, *Über die Theorie des Kreisels* (1965).

57. Note added by Klein to Felix Klein, "On the Stability of the Sleeping Top," in *Felix Klein: Gesammelte Mathematische Abhandlungen*, ed. R. Fricke and H. Vermeil (1922), 658–9.

58. Felix Klein, *Ausgewählte Kapitel der Zahlentheorie* (1896).

59. The third volume, published in 1903 (after Sommerfeld left Göttingen), dealt largely with astronomical and geophysical applications, which Klein acknowledged Sommerfeld had essentially worked on alone (Klein, "On the Stability of the Sleeping Top," 659). The fourth volume appeared in 1910, after Sommerfeld had taken up the position in Munich and was compiled and edited by Fritz Noether. The effective change in editorial control can be tracked by the shift in the mathematical topics. To quote the *Vorwort* of the fourth volume: "Therefore, for example, the parameters $\alpha, \beta, \chi, \delta\ldots$ whose geometrical meaning in the first, and whose analytical importance in the second volume were worked out, were let slide in the third and fourth, naturally with the full concurrence of Klein himself, whose interests had turned, in any case, more and more toward applications." Klein and Sommerfeld, *Über die Theorie des Kreisels*, volume 4, iii.

60. Klein and Sommerfeld, *Über die Theorie des Kreisels*, volume 1, 4.

61. Ibid.

62. Ibid., volume 1, 5. For a detailed history of the theory and use of Gyroscopes, see Jobst Broelmann, *Intuition und Wissenschaft in der Kreiseltechnik, 1750 bis 1930* (2002). This discusses the text by Sommerfeld and Klein on 132–5.

63. Hermann ("Der Brückenschlag zwischen Mathematik und Technik") characterizes the difference between Sommerfeld and Klein as one of interest and ability. Sommerfeld possessed not only an exceptional mathematical ability, but also the ability for visualisable or intuitive thought [*anschaulich*]. The coming together of Klein and Sommerfeld was thus "a lucky chance that was essential for the fusion [*Brückenschlag*] of technology and mathematics." Klein's usual failure to deal with problems of industry, however, was deeper than that, and represented a more fundamental difference between him and his young student. I concur more with the view of Einstein, who once said bluntly to Klein: "It nonetheless appears to me that you vastly overrate the value of the formal view-point." Both quotations are from Hermann, "Der Brückenschlag zwischen Mathematik und Technik," 445.

64. Klein and Sommerfeld, *Über die Theorie des Kreisels*, Vorwort, iii. Given this clear statement of a difference of interests, it is somewhat peculiar that Sommerfeld scholars have tended to see *The Theory of Gyroscopes* as an unproblematic co-authored book, one that showed the disciple Sommerfeld in full accord with the man he called "The Great Felix." Part of the problem is that the book is usually discussed as part of Sommerfeld's early years in Göttingen, and is rarely mentioned in discussions of the Munich period. See, e.g., Eckert, *Die Atomphysiker*, 25–6, where, after a discussion of the Sommerfeld's assistantship and the *Gyroscope* project, Eckert notes that "Klein could hardly have hoped for a more eager interpreter of his efforts." The book is not mentioned again. More attention is paid in Eckert's introductory article in Eckert and Märker, eds., *ASWB I*, 140–3. There Eckert quotes from the joint introduction of the fourth volume, where the gyroscope is referred to as a "philosophical instrument." Although Sommerfeld signs his name to this introduction as well, that point of view seems much more like Klein's. Sommerfeld treated the gyroscope as an instrument for war, for industry, and for science, not just for philosophy. The point to be emphasized is that Sommerfeld's subordination to Klein clearly changed after he left Göttingen, even if he remained in agreement with the larger aims of his highly respected mentor.

65. Whether his analyses were deemed useful by practitioners was another question, however. For a discussion of Sommerfeld's exchanges with the torpedo engineer Carl Diegel on the technical application of gyroscopes in the guidance of weapons, see Broelmann, *Intuition und Wissenschaft in der Kreiseltechnik, 1750 bis 1930*, 136–8.

66. Arnold Sommerfeld, ed., *Die Enzyklopädie der mathematischen Wissenschaften mit Einschluss ihrer Anwendungen* (1904–26). This can be taken as support for my earlier assertion about Klein. Note that most of the applications are to natural sciences, not to engineering.

67. This applies, of course, principally to academics in subjects related to the physical sciences. It is this, no doubt, along with his efforts in teaching, that makes the Sommerfeld correspondence so voluminous and varied.

68. Paul Ewald would recall that it was Sommerfeld's lectures on Hydrodynamics (to which he was dragged by another student, Demetrios Hondros) that converted him to the study of theoretical physics.

69. Arnold Sommerfeld, "Ein Beitrag zur hydrodynamischen Erklärung der turbulenten Flüssigkeitsbewegung," in *Arnold Sommerfeld: Gesammelte Schriften*, ed. F. Sauter (1968).

70. Ibid. For an excellent discussion of the history of the turbulence problem, see Olivier Darrigol, "Turbulence in 19th-Century Hydrodynamics," *Historical Studies in the Physical and Biological Sciences* 32 (2002).

71. Illustrating the ongoing connection between Sommerfeld's research and teaching, Sommerfeld made the same point (in almost the same words) in his lectures. Arnold Sommerfeld, *Mechanics of Deformable Bodies*, volume II: *Lectures on Theoretical Physics* (1950), 113–4.

72. Ibid.

73. Eckert, "Mathematics, Experiments, and Theoretical Physics," 244–5.

74. Ludwig Hopf, *Hydrodynamische Untersuchungen: Turbulenz bei einem Flusse. Über Schiffswellen. Dissertation, München 1909* (1910), 5.

75. Ibid., 6.

76. Arnold Sommerfeld, "Report to the Faculty on the Dissertation of Ludwig Hopf" (5 July 1909).

77. Hopf, *Hydrodynamische Untersuchungen*, 47.

78. Ewald, "Erinnerungen an die Anfänge des Münchener Physikalischen Kolloquiums," 539.

79. Some material in this section has been published as Suman Seth, "Quantum Theory and the Electromagnetic Worldview," *Historical Studies in the Physical and Biological Sciences* 35 (2004). See also Shaul Katzir, "On 'The Electromagnetic Worldview': A Comment on an Article by Suman Seth," *Historical Studies in the Physical and Biological Sciences* 36 (2005); Suman Seth, "Response to Shaul Katzir: 'On the Electromagnetic Worldview,'" *Historical Studies in the Physical and Biological Sciences* 36 (2005).

80. Sommerfeld to Lorentz, 12 December 1906 (document 103 in Eckert and Märker, eds., *ASWB I*, 257–8).

81. Arnold Sommerfeld, "Lectures, Maxwell'sche Theorie u. Elektronenth." (DM: NL 089 (028), Winter 1906/07 and Winter 1908/09). "Freilich gibt das alles nur von der negativen Elektr. u. der mit ihr verbundenen scheinbaren Masse. Über die positive Elektr. u. die mit ihr scheinbar untrennbar verbundene Materie wissen wir nichts. Auch sind noch ernstl. Schwierigkeiten zu überwinden bei den elektroopt. Erscheinungen, die den Einfluss der Erdbewegung nach der Elektronenth. zeigen sollten. Lorentz sagte kurzl. *auf meine Frage, wie es den Elektronen geht: schlecht*. Die Schwierigkeiten auf elektroopt. Gebiet lassen sich angesichts der neuesten Messungen von Kaufmann nicht überwinden. So war auch Planck pessimist." The italicized sentence was written in shorthand. I am indebted to the Bonner Steno-Club for translating it for me. Many

of Sommerfeld's lectures have been microfilmed as part of the project carried out by Thomas S. Kuhn, John L. Heilbron, and Paul Forman for the Archive for the History of Quantum Physics. All references here are to the original documents, located in the Deutsches Museum.

82. Toward the end of 1905, Sommerfeld wrote to Wilhelm Wien, asking whether he already knew that Kaufmann had completed his measurements and had determined that the rigid electron had "won brilliantly." Kaufmann's measurements, Sommerfeld claimed, indicated that the "Lorentz formula for the deformable electron lay completely outside the [range of] possible observational errors." (Sommerfeld to Wien, 5 November 1905, document 96 in Eckert and Märker, eds., *ASWB I*, 250) This initial affirmation of Kaufmann's results was soon brought into question by the fact that Wilhelm Röntgen (Sommerfeld's senior colleague, and professor of experimental physics at Munich) did not believe that Kaufmann's measurements had been exact enough to decisively rule out the Lorentz theory. See Sommerfeld to Wien, 23 November 1906, document 102 in Eckert and Märker, eds., *ASWB I*, 255–7; Sommerfeld to Lorentz, 12 December 1906, document 103 in Eckert and Märker, eds., *ASWB I*, 257–58.

83. Max Planck, "Die Kaufmannschen Messungen der Ablenkbarkeit der γ-Strahlen für die Dynamik der Elektronen," *Physikalische Zeitschrift* 7 (1906): 758.

84. Sommerfeld to Lorentz, 12 December 1906, document 103 in Eckert and Märker, eds., *ASWB I*, 257–8.

85. "Diskussion" following Planck, "Die Kaufmannschen Messungen der Ablenkbarkeit der γ-Strahlen für die Dynamik der Elektronen," 759–61, on 761.

86. Russell McCormmach, "H. A. Lorentz and the Electromagnetic View of Nature," *Isis* 61 (1970): 489–90. While I concur with this position, and assume it in the argument that follows, it should be noted that the anti-Semitism that was so prevalent in German academia at this time also seems to have contributed to Sommerfeld's initial hostility toward Einstein's theory. On 26 December 1907, for example, Sommerfeld wrote to Lorentz (document 115 in Eckert and Märker, eds., *ASWB I*, 318–20), telling his colleague that he had been immersed in a study of Einstein's papers. "As brilliant at they are," Sommerfeld wrote, "there appears nonetheless to lie in this unconstructable, non-intuitive dogmatism almost something unhealthy. An Englishman would have found it difficult to produce such a theory. Perhaps there is something here that corresponds, as with Cohn, to the abstract-conceptual style of the Semite."

87. McCormmach, "H. A. Lorentz and the Electromagnetic View of Nature," 459. See also Olivier Darrigol, *Electrodynamics from Ampere to Einstein* (2000), esp. 351–94. Darrigol has argued that Lorentz should not be considered as strong a supporter of the electromagnetic view as Sommerfeld or Wien.

88. That Sommerfeld believed in an electromagnetic origin for the electron's (and other particles') mass can be seen from the text of the lectures he delivered on Maxwell's Theory and Electron Theory: "Cathode Rays. Microatoms. Their mass is alterable; an electromagnetic action. The mechanics of the smallest and simplest masses is really electromagnetically based." Sommerfeld, "Lectures, Maxwell'sche Theorie u. Elektronenth.," 2.

89. Sommerfeld's students would gently satirize his position in a poem written for one of the "Semestereinladungen" held at his house in Munich. Soon after his 1908 course on Maxwell's Theory, in which Sommerfeld introduced the concepts of special relativity, his students put on a play in which a "primed" and an "unprimed" observer ran to and fro delivering lines, including the following: "Die Mechanik, die ist ein gefügiges Ding / Die Geschwindigkeiten sind so lumpig gering, / Und widerspricht was, behauptete keck ich / Der Aether ist rein, die Materie dreckig." (P. P. Ewald, "Arnold Sommerfeld als Mensch, Lehrer und Freund: Rede, gehalten zur Feier der 100sten Wiederkehr seiner Geburt," in *Physics of the One- and Two-Electron Atoms: Proceedings of the Arnold Sommerfeld Centennial Meeting and of the International Symposium on the Physics of the One- and Two-Electron Atoms, Munich, 10–14 September 1968*, ed. F. Bopp and H. Kleinpoppen (1969), 12)

90. McCormmach, "H. A. Lorentz and the Electromagnetic View of Nature," 489.

91. James T. Cushing, *Quantum Mechanics: Historical Contingency and the Copenhagen Hegemony* (1994).

92. Sommerfeld, "Lectures, Maxwell'sche Theorie u. Elektronenth.," 2. "Gewiss ist die electromagn. Begründung der Mechanik Zukunftsmusik. Aber ich bin überzeugt, dass es hier ebenso gehen wird wie mit der Musik, die vor 30 Jahren den Namen Zukunftsmusik erhielt." Sommerfeld's reference here is a dual one. *Zukunftsmusik* may be translated colloquially as "dreams of the future." It was used in the late nineteenth century as a description of the music of Richard Wagner and Franz Liszt. Sommerfeld would appear to be playing on both meanings here, and noting that Wagner's music was part of the German canon by the early twentieth century, when he was writing.

93. Max Planck, *Vorlesungen über die Theorie der Wärmestrahlung* (1906).

94. Sommerfeld makes a point of noting that he is doing so: "Bezeichnungen wie bei Planck. Nur in der Wahl der elektr. Einheiten bediene ich mich der 'rationellen.'" "Zusammenfassung aller drei Methoden bei Planck." Sommerfeld, "Lectures, Theorie der Strahlung" (quotations on backs of 5 and on 4 respectively)

95. Ibid. The original is "Strahlung ein Brennpunkt moderner Forschung. Drei Strahlen kommen darin zusammen: Thermod., Electrod., Statistische Methoden." Planck did not name the statistical methods in his subject headings. Nonetheless, he treated the questions of probability and entropy in detail.

96. Source: Suman Seth, "Quantum Theory and the Electromagnetic World-View," *Historical Studies in the Physical and Biological Sciences* 35, no. 1 (2004): 67–93.

97. Paul Ehrenfest, "Zur Planckschen Strahlungstheorie," *Physikalische Zeitschrift* 7 (1906). The "impotence of resonators" and Ehrenfest's paper on it are addressed in Kuhn, *Black-Body Theory*, 158–66.

98. Quoted in Kuhn, *Black-Body Theory*, 159–60.

99. Planck (*Vorlesungen über die Theorie der Wärmestrahlung*, 220) cites an earlier paper by Paul Ehrenfest: "Über die physikalischen Voraussetzungen der Planck'schen Theorie der irreversiblen Strahlungsvorgänge," *Wiener Berichte* 114 (1905).

100. Sommerfeld, "Lectures, Theorie der Strahlung," in §9, "Planck's Theorie." The reference to the section on Jeans does not actually have a great deal of relevance to this particular point, though that section does end with a comparison of Planck's and Jeans's methods.

101. Ehrenfest, "Zur Planckschen Strahlungstheorie," 532. Quotations are from Sommerfeld, "Lectures, Theorie der Strahlung." §9 "Planck's Theorie," "Kritische Bemerkungen," point 4.

102. Sommerfeld, "Lectures, Theorie der Strahlung." §9 "Planck's Theorie," "Kritische Bemerkungen," point 3.

103. For a discussion of the experimental side of the black-body problem, see Hans Kangro, *Vorgeschichte des Plankschen Strahlungsgesetzes* (1970).

104. Sommerfeld, "Lectures, Theorie der Strahlung." §6, emphasis added. "Die interessanteste Frage ist nun die: Warum erhalten wir hier nur eine Näherungsformel?
1) Der Satz der gleichen Energieverteilung gilt nicht für den Äther allgemein, ist mechanisch abgeleitet. Es ist sozusagen Zufall, dass es noch fur ~~kurze~~ lange Wellen gilt. Lang heisst dabei nicht: gross gegen l, den l fällt heraus.
2) Standp. von Planck. Die Grosse h ist das Wirkungsquantum der Energie. Die Energie kann nicht beliebig unterteilt werden. Wäre die kleinste Energiemenge $h = 0$, so würde ~~sich~~ auch ~~aus~~ die Planck'sche Formel in die Jeans'sche degenerieren.
3) ~~Standp. von Jeans.~~

105. Ibid., under the heading "Allgemeine Bemerkungen dazu" [strikethroughs and superscripts in original]: "Die Thermod. ist die sicherste Grundlage aber die am wenigsten befriedigende. ^(Im Gegensatz zur Energetik verlangt man) Verständnis des Mechanismus oder Elektrodynamismus. In der Gastheorie hat man die Thermod. eliminirt, mechanisch-statistisch erklärt. Das Program ~~von Planck lautete~~ der Strahlungsth. sollte lauten; die Thermod. elektr.-statistisch zu erklären."

106. Planck, *Vorlesungen über die Theorie der Wärmestrahlung*, 132.

107. Sommerfeld, "Lectures, Theorie der Strahlung," under "Allgemeine Bemerkungen dazu": "Die Planck'sche Th. ist also nicht ideal; die Theorien von Jeans u. Lorentz principiell besser."

108. Ibid. "Die Elektrodyn. schafft auch hier die höchste Einheit. Wärme (gestrahlt) ist Licht, also Elektr., aber Wärme ist andererseits Molekularbewegung. ~~Wie soll sich~~ ^(Es muss sich) Elektr. Wirkung in Tragheitswirkung umsetzen; wie sie das tut, zeigt die Theorie der scheinbaren Masse, wonach kinetische Energie tatsächlich elektromagn. Energie der geladener Materie sein soll. Also kurz: Aus der Ident. von Licht u. Wärme ^(Leslie Prevost Rumford 18 Jahrh.), der Id. von Licht u. Elektr. ^(Maxwell Hertz Ende d. 19.) Jahrh. und der Id. von Wärme u Molekular mechanik ^(Clausius Maxwell Boltzmann 19 Jahrh.) folgt mit Notwendigkeit die Id. von Molekularmech. u. Elektrodynamik (20. Jahrh.)"

109. Ibid.: "Daraufhin werden wir ^(zu) den Planck'schen Einleitungsthesen gerade das Gegenteil aussagen können: 1) Wärme pflanzt sich auf 2 versch. Artens fort, Leitung u. Strahlung. 1a) Wärme pflanzt sich nur auf eine Art fort, elektrod., bei der Leitung sind die elektr. Felder an Ladungen gebunden, bei der Strahlung breiten si sich frei im Äther aus. 2) Die Wärmestr. ist viel compl. wie die Wärmeleitung, weil sich dort der Zustand nicht durch einen Vektor charakterisieren lasst. 2a) Die Wärmestr. ist viel einfacher wie die Wärmeleitung, weil die Besonderheiten

der Ladungsverteilung (Materie) nicht mitspielen. Im Äther allein die Strahlungsrichtung u. Intensitäten, im Wärmeleiter ausserdem die Bewegungsrichtung der Molekule."

110. Thomas S. Kuhn, *Black-Body Theory and the Quantum Discontinuity, 1894–1912*, second edition (1987), 189.

111. Lorentz wrote: "I admit that, when Jeans published his theory, I hoped that by examining it more closely one would be able to demonstrate the inapplicability to the ether of the theorem of "equipartition of energy" on which it is based.... The preceding considerations seem to me to prove that that is not the case and that one cannot escape Jeans's conclusion, at least not without profoundly modifying the fundamental hypotheses of the theory." ("Le partage de l'énergie entre la matière pondérable et l'éther," *Atti del IV Congresso Internazionale dei Matematici (Roma, 6–11, Aprile 1908)* (1909), volume I, 145–65, reproduced in H. A. Lorentz, *Collected Papers*, volume VII (1934), 317–43. Quotation from Kuhn, *Black-Body Theory*, 191.

112. Kuhn, *Black-Body Theory*, 192.

113. H. A. Lorentz, "Zur Strahlungstheorie," *Physikalische Zeitschrift* 9 (1908), reproduced in Lorentz, *Collected Papers*, 344–6.

114. Kuhn, *Black-Body Theory*, 195.

115. Kuhn writes: "Lummer and Pringsheim concluded, that the Jeans law is experimentally impossible." The quotation from the pair that he gives at the top of the page, however, is a criticism of the "Jeans-Lorentz formula," and hence presumably of both derivations. (ibid., 193)

116. Lorentz to Wien, June 6 1908, quoted in ibid., 194.

117. Lorentz, *Collected Papers*, 345.

118. Wien to Sommerfeld, 18 May 1908, document 132 in Eckert and Märker, eds., *ASWB I*, 338–39. Cf. Kuhn, *Black-Body Theory*, 192.

119. Wien to Sommerfeld, 15 June 1908, document 134 in Eckert and Märker, eds., *ASWB I*. Cf. Kuhn, *Black-Body Theory*, 203.

120. Sommerfeld to Wien, 20 June 1908, document 135 in Eckert and Märker, eds., *ASWB I*, 341–3. Sommerfeld claimed that Lorentz's assumption that one could treat electron motions as quasi-stationary, and hence restrict the number of degrees of freedom that needed to be considered to 6, was "definitely not acceptable" [*gewiss unzulässig*]. One needed, he argued, to take account of the existence of "free electron vibrations," which provided an infinite spectrum of free periods of vibration. "In the distribution of energy one must pay exactly as much attention to these infinitely many degrees of freedom as to the Eigen-vibrations of the box. The possibility of an equilibrium of energy appears thereby to be given, and a part of the energy for small λ [wavelength], which Lorentz gives to the ether alone, must go over to the electrons." Summarizing, Sommerfeld argues that "Lorentz only actually considers the equilibrium between the ether and uncharged atoms. This is not allowed, because the electrons mediate the energy exchange in the first place. In this energy exchange a part of the energy transferred remains stored up in the electrons."

121. A section of Sommerfeld's letter to Lorentz that repeats much of the material of the Wien letter is reproduced in note 2 to document 135 in ibid., 342: "Als ich einmal über die Theorie der Strahlung vortrug, glaubte ich dem Jeans'schen Paradoxon dadurch entgehen zu können, dass ich sagte, die Elektrodynamik ist nicht den mechanischen Gesetzen unterworfen. Ihre jetzigen Ausführungen scheinen mir ein vorzügliches Fundament zur Entscheidung dieser Frage."

122. Kuhn (*Black Body Theory*, 188–94, 202–4) locates Lorentz's and Wien's public espousals of the quantum discontinuity within a few months of each other during 1909.

123. In a letter dated 16 November 1908 (document 140 in Eckert and Märker, eds., *ASWB I*, 348–9), Sommerfeld urged Lorentz not to spend too much time in answering his letter: "You will respond that the small wavelengths—ultra-ultra-violet—can not play a role in the region of heat [*Wärmegebiet*]. But I am lacking the proof for that." In this letter Sommerfeld also congratulates Lorentz for the "victory" of the relativity theory, which Bucherer's experiments had recently supported, but goes on to lament that "a great deal of the clarity and causality of the physical foundations of your original theory is lost."

124. The language of conversion is Sommerfeld's: "Ich bin jetzt auch zur Relativtheorie bekehrt; besonders die systematische Form und Auffassung Minkowski's hat mir das Verstaendnis erleichtert." (Sommerfeld to Lorentz, 9 January [1910], document 163 in ibid., 375–6) Minkowski's enthusiasm for the electromagnetic worldview—an enthusiasm that may even be found in the text of his lecture on *Raum und Zeit*—may have appealed to Sommerfeld: "The validity without exception of the world-postulate, I like to think," Minkowski wrote, "is the true nucleus of an electromagnetic image of the world...." (quoted in Peter Galison, "Minkowski's Space-Time: From Visual Thinking to the Absolute World," *Historical Studies in the Physical Sciences* 10 (1979): 93)

125. Sommerfeld in fact taught two classes in the winter of 1909 that dealt with relativity theory: one on the thermodynamics of moving systems and the one quoted here, titled "Elektronentheorie, II Teil." (*Deutsches Museum NL 089 (028)*)

126. On Lorentz's comparatively "soft" support for the electromagnetic worldview, however, see Darrigol, *Electrodynamics from Ampere to Einstein*, 361.

127. J. H. Jeans, "Non-Newtonian Mechanical Systems, and Planck's Theory of Radiation," *Philosophical Magazine* 20 (1910), quoted in Kuhn, *Black-Body Theory*, 204.

128. "One final aspect of the problem-solving approach to mathematics that was of special significance to wrangler research," Warwick writes, "is the technical unity lent to mathematical physics by the analogical application of common mathematical methods.... Undergraduate training thus encouraged students to approach physical phenomena as illustrative exercises in the application of common physical principles...and mathematical methods." (Warwick, *Masters of Theory*, 278)

129. Arnold Sommerfeld, "Report to the Faculty on the Dissertation of Hermann von Hoerschelmann" (January 1911).

130. Arnold Sommerfeld, "Report to the Faculty on the Dissertation of Hermann W. March" (13 June 1911). March's dissertation was titled Die Ausbreitung der Wellen des Drahtlosen Telegraphie auf der Erdkugel.

131. Ibid.

132. Ewald, "The Setting for the Discovery of X-Ray Diffraction by Crystals. Speech Given at the First General Assembly of the International Union of Crystallography at Harvard University, 2 August 1948," 18.

Chapter 2

1. Thomas Mann, "Gladius Dei," in *Tonio Kröger and Other Stories* (1970), 81.

2. Ewald, "The Setting for the Discovery of X-Ray Diffraction by Crystals. Speech Given at the First General Assembly of the International Union of Crystallography at Harvard University, 2 August 1948," 5–6.

3. Mann, "Gladius Dei," 84.

4. Ewald, "Arnold Sommerfeld als Mensch, Lehrer und Freund," 14. Although clearly tinted with a nostalgic glow, Ewald's description is not without validity. As Pyenson and Skopp have shown, physics in Germany around 1900 was the domain of the upper classes: "From the point of view of demography, physics had more in common with classical philology than with organic chemistry.... Although the success of academic physics was closely related to the rise of academic chemistry, these statistics suggest that, by 1900, physics had evolved as a distinct scientific enterprise appealing for the most part to the highest social classes." Physics students, in other words, could afford the luxury of the pursuit of "pure" knowledge. Pyenson and Skopp also suggest that Sommerfeld's timing may have contributed as much to his students' success as any of his personal or pedagogical attributes: "[P]roductive physicist doctor-fathers...exercised enormous power through their students during the late Wilhelmian and Weimar periods. Part of their influence is an accident of history. German tertiary education expanded at a rapid rate in the period before the First World War, and many new posts had to be filled. Having been promoted in 1900 meant that one would be around 45 years old in 1920. If one had maintained visibility and came from the right school, the chances were good of receiving a professorship. Nearly 20 percent of the physicists in our sample held a chair by the middle of the Weimar period; in contrast, by 1927 only around 10 percent of the PhDs in the other disciplines of our sample had been called to a chair, or to direct a university institute, library, or museum." ("Educating Physicists in Germany circa 1900," 338, 356; see also Paul Forman, John L. Heilbron, and Spencer Weart, "Physics circa 1900: Personnel, Funding, and Productivity of the Academic Establishments," *Historical Studies in the Physical Sciences* 5 (1975)) Since Ewald had held a position as chair of the physics department at the Brooklyn Polytechnic Institute from 1949 to 1957, he may also have been drawing on a trope that was common in US universities in the postwar years, one that longed for a time when physicists did not undertake their studies in search of financial remuneration. See David Kaiser, "The Postwar Suburbanization of American Physics," *American Quarterly* 56 (2004).

5. Albert Einstein to Arnold Sommerfeld, 14 January 1908, in Eckert and Märker, eds., *ASWB I*, 321–3.

6. Einstein to Sommerfeld, between 11 and 28 January. Eckert and Märker, eds., *ASWB 2*, 111–113, on 113.

7. Born, "Sommerfeld als Begründer einer Schule," 1035–6.

8. Paul Kirkpatrick, "Address of Recommendation to Arnold Sommerfeld upon the award of the 1948 Oersted Medal for Notable Contributions to the Teaching of Physics," *American Journal of Physics* 12 (1949): 313. Kirkpatrick lists Wentzel, Heitler, Houston, Stueckelberg, Ewald, Morse, Pauling, Laporte, Herzfeld, Teller, Brillouin, Eckart, Pauli, Condon, Landé, London, Bethe, Debye, Heisenberg, Laue, and Bragg as having "spent one, two, or several years at Sommerfeld's famous institute."

9. Warwick, *Masters of Theory*.

10. Jungnickel and McCormmach, *Intellectual Mastery of Nature*, volume 2, 153.

11. Sommerfeld seems to have restored the instruments quickly, for by the winter of 1908 he was offering a class for students wishing to undertake "independent work in the mathematical-physical collection."

12. Ewald, "Erinnerungen an die Anfänge des Münchener Physikalischen Kolloquiums."

13. Arnold Sommerfeld, "Das Institut für Theoretische Physik," in *Die Wissenschaftlichen Anstalten der Ludwig-Maximilians-Universität zu München: Chronik zur Jahrhundertfeier im Auftrag des Akademischen Senats*, ed. K. von Müller (1926).

14. Jungnickel and McCormmach, *Intellectual Mastery of Nature*, volume 2, 152–3.

15. Bavarian Minister of the Interior to Munich University Senate, 2 August 1886. Quoted in ibid., 38.

16. On Graetz at Munich, and Röntgen's attitude toward him, see ibid., 276–81.

17. K. Kuhn, "Erinnerungen an die Vorlesungen von W. C. Röntgen und L. Grätz," *Physikalische Blätter* 18 (1962): 314.

18. Ibid. Kuhn suggests (315) that Röntgen's lectures (but not, one assumes, Graetz's) may have been a little too much for students of medicine and pharmacy. Walther Friedrich, an early student of Sommerfeld's and Röntgen's, said in an interview (Thomas S. Kuhn, "Interview with Walther Friedrich, 05/15/1963" (1963), 7–8) that Sommerfeld "was the first that got students in theoretical physics, because before him was Leo Graetz, and...he dealt with theoretical physics the old way with all the variables, writing formulas that spilled over not only one row of the board, but needed several rows. The first actual one, e. g. that introduced the modern theories of the mathematical tools, was Sommerfeld."

19. Thomas S. Kuhn, "Interview with Werner Heisenberg, 02/07/1963" (1963), 8.

20. Arnold Sommerfeld, *Mechanics*, volume I: *Lectures on Theoretical Physics* (1952), ix.

21. Kuhn, "Interview with Werner Heisenberg, 02/07/1963," 7. This logic did not always apply. Pauli and Heisenberg, for example, both finished after only six semesters.

22. Thomas S. Kuhn and G. Uhlenbeck, "Interview with Peter Debye, 05/03/1962" (1962), 7. According to Debye, "they were all experimentalists, or preparing for their high school teaching exams."

23. Kuhn, "Interview with Werner Heisenberg, 02/07/1963," 1, 8. Heisenberg gives the figure of 80 to 100 for the lectures in general at first, and later suggests that there were about 60 people in the hydrodynamics lectures he attended.

24. Sommerfeld's lecture notes, with their regular emendations and modifications, provide a rich record of his thinking over time—similar in some ways to the examination scripts that Warwick uses to capture "mathematical physics in the making." It is much harder to garner from them what was seen as important and significant for his listeners (Warwick, *Masters of Theory*, 18). I have been able to find few contemporary accounts of Sommerfeld's lecturing and pedagogical style, and the sources used below are often drawn from interviews conducted considerably after the fact. These present obvious problems, although as a point of reference I note the largely post-facto source base of Warwick's superb reconstruction of Routh's pedagogical method to make clear that the problems are not intrinsically insurmountable. I have tried to provide multiple sources for the claims made in participants' reminiscences.

25. Quoted in Kirkpatrick, "Oersted Medal." Sommerfeld was critiqued by Born for not emphasizing enough the uncertain areas in atomic spectroscopy in the first edition of *Atombau und Spektrallinien*.

26. Sommerfeld, "Die naturwissenschaftlichen Ergebnissen und die Ziele der modernen technischen Mechanik." In Sauter, ed., *GS I*, 446–55.

27. Quoted in Kirkpatrick, "Oersted Medal," 313. According to Hönl, however, the smoothness of Sommerfeld's presentation could sometimes lead older students to gripe at the fact that difficulties were thereby being dodged (H. Hönl, *Memoirs of Research on Zeeman Effect at Munich in Early 1920s* (AHQP), 1).

28. Kuhn, "Interview with Werner Heisenberg, 02/07/1963," 10.

29. Sommerfeld, *Mechanics*. ix–x.

30. Arnold Sommerfeld *Thermodynamics and Statistical Mechanics*, Lectures on Theoretical Physics, volume V, ed. F. Bopp and J. Meixner (1956), v.

31. In contrast, the preface to the second edition of Planck's classic *Vorlesungen über Thermodynamik* noted: "Since this book is chiefly concerned with the exposition of these fundamental [thermodynamic] principles...the applications are given more as illustrative examples...."

32. Debye noted that in the beginning Sommerfeld offered four "fundamental" courses: theoretical mechanics, thermodynamics, electromagnetism, and optics (Kuhn and Uhlenbeck, "Interview with Peter Debye, 05/03/1962," 7; Sommerfeld's lecture schedule for 1906–1930 is included in appendix A).

33. Sommerfeld, *Mechanics*. ix.

34. Ewald, "Foreword to Sommerfeld's Course," in ibid., vi–vii. See also David C. Cassidy, *Uncertainty: The Life and Science of Werner Heisenberg* (1992), esp. 91–109.

35. Sommerfeld, *Mechanics*. x. Cf. Sommerfeld, "Das Institut für theoretische Physik," 291-2: "[W]eekly exercises in connection to the lectures…are held, in which written tasks are posed and are worked through by the students."

36. Kuhn and Uhlenbeck, "Interview with Peter Debye, 05/03/1962," 3, 7.

37. Kuhn, "Interview with Werner Heisenberg, 02/07/1963," 1.

38. Ibid., 1–2.

39. Kuhn, "Interview with Walther Friedrich, 05/15/1963," 11–12.

40. Quoted in Warwick, *Masters of Theory*, 276. This is not to suggest, of course, that problem solving was entirely absent from the German pedagogical tradition, but rather that—in comparison to Britain—it was not characteristic of teaching in mathematical or theoretical physics.

41. Many students, it is true, sat for an examination that enabled them to teach in secondary schools, but the level for such an examination was nowhere near as high as that of the Tripos and was significantly lower than the training problems students solved in Sommerfeld's classes. One suspects that many of the students in Graetz's lectures were preparing for careers as teachers in Realschulen and Gymnasia.

42. "Vorwort" by Felix Klein to Edward John Routh, *Die Dynamik der Systeme starrer Körper* (1898). Cf. Rowe, "Klein, Hilbert, and the Göttingen Mathematical Tradition"; Warwick, *Masters of Theory*, 252–3.

43. Note, however, the distinction between applications to the other sciences and applications to engineering problems, discussed in the preceding chapter.

44. Klein and Sommerfeld, *Über die Theorie des Kreisels*, volume 1: *Einführung in Kinematik und Kinetik des Kreisels (1897)*; volume 2: *Durchführung der Theorie im Falle des Schweren symmetrischen Kreisels (1898)*; volume 3: *Die störende Einflüsse, Astronomische und geophysikalische Anwendungen (1903)*; volume 4: *Die technische Anwendungen der Kreiseltheorie*, 5. The introduction to the first volume mentions Routh's book as currently in press.

45. Sommerfeld, *Mechanics*. x.

46. Sommerfeld's more direct and instrumental enthusiasm for Routh's work lasted throughout his lifetime, the mechanics volume of his *Lectures* recommending the book Klein had introduced as "a collection of problems of unique variety and richness" (ibid., 223, note 3).

47. Kuhn, "Interview with Werner Heisenberg, 02/07/1963," 2.

48. Kuhn and Uhlenbeck, "Interview with Peter Debye, 05/03/1962," 15.

49. Marvin L. Goldberger, "Enrico Fermi (1901–1954): The Complete Physicist," in James W. Cronin, ed., *Fermi Remembered* (2004), on 155.

50. Thomas S. Kuhn, "Interview with Werner Heisenberg, 02/15/1963" (1963), 7.

51. Kuhn, "Interview with Werner Heisenberg, 02/07/1963," 2.

52. Bethe, for example, was admitted to the seminar on the basis of a recommendation from Meissner (Thomas S. Kuhn, "Interview with Hans Bethe, 01/17/1964" (1964), 7). Sommerfeld admitted Heisenberg after hearing that the young man had worked his way, on his own, through Hermann Weyl's *Raum, Zeit, Materie*—a point that had condemned Heisenberg in the eyes of Lindemann, a more conservative mathematician. In Heisenberg's memory, Sommerfeld replied to his rather presumptuous request in a friendly manner: "All right, you have an interest in mathematics; it may be that you know something; it may be that you know nothing; we will see. All right, you come to the seminar, and we'll see what you can do." (Thomas S. Kuhn, "Interview with Werner Heisenberg, 11/30/1962" (1962), 5)

53. From Linus Pauling's memory, Epstein's comparison to American graduate schools may have been overly generous, since it was apparently assumed that visiting American students were rather weak in relation to their German counterparts (John L. Heilbron, "Interview with Linus Pauling, 03/27/1964" (1964), 3a, 26).

54. This followed Sommerfeld's lectures on "selected parts of electrodynamics and mechanics from the standpoint of the relativity principle." It constituted, as Sommerfeld was later boasted, the first undergraduate classes on relativity given anywhere in the world.

55. Kuhn, "Interview with Werner Heisenberg, 02/07/1963," 3.

56. Kuhn, "Interview with Hans Bethe, 01/17/1964," 7.

57. Ibid., 10. This insistence on including all the algebra constituted a more general principle of Sommerfeld's work. As Debye noted, Sommerfeld always wanted publications to be longer than editors usually desired, since he insisted on including all the calculations in the manuscript. The idea was that one could thus start from the beginning without taking on anyone else's assumptions. One might also add that it made it much easier to learn a new theory this way. Kuhn and Uhlenbeck, "Interview with Peter Debye, 05/03/1962," 11. See also chapter 5 of the present volume.

58. Kuhn and Uhlenbeck, "Interview with Peter Debye, 05/03/1962," 6.

59. Ewald, "Erinnerungen an die Anfänge des Münchener Physikalischen Kolloquiums," 541. This "Munich Wednesday Colloquium" ran for more than 30 years (1908–1939). Sommerfeld soon became a regular participant. At the end of a semester, students and professors would gather at a nearby tavern to eat roast pork and play skittles. Zenneck, one of Sommerfeld's colleagues, claimed later that he no longer could remember whether Sommerfeld was any good at the game, but assumed he had to be: the co-author of a large book on gyroscopic motion should, he suspected, be able to deal with this special case of the rolling sphere (Zenneck, quoted in Benz, *Arnold Sommerfeld*, 65). More formal lectures were heard on Mondays, at the Sohnke Colloquium.

60. Kuhn, "Interview with Werner Heisenberg, 02/07/1963," 3–4.

61. Thomas S. Kuhn, "Interview with Werner Heisenberg, 02/19/1963" (1963), 8.

62. Kuhn, "Interview with Werner Heisenberg, 02/15/1963," 23. Heisenberg's recollection was that, although he had fewer scientific conversations with Born, he had more social interaction with Born than with the more formal Sommerfeld: "Well, Born would not see the young students as much as that, so I had fewer scientific discussions with Born. On the other hand, Born had more private connections with the students; we had many parties at his house, and we had excursions to the [Heinberg]. The whole style in Göttingen was less the traditional style; I mean Sommerfeld was still the Geheimrat and was a bit traditional, while Born tried to be everything else, but no Geheimrat.... The whole thing was more intended as a kind of family life among the young people with the Borns." (Kuhn, "Interview with Werner Heisenberg, 02/15/1963," 10)

63. Thomas S. Kuhn, "Interview with Paul Epstein, 05/25/1962" (1962), 6.

64. Kuhn, "Interview with Werner Heisenberg, 02/07/1963," 3.

65. Einstein to Sommerfeld, 29 September 1909, in Eckert and Märker, eds., *ASWB I*, 362–3.

66. Interview with Heisenberg.

67. Thomas S. Kuhn, "Interview with G. Uhlenbeck, 03/30/1962" (1962).

68. Quoted in Jeremy Bernstein, *Hans Bethe, Prophet of Energy* (1980), 15. Cf. Eckert, "Mathematics, Experiments, and Theoretical Physics," 252.

69. Kuhn, "Interview with G. Uhlenbeck, 03/30/1962," 10.

70. Kuhn and Uhlenbeck, "Interview with Peter Debye, 05/03/1962," 3–4. Cf. Armin Hermann, *The Genesis of the Quantum Theory, 1899–1913* (1971), 103.

71. Arnold Sommerfeld, "Some Reminiscences of My Teaching Career," *American Journal of Physics* 12 (1949): 315.

72. Ewald wrote: "I know of no other Ordinarius in the Ludwigs-Maximilians-University at that time, who would have offered such an intimate gathering of his seminar members and doctoral students...." ("Arnold Sommerfeld als Mensch, Lehrer und Freund," 10). Cf. Benz, *Arnold Sommerfeld*, 66–8; Eckert, *Die Atomphysiker*, 38–9. Sommerfeld's love of skiing was gently satirized by his students at the dinner in Sommerfeld's house. Soon after Sommerfeld's first lectures on the relativity theory, two students performed as a "primed observer" and an "unprimed observer." Near the close, the unprimed observer declaimed: "Es lockt ihn von der Arbeit fort / Einzig und allein der Wintersport / Drum strahlt im Sonnenschein die Welt / Zieht ins Gebirg Professor Sommerfeld / Und fühlt in Eis und Schnee sich frei / Von seiner Schüler dummer Rederei. / Denkt nicht der Faulen die Kollege schwänzen / Denkt nicht an Hopf und seine Turbulenzen / An Wellen nicht, von Hondros wohlerwogen / Nicht an Debyes schönen Regenbogen, / Ja, er vergisst sogar auf seinen Paar Ski / Das Mach'sche Phänomen und Fräulein Kucharski." (Ewald, "Arnold Sommerfeld als Mensch, Lehrer und Freund," 12)

73. Ewald, "Arnold Sommerfeld als Mensch, Lehrer und Freund," 10.

74. Ewald, "The Setting for the Discovery of X-Ray Diffraction by Crystals. Speech Given at the First General Assembly of the International Union of Crystallography at Harvard University, 2 August 1948," 21–2. Cf. Eckert, "Mathematics, Experiments, and Theoretical Physics," 247. In the end, Ewald chose yet another topic, one in crystal optics, which Sommerfeld also supported.

75. For early discussions of research schools and the insistence on the existence of a "reasonably coherent program of research," see J. B. Morrell, "The Chemist Breeders: The Research Schools of Liebig and Thomas Thomson," *Ambix* 19 (1972); Gerald Geison, *Michael Foster and the Cambridge School of Physiology: The Scientific Enterprise in late Victorian Society* (1978); Geison, "Scientific Change, Emerging Specialties, and Research Schools," *History of Science* 19 (1981). A more recent collection is *Research Schools: Historical Reappraisals*, volume 8, ed. G. Geison and F. Holmes (1993). For a critique similar to that offered here, see Olesko, *Physics as a Calling*, 8–9.

76. As Warwick and Kaiser note, Kuhn, having pointed out that students who learn physics from a textbook often have no idea how to solve the problems at the end of the section, nonetheless offers textbooks as a principal means by which a community of researchers is produced. The implication is that "worked problems" provide a template that students can imitate, yet such an answer ignores the basic fact that some pedagogical environments or "training regimes" are vastly more successful than others and almost always more successful than solitary study.

77. Andrew Warwick and David Kaiser, "Kuhn, Foucault and the Power of Pedagogy," in *Pedagogy and the Practice of Science: Historical and Contemporary Perspectives*, ed. D. Kaiser (2005), 403.

78. Michel Foucault, *Discipline and Punish: The Birth of the Prison*, second edition (1995), 136.

79. Ibid. And Foucault writes: "By the late eighteenth century, the soldier has become something that can be made; out of a formless clay, an inapt body, the machine required can be constructed; posture is gradually corrected; a calculated restraint runs slowly through each part of the body, mastering it, making it pliable, ready at all times, turning silently into the automatism of habit...." (*Discipline and Punish*, 135)

80. Warwick and Kaiser, "Power of Pedagogy," 401.

81. Foucault, *Discipline and Punish*, 154; Warwick and Kaiser, "Power of Pedagogy," 399.

82. Foucault, *Discipline and Punish*, 138. See also Michel Foucault, *The History of Sexuality*, volume 1: *An Introduction* (1979), 139. Discussing the evolution of the "power over life," starting in the seventeenth century, Foucalt writes of "the body as machine: its disciplining, the optimization of its capabilities, the extortion of its forces, *the parallel increase of its usefulness and its docility* [emphasis added], its integration into systems of efficient and economic controls, all this was ensured by the procedures of power that characterized the *disciplines*: an *anatomo-politics of the human body*."

83. Warwick, *Masters of Theory*, 273.

84. Sharon Traweek, *Beamtimes and Lifetimes: The World of High Energy Physicists* (1988), 88. This, of course, is an example of what Thomas Kuhn termed "the essential tension" between tradition

and innovation inherent in scientific research ("The Essential Tension: Tradition and Innovation in Scientific Research," in *The Essential Tension: Selected Studies in Scientific Tradition and Change* (1977)).

85. Note that for Foucault's main examples (students, prisoners, soldiers, the ill, the insane) none of those exercising disciplinary power desire independent action for those being disciplined. Foucault writes: "What is specific to the disciplinary penalty is non observance, that which does not measure up to the rule, that departs from it. The whole indefinite domain of the non-conforming is punishable: the soldier commits an 'offence' whenever he does not reach the level required; a pupil's 'offence' is not only a minor infraction, but also an inability to carry out his tasks. The regulations for the Prussian infantry ordered that a soldier who had not correctly learned to handle his rifle should be treated with the 'greatest severity.' Similarly, 'when a pupil has not retained the catechism from the previous day, he must be forced to learn it, without making any mistake, and repeat it the following day; either he will be forced to hear it standing or kneeling, his hands joined, or he will be given some other penance." (*Discipline and Punish*, 178–9)

86. Warwick, *Masters of Theory*, 303–6.

87. Foucault, *Discipline and Punish*, 159.

88. For studies of the creative possibilities available to "discipliners," see Jan Goldstein, "Foucault Among the Sociologists: The "Disciplines" and the History of the Professions," *History and Theory* 23 (1984); Timothy Lenoir, "The Discipline of Nature and the Nature of Disciplines," in *Instituting Science: The Cultural Production of Scientific Disciplines* (1997).

89. Goldstein, quoted in Olesko, *Physics as a Calling*, 14.

90. Ibid., 15.

91. Ibid., 17.

Chapter 3

1. Ernst Jünger, *The Storm of Steel: From the Diary of a German Storm-Troop Officer on the Western Front* (1996), 1.

2. Ludwig Hopf to Arnold Sommerfeld, 30 Sept. 1916. DM: NL 089 (059).

3. Ludwig Hopf to Arnold Sommerfeld, 31 May 1915. DM: NL 089 (059).

4. Hopf to Sommerfeld, 30 Sept. 1916. DM: NL 089 (059).

5. This is particularly true for the German case. Michael Eckert has considered some of the war work of Arnold Sommerfeld in *Die Atomphysiker*. See also the essay in Eckert and Märker, eds., *ASWB I*, 429–66, esp. 445–55. Stefan L. Wolff ("Physicists in the "Krieg der Geister": Wilhelm Wien's Proclamation," *Historical Studies in the Physical and Biological Sciences* 33 (2003)) supplies a history of Wilhelm Wien's war work. For the British side of this "war of words," see Andrew

Hull, "War of Words: The Public Science of the British Scientific Community and the Origins of the Department of Scientific and Industrial Research, 1914–16," *British Journal for the History of Science* 32 (1999). England and the United States are both discussed in more detail in the literature. Even so, specific studies of the physics community in any country are rare. The work of Daniel Kevles for the case of the US stands out. See Daniel J. Kevles, "George Ellery Hale, the First World War, and the Advancement of Science in America," *Isis* 59 (1968); "Into Hostile Political Camps: The Reorganization of International Science After World War I," *Isis* 62 (1971); *The Physicists: The History of a Scientific Community in Modern America* (1995). On the British case but with less of a focus on a community history, see the extensive work of Roy Macleod—e.g., R. Macleod and K. Macleod, "War and Economic Development: Government and the Optical Industry in Britain, 1914–1918," in *War and Economic Development*, ed. J. Winter (1975); R. Macleod and E. K. Andrews, "Scientific Advice for the War at Sea, 1915–1917: The Board of Invention and Research," *Journal of Contemporary History* 6 (1971). On chemists, see Roy Macleod and Jeffrey Allan Johnson, eds., *Frontline and Factory: Comparative Perspectives on the Chemical Industry at War, 1914–1924* (2006); Jeffrey A. Johnson, *The Kaiser's Chemists: Science and Modernization in Imperial Germany* (1990); Roy Macleod, "The Chemists Go to War: The Mobilization of Civilian Chemists and the British War Effort, 1914–1918," *Annals of Science* 50 (1993). On the "War of Invention," see G. Hartcup, *The War of Invention: Scientific Developments, 1914–1918* (1988); Michael Pattison, "Scientists, Inventors and the Military in Britain, 1915–19: The Munitions Inventions Department," *Social Studies of Science* 13 (1983). For an excellent and concise English-language overview of the French context, see George K. Burgess, "Applications of Science to Warfare in France," *Scientific Monthly* 5 (1917).

6. Johnson, *The Kaiser's Chemists*. It is, of course, from Johnson that I draw this chapter's title.

7. Wolff, "Physicists in the 'Krieg der Geister.'" Wolff takes the title from H. Kellermann, *Der Krieg der Geister* (1915).

8. This "Aufruf" is reproduced in Eckert and Märker, eds., *ASWB I*, 488–9.

9. Karl Herzfeld to Arnold Sommerfeld, 14 Nov. 1916. DM: NL 089 (059)

10. Wilhelm Wien, *Aus dem Leben und Wirken eines Physikers* (1930), 31. "[D]ie Deutschen waren zu ihren Fahnen geeilt."

11. Max Born to Arnold Sommerfeld, 2 Feb. 1915. DM: NL 089 (059).

12. Redaktion, "Übersicht über die Kriegsbeteiligung der Deutschen Physiker," *Physikalische Zeitschrift* 16 (1915). Corrections and addenda were included in subsequent editions of the journal until the end of 1916. The original list comprised 163 names. In successive months, 35 names were added. A similar list of members of the Society of German Engineers with diplomas was published in the society's journal through 1917. The first listing is Redaktion, "Vom Kriegsschauplatze" Verbandsmitglieder im Felde," *Zeitschrift des Verbandes deutscher Diplom-Ingenieure* 5 (1914). Note that the "Übersicht" provides information concerning the military rank and posting of physicists but not of their activities. It is thus probable that some of those listed under "combat service" (a category that includes artillery officers) may have been calculating the trajectories of

shells, for example. However, the general point—that most physicists were serving as soldiers rather than as scientists—would seem to hold.

13. That said, it is worth noting that the area in which physicists found their skills most applicable—telegraphy—was a central problem of the Sommerfeld School. When this concentration is coupled with the growing utilization of X-rays (another area of specialization for the Munich School) in military medicine, one can begin to understand the comparatively large number of *Sommerfeld's* students who would ultimately turn their background in physics toward problems associated with the war.

14. Robert Wohl, *The Generation of 1914* (1979), esp. 42–84.

15. Jünger, *The Storm of Steel: From the Diary of a German Storm-Troop Officer on the Western Front*, 1.

16. Wien, *Aus dem Leben und Wirken eines Physikers*, 32; Michael Eckert, "Sommerfeld in World War I: Military Research and Political Attitude (Conference talk delivered at the EASST-4S-Meeting at the University of Bielefeld, October 1996)" (1996), 10.

17. Johnson, *The Kaiser's Chemists*, 184.

18. Max Born, *My Life: Recollections of a Nobel Laureate* (1978), 169–70.

19. Roger Chickering, *Imperial Germany and the Great War, 1914–1918* (1998), 35–40. See also Gerald D. Feldman, *Army, Industry and Labor in Germany, 1914–1918* (1966).

20. Max Wien to Arnold Sommerfeld, 4 Jan. 1916. DM: NL 089 (059).

21. Paul Ewald to Arnold Sommerfeld, 5 Sept. 1915. DM: NL 089 (059).

22. Born, *My Life*, 170–3.

23. Quoted in Ken Beauchamp, *History of Telegraphy* (2001), 364.

24. Wien, *Aus dem Leben und Wirken eines Physikers*, 37: "English physicists had it much better in this, since they could participate in a more significant fashion through the determination of more suitable methods for the discovery of U-boats. The task of working against the English methods was not, unfortunately, assigned to us."

25. The question of the military-industrial connections to physics in World War I has, as far as I know, gone almost completely unanswered. However, a few hints in Sommerfeld's correspondence suggest that industrial concerns, like those of Rathenau and Siemens, had the capability of pulling soldiers away from the front and bringing them to work on industrial problems. Thus, Max Abraham, who had worked on electromagnetic theory before the war, was "reclaimed by the *Gesellschaft Telefunken* (the Radiotelegraphic Society, a concern founded by AEG and Siemens in 1904 for wireless telegraphy). See letter from Sommerfeld to Dr Eilbott 23 July 1923 DM: NL 089 (004). Otto Blumenthal, one of Sommerfeld's students, was also reclaimed by Siemens. See letter from Paul Ewald to Sommerfeld, 24 Nov. 1917 DM: NL 089 (007). On chemistry, see Macleod and Johnson, eds., *Frontline and Factory: Comparative Perspectives on the Chemical Industry at War, 1914–1924*.

26. Born, *My Life*, 168. Born was already a well-respected scholar. One assumes that the option of simply reporting to a unit so directly was not available to everyone. Nonetheless, many did volunteer for particular units in the army. Given the match between assumed research interests and postings, one can assume that these requests were often acceded to.

27. Ibid., 171.

28. Arnold Sommerfeld, "Zu Röntgens siebzigstem Geburtstag," *Deutsche Revue* (1915): 89.

29. Sommerfeld to Paul Ehrenfest, 26 January 1915. Quoted in Eckert and Märker, eds., *ASWB I*, 448.

30. Paul Ewald to Sommerfeld, 9 November 1915. DM: NL 089 (059).

31. For contemporary histories on this topic, see Koelzer, "Der Militärische Wetterdienst," in *Die Technik im Weltkriege*, ed. M. Schwarte (1920); Siegmund Günther, "Die Meteorologie im Kriege," in *Deutsche Naturwissenschaft, Technik und Erfindung im Weltkriege*, ed. B. Schmid (1919).

32. Quoted in Johnson, *The Kaiser's Chemists*, 190. A similar meteorological phenomenon was behind Robert Millikan's suggested use of "balloon bombers." See Charles A. Ziegler, "Weapons Development in Context: The Case of the World War I Balloon Bomber," *Technology and Culture* 35 (1994).

33. Rudolf Seeliger to Arnold Sommerfeld, undated letter (internal evidence points to a date in 1915), DM: NL 089 (059).

34. Manfred Rasch, "Wissenschaft und Militär: Die Kaiser Wilhelm Stiftung für kriegstechnische Wissenschaft," *Militärgeschichtliche Mitteilungen* 49 (1991).

35. This incomplete letter to the Kaiser Wilhelm Foundation, titled "Für den Jahresbericht der KWKW," is reproduced on pp. 586–7 of Eckert and Märker, eds., *ASWB I*.

36. Albert Einstein, "Wahlvorschalg für A. Sommerfeld und P. Debye zur Aufnahme als korrespondierende Mitglieder in die Akademie d. Wiss.," in *Albert Einstein in Berlin, 1913–1933*, ed. C. Kirsten and H. Treder (1979).

37. An excellent recent history of wireless telegraphy, covering the war years, is Chen-Pang Yeang, "The Study of Long-Distance Radio-Wave Propagation, 1900–1919," *Historical Studies in the Physical and Biological Sciences* 33 (2003).

38. Sommerfeld, letter to KWKW, in Eckert and Märker, eds., *ASWB I*, 588.

39. John Ambrose Fleming, *An Elementary Manual of Radiotelegraphy and Radiotelephony for Students and Operators*, first edition (1908), 168, note 7.

40. John Ambrose Fleming, *An Elementary Manual of Radiotelegraphy and Radiotelephony for Students and Operators*, third edition (1916).

41. On the relationship between Sommerfeld and Zenneck, and on their work on wireless telegraphy, see Eckert and Märker, eds., *ASWB I*, 284–9.

42. Arnold Sommerfeld, "Über die Ausbreitung elektrischer Wellen in der drahtlosen Telegraphie," in *Arnold Sommerfeld: Gesammelte Schriften*, ed. F. Sauter (1968).

43. Ibid., 692. This theory met with some controversy, even within Germany. See Michael Eckert and Walther Kaiser, "An der Nahstelle von Theorie und Praxis: Arnold Sommerfeld und der Streit um die Wellenausbreitung in der drahtlosen Telegraphie," in *Chemie-Kultur-Geschichte*, ed. A. Schurmann and B. Weiss (2002).

44. Lenz to Sommerfeld, 28 Nov. 1916. DM: NL 089 (059).

45. Fritz Noether, "Über analytische Berechnung der Geschosspendelungen," *Artillerist. Monatsheft* 149/50 (1919). Also Fritz Noether, "Über analytische Berechnung der Geschosspendelungen," *Nachrichten der K. Gesellschaft der Wissenschaften zu Göttingen (Mathematisch-physikalische Klasse)* (1919).

46. Cranz cited in Eckert, "Sommerfeld in World War I," 7–8.

47. Ewald to Sommerfeld 21 Oct. 1916. DM: NL 089 (059). Ewald also noted that there were currently four of Sommerfeld's students in close proximity and that he hoped to gather them all for an upcoming "physicists' evening."

48. Noether to Sommerfeld, 23 May 1915. DM: NL 089 (059).

49. Lenz to Sommerfeld, 28 Jan. 1916. DM: NL 089 (059).

50. Lenz to Sommerfeld, 16 Jan 1915. DM: NL 089 (059).

51. Lenz to Sommerfeld, 28 January 1916. DM: NL 089 (059).

52. John Howard Morrow, *German Air Power in World War I* (Lincoln, 1982).

53. Editor, "Zur Entstehung der Technischen Berichte der Flugzeugmeisterei," *Technische Berichte: Herausgegeben von der Flugzeugmeisterei der Inspektion der Fliegertruppen Charlottenburg* I (1917), iii.

54. Hopf to Sommerfeld, 30 Sept. 1916. DM: NL 089 (059).

55. Hopf to Sommerfeld, 20 Oct. 1916. DM: NL 089 (059).

56. Hopf to Sommerfeld, 16 Sept. 1914. DM: NL 089 (059).

57. Hopf to Sommerfeld, 31 May 1915. DM: NL 089 (059).

58. Richard Fuchs and Ludwig Hopf, *Aerodynamik* (1922).

59. Ibid., v.

60. Ibid., 30.

61. Eddies will last in such fluids, but will not arise spontaneously.

62. Fuchs and Hopf, *Aerodynamik*, 30.

63. Ibid.

64. Ibid.

65. Pages denoted by Roman Numerals are those of the "Vorwort."

66. R. Fuchs and L. Hopf, "Die allgemeine Längsbewegung des Flugzeuges," *Technische Berichte: Herausgegeben von der Flugzeugmeisterei der Inspektion der Fliegertruppen Charlottenburg* III (1918): 317.

67. Hopf to Sommerfeld, 30 Sept. 1916. DM: NL 089 (059).

68. Lenz to Sommerfeld, 18 May 1916. DM: NL 089 (059).

69. Lenz To Sommerfeld, No date, presumably 1916. DM: NL 089 (059).

70. Sommerfeld and Wilhelm Wien are clear exceptions.

Chapter 4

1. Christa Kirsten and Hans-Günther Körber, eds., *Physiker über Physiker: Wahlvorschläge zur Aufnahme von Physikern in die Berliner Akademie 1870 bis 1929, von Hermann v. Helmholtz bis Erwin Schrödinger* (1975). Nr. 23, "Wahlvorschlag für Max Planck (1858–1947) zum OM."

2. Heilbron, *The Dilemmas of an Upright Man*, vii.

3. Max Planck, "Antrittsrede zur Aufnahme in die Akademie vom 28. Juni 1894," in *Physikalische Abhandlungen und Vorträge* (1958), 1.

4. Ibid., 3.

5. Kirsten and Körber, eds., *Physiker über Physiker*. Nr. 23, "Wahlvorschlag für Max Planck (1858–1947) zum OM."

6. Max Planck, ed., *Abhandlungen zur Thermodynamik, von H. Helmholtz*, volume 124 (1902). Five years earlier, however, Planck cast Helmholtz's approach to thermodynamics as one midway between a micro-mechanical one and his own, which did not place the mechanical nature of heat in the foreground of his analysis, "proceeding instead directly from a few very general facts of experience, mainly from both of the so-called fundamental laws of heat theory." Helmholtz's method, Planck noted, restricted itself to the most important assumption of the mechanical heat theory, the notion that heat was a form of motion, but avoided any hypothesizing about the nature of this motion. Max Planck, *Vorlesungen über Thermodynamik* (1897).

7. Planck, quoted in Kuhn, *Black-Body Theory*, 27.

8. Hiebert, "Conception of Thermodynamics."

9. Writing of Planck's work on cavity radiation, Darrigol has rightly noted—presumably with Henri Poincaré in mind—that "it is typical of a style of theoretical physics, namely, concentrating on the features of physical systems which can be determined on the basis of general principles only, without recourse to detailed microscopic assumptions." Olivier Darrigol, *From c-Numbers to q-Numbers: The Classical Analogy in the History of Quantum Theory* (1992), 31; Darrigol, "Henri

Poincaré's Criticism of Fin-de-siècle Electrodynamics"; Darrigol, "The Electrodynamic Origins of Relativity Theory," *Historical Studies in the Physical and Biological Sciences* 26 (1996).

10. Holton, "Einstein and the Cultural Roots of Modern Science."

11. On Planck and the energetics debate, see Robert J. Deltete, "The Energetics Controversy in Late Nineteenth-Century Germany: Helm, Ostwald, and Their Critics" (1983); Robert J. Deltete, "Gibbs and the Energeticists," in *No Truth Except in the Details*, ed. A. Kox and D. Siegel (1995); C. Hakfoort, "Wilhelm Ostwald's Energeticist Worldview and the History of Scientism," *Annals of Science* 49 (1992); Erwin N. Hiebert, "The Energetics Controversy and the New Thermodynamics," in *Perspectives in the History of Science and Technology*, ed. D. Roller (1971), Arie Leegwater, "The Development of Wilhelm Ostwald's Chemical Energetics," *Centaurus* 29 (1986).

12. In 1910, told of plans to create a *Zeitschrift für Theoretische Physik,* Planck wrote to Wien, noting that "on the other hand I do not at all like a stricter separation of theoretical from experimental research." Quoted in Hans Kangro, *Early History of Planck's Radiation Law* (1976), 121.

13. Cf. Helmholtz's physics, in which bodies were built "out of small pieces that were entirely similar to the bodies proper that it took them to be, that is, essentially as they were in the *laboratory* world." Jed Z. Buchwald, "Electrodynamics in Context: Object States, Laboratory Practice, and Anti-Romanticism," in *Hermann von Helmholtz and the Foundations of Nineteenth-Century Science*, ed. D. Cahan (1993).

14. J. H. Van 't Hoff, *Lectures on Theoretical and Physical Chemistry: Chemical Statics,* volume II (1899), 32.

15. Arne Schirrmacher, "Experimenting Theory: The Proofs of Kirchhoff's Radiation Law Before and After Planck," *Historical Studies in the Physical and Biological Sciences* 33 (2003).

16. Svante Arrhenius, *Theories of Solutions* (New Haven, 1912), 89–90. The debate is discussed in detail by Robert Scott Root-Bernstein, "The Ionists: Founding Physical Chemistry, 1872–1890" (Princeton University, 1980). My analysis of the debate between Arrhenius and Planck is indebted to Root-Bernstein's pioneering work. A more sympathetic portrayal of Planck is offered in Elisabeth Crawford, *Arrhenius: From Ionic Theory to the Greenhouse Effect* (1996), 88–91.

17. Max Planck, "Wissenschaftliche Selbstbiographie," in *Max Planck: Physikalische Abhandlungen und Vorträge III* (1958), 381.

18. Max Planck, "Über das Prinzip der Vermehrung der Entropie: 1. Abhandlung," *Wiedemanns Annalen* 30 (1887). Reproduced in Planck, *Physikalische Abhandlungen und Vorträge* (hereafter cited as *PAV*) I, 196–216, on 216.

19. Max Planck, "Zur Geschichte der Auffindung des physikalischen Wirkungsquantums," *Naturwissenschaften* 31 (1943). Reproduced in Planck, *PAV III*, 255–67, on 255.

20. Max Planck, *Gleichgewichtszustände isotroper Körper in verschiedenen Temperaturen* (1880). Reproduced in Planck, *PAV I*, 62–124, on 62.

21. Max Planck, "Verdampfen, Schmelzen, und Sublimiren," *Wiedemanns Annalen* 13 (1882). Reproduced in Planck, *PAV I*, 134–63, on 140.

22. Planck, "Verdampfen, Schmelzen, und Sublimiren," 162–3.

23. Planck, "Über das Prinzip der Vermehrung der Entropie: 1. Abhandlung," 196.

24. Max Planck, "Über das Prinzip der Vermehrung der Entropie. 3. Abhandlung," *Wiedemanns Annalen* 32 (1887). Reproduced in Planck, *PAV I*, 232–73, on 257–8. Italics mine.

25. Root-Bernstein calls Planck's move here a "chemical slip," one of many Planck would make. "In the briefest terms, then, Planck did not demonstrate any understanding of the wider *chemical* implications of his results. They stood as a monument to logic and reductionism not as a representation of nature." Root-Bernstein, "The Ionists." 465, 471. In the case of his "ideal process," however, Planck was clearly aware of the problem and would later defend it. No simple mistake, Planck believed that the process was justified on good, theoretical grounds. It followed, Planck suggested, from his earlier analysis, from a "mixed gas" of arbitrarily many substances to arbitrarily many solutes in a solvent.

26. Here the constants k are assumed to be independent of the number of molecules, n.

27. Planck does not make the assumption that the salt is vaporized along with the water, so he was well aware of the normal chemical processes involved in the change of phase from solution to gas.

28. Root-Bernstein, "The Ionists." 454.

29. Planck, "Über das Prinzip der Vermehrung der Entropie. 3. Abhandlung," 271.

30. Ibid., 273.

31. Root-Bernstein, "The Ionists." 417.

32. Max Planck, "Das chemische Gleichgewicht in verdünnten Lösungen," *Wiedemanns Annalen* 34 (1888). Reproduced in Planck, *PAV I*, 280–95, on 280–1.

33. Arrhenius to Ostwald, February 2, 1888. Quoted in Root-Bernstein, "The Ionists." 498.

34. Quoted in Leo Koenigsberger, *Hermann von Helmholtz* (1906), 340.

35. Planck, "Allgemeines zur neuren Entwicklung der Wärmetheorie." Reproduced in Planck, *PAV I*, 372–81.

36. Planck, "Allgemeines zur neuren Entwicklung der Wärmetheorie," 373. This comment, of questionable accuracy, would understandably draw Boltzmann's ire, leading to a frosty relationship between the two men that would last for several years.

37. Ibid.

38. Ibid., 375.

39. In fact, these too were also examples of abstractions. For the case of the semi-permeable membrane, Planck argued that "In a strong sense there are in general perhaps no semi-permeable membranes; because the property that allows a substance to appear suitable for this [task] probably only consists in a particularly large difference in the velocities of the various substances that diffuse through it." Ibid., 378.

40. See, e.g., J. H. Van 't Hoff, "Die Rolle des osmotischen Druckes in der Analogie zwischen Lösungen und Gasen," *Zeitschrift für physikalische Chemie* 1 (1887): 482.

41. Planck, "Allgemeines zur neuren Entwicklung der Wärmetheorie," 377–8.

42. Ibid., 378.

43. Ibid.

44. Ibid., 375.

45. Ibid., 379.

46. Ibid.

47. Ibid. (emphasis added).

48. Ibid.

49. Svante Arrhenius, "Über die Gültigkeit des Beweises von Herrn Planck für das van 't Hoffsche Gesetz," *Zeitschrift für physikalische Chemie* 9 (1892): 332.

50. The use of "forget" is Arrhenius's (ibid., 330).

51. Ibid., 332.

52. Max Planck, "Erwiderung auf einen von Herrn Arrhenius erhobenen Einwand," *Zeitschrift für physikalische Chemie* 9 (1892). Reproduced in Planck, *PAV I*, 433–4.

53. On Arrhenius's turn away from physical chemistry after the early 1890s, see Crawford, *Arrhenius: From Ionic Theory to the Greenhouse Effect*.

54. Planck published a short book that included one of the three additional articles he wrote in this three-year period. Neither of the other two papers concerned thermochemistry. Max Planck, *Grundriss der allgemeinen Thermochemie* (Breslau, 1893). In 1897, Planck published his classic text on thermodynamics, one that summarized and clarified his positions but did not substantially add to them. The book was in its eleventh edition by the early 1960s. Planck, *Vorlesungen über Thermodynamik*. See Hiebert, "Conception of Thermodynamics," 65–72.

55. "Wie verdünnt eine Lösung anzunehmen ist, damit sie den Bedingungen der idealen Verdünnung Genüge leistet, kann man ebensowenig allgemein sagen, wie es auch unmöglish ist, ganz allgemein die Dichte anzugeben, welche ein Gas höchstens haben darf, damit es sich im idealen Gaszustand befindet." Max Planck, "Zur Thermodynamik und Dissoziationstheorie binärer Elektrolyte," *Zeitschrift für physikalische Chemie* 41 (1902). Reproduced in Planck, *PAV II*, 26–37, on 31.

56. Mathias Cantor, "Über die Grundlage der Lösungstheorie," *Annalen der Physik* 10 (1903).

57. Ibid., 205.

58. Ibid., 206.

59. Planck, "Über die Grundlage der Lösungstheorie, eine Erwiderung." Reproduced in Planck, *PAV II*, 45–54.

60. Planck, "Über die Grundlage der Lösungstheorie, eine Erwiderung," 47.

61. Ibid.

62. Ibid., 50.

63. Ibid., 49.

64. Darrigol, *From c-Numbers to q-Numbers*; Heilbron, *The Dilemmas of an Upright Man*; Kangro, *Vorgeschichte des Planckschen Strahlungsgesetzes*; Klein, "Max Planck and the Beginnings of the Quantum Theory"; Kuhn, *Black-Body Theory*; Needell, "Irreversibility and the Failure of Classical Dynamics."

65. Planck, "Zur Geschichte der Auffindung des physikalischen Wirkungsquantums." Reproduced in Planck, *PAV III*, 255–67, on 258; emphasis added. Cf. Klein, "Max Planck and the Beginnings of the Quantum Theory."

66. Max Planck, "Absorption und Emission elektrischer Wellen durch Resonanz," *Wiedemanns Annalen* 57 (1896). Reproduced in Planck, *PAV I*, 445–58. Quote on 458.

67. Buchwald, "Electrodynamics in Context"; Edward Jurkowitz, "Helmholtz and the Liberal Unification of Science," *Historical Studies in the Physical Sciences* 32 (2002).

68. Jed Z. Buchwald, "Reflections on Hertz and the Hertzian Dipole," in *Heinrich Hertz: Classical Physicist, Modern Philosopher*, ed. D. Baird, R. Hughes, and A. Nordmann (1998), 275. See, more generally, Jed Z. Buchwald, *The Creation of Scientific Effects: Heinrich Hertz and Electric Waves* (1994).

69. Max Planck, "Über elektrische Schwingungen, welche durch Resonanz erregt und durch Strahlung gedämpft werden," *Wiedemanns Annalen* 60 (1897).

70. Planck, *PAV I*, 481. Planck tended to use the logarithmic decrement of the damping term, given by $\sigma = 2\pi/3c^3 \sqrt{K/L^3}$

71. For a detailed discussion of Planck's route to this term, see Darrigol, *From c-Numbers to q-Numbers*, 29–36.

72. Max Planck, "Über irreversible Strahlungsvorgänge, 1. Mitteilung," *Sitzungsberichte der Preussischen Akademie der Wissenschaften* (1897). Reproduced in Planck, *PAV I*, 493–504.

73. Ludwig Boltzmann, "Über irreversible Strahlungsvorgänge I," *Berliner Berichte* (1897). Reproduced in F. Hasenöhrl, ed., *Wissenschaftliche Abhandlungen von Ludwig Boltzmann*, volume III: *1882–1905* (1909), 615–7.

74. Max Planck, "Über irreversible Strahlungsvorgänge, 2. Mitteilung," *Sitzungsberichte der Preussischen Akademie der Wissenschaften* (1897). Reproduced in Planck, *PAV I*, 505–7. Emphasis added.

75. See Buchwald, "Reflections on Hertz and the Hertzian Dipole."

76. Heinrich Hertz, "The Forces of Electric Oscillations, Treated According to Maxwell's Theory [Translation]," *Wiedemanns Annalen* 36 (1889): 146.

77. Planck, "Absorption und Emission elektrischer Wellen durch Resonanz." Reproduced in Planck, *PAV I*, 445–58. Quote on 446 (emphasis added)

78. Max Planck, "Über irreversible Strahlungsvorgänge. 3. Mitteilung," *Sitzungsberichte der Preussischen Akademie der Wissenschaften* (1897). Reproduced in Planck, *PAV I*, 508–31. Cf. Darrigol, *From c-Numbers to q-Numbers*, 39.

79. Planck, *PAV I*. See 518–19, 529–31.

80. Ibid., 511.

81. Ludwig Boltzmann, "Über irreversible Strahlungsvorgänge II," *Berliner Berichte* (1897). Reproduced in Hasenöhrl, ed., *Wissenschaftliche Abhandlungen von Ludwig Boltzmann*, 618–21.

82. Hasenöhrl, ed., *Wissenschaftliche Abhandlungen von Ludwig Boltzmann*, 619.

83. Max Planck, "Über irreversible Strahlungsvorgänge, 4. Mitteilung," *Sitzungsberichte der Preussischen Akademie der Wissenschaften* (1898). Reproduced in Planck, *PAV I*, 532–59. Quote on 559.

84. One could exclude, in Boltzmann's examples, cases where each molecule moved toward its nearest neighbour, or each molecule within a certain velocity range was surrounded by ten slower-moving molecules. Ludwig Boltzmann, *Lectures on Gas Theory* (1995), 40–2.

85. Max Planck, "Über irreversible Strahlungsvorgänge," *Annalen der Physik* 1 (1900). Reproduced in Planck, *PAV I*, 614–67, quote on 618.

86. Kuhn, *Black-Body Theory*, 77. Darrigol also accepts that a certain "procedural parallelism" existed between Planck and Boltzmann's derivation, but notes that it is vague in its details: "…the relevant analogies concerned general concepts or categories rather than specific mathematical expressions.…This should not surprise us, because the original dynamical systems, the molecular gas on the one hand, the resonator in cavity radiation on the other, exhibited strong qualitative differences. As vague as it was, the idea of a selection of disordered states for which the evolution of all measurable quantities would not depend on finer uncontrollable details was all Planck needed to establish his "fundamental equation." The rest…proceeded from the autonomous development of Planck's program, with the exception of the entropy formula." Darrigol, *From c-Numbers to q-Numbers*, 52.

87. Kuhn, *Black-Body Theory*, 73.

88. Planck, *PAV I*, 536.

89. Ibid., 559.

90. Planck, "Über irreversible Strahlungsvorgänge IV." Reproduced in Planck, *PAV I*, 532–59, on 551.

91. Planck, *PAV I*, 551–2. Emphasis added.

92. See, e.g., Planck's dissertation: Max Planck, *Über den zweiten Hauptsatz der mechanischen Wärmetheorie* (1879). Reproduced in Planck, *PAV I*, 1–61.

93. Planck, *PAV I*, 533.

94. Ibid., 557. See Darrigol, *From c-Numbers to q-Numbers*, 52–3.

95. Quoted in Kuhn, *Black-Body Theory*, 82.

96. Planck, *PAV I*, 532.

97. Planck, "Zur Geschichte der Auffindung des physikalischen Wirkungsquantums," 259, 264.

98. Ibid., 264.

99. See Needell, "Irreversibility and the Failure of Classical Dynamics." Darrigol, *From c-Numbers to q-Numbers*, 52–3. For Planck's explicit acknowledgment of this point, see Planck, "Das Einheit des physikalischen Weltbildes (Lecture, 9 December 1908, at the University of Leiden)." Reproduced in Planck, *PAV III*, 20.

100. Planck, *PAV I*, 633.

101. Here a and b are constants, and e is the inverse of the natural logarithm.

102. Max Planck, "Über irreversible Strahlungsvorgänge. 5. Mitteilung," *Sitzungsberichte der Preussischen Akademie der Wissenschaften* (1899). Reproduced in Planck, *PAV I*, 560–600.

103. Planck, *PAV I*, 593.

104. Klein, "Max Planck and the Beginnings of the Quantum Theory," 463, Kuhn, *Black-Body Theory*, 89–90.

105. Planck, *PAV I*, 599–600. See also Kuhn, *Black-Body Theory*, 89–90.

106. On Planck's faith in the absolute, see in particular Goldberg, "Planck's Philosophy of Nature"; Heilbron, *The Dilemmas of an Upright Man*; Klein, "Max Planck and the Beginnings of the Quantum Theory."

107. See, e.g., Olivier Darrigol, "The Historians' Disagreement over the Meaning of Planck's Quantum," *Centaurus* 43 (2001). Clayton A. Gearhart, "Planck, the Quantum, and the Historians," *Phys. Persp.* 4 (2002).

108. Wien's law of displacement, not to be confused with his expression for the black-body curve, was a general functional expression relating the density of radiant energy at a given wavelength to temperature. On the early history of such laws, see Kangro, *Early History*, 37–47. Wien stated his law in 1893 as follows: " In the normal emission spectrum from a black body each wavelength is displaced with change of temperature in such a way that the product of temperature and wavelength remains constant." Quoted in Kangro, *Early History*, 45. The form most familiar today, $E_\lambda = \lambda^{-5}\phi(\lambda T)$ was first derived by Joseph Larmor in 1900. Planck derived the distribution law in the form given above in January 1901.

109. Max Planck, "Zur Theorie des Gesetzes der Energieverteilung im Normalspektrum," *Verhandlungen der deutschen physikalischen Gesellschaft* 2 (1900). Reproduced in Planck, *PAV I*, 698–706, on 699 and 698. Note the similarity of this definition of disorder to that used in the third paper on irreversible radiative processes.

110. Max Planck, "Über das Gesetz der Energieverteilung im Normalspektrum," *Annalen der Physik* 4 (1901). Reproduced in Planck, *PAV I*, 717–27. Kuhn, *Black-Body Theory*, 108.

111. The numerical factor enters here from a substitution of the energy density u for the energy, U, where, from an earlier calculation, $u/U = 8\pi v^2/c^2$.

112. See Martin J. Klein, "Planck, Entropy, and Quanta," *The Natural Philosopher* 1 (1963).

113. Max Planck, "Über die Verteilung der Energie zwischen Äther und Materie," *Annalen der Physik* 9 (1902). Reproduced in Planck, *PAV I*, 731–43, on 732.

114. Planck, *PAV I*, 734. Klein, "Planck, Entropy, and Quanta," 94.

115. Pockels and Voigt, quoted in Schirrmacher, "Experimenting Theory," 311 and 309. Emphasis added.

116. John L. Heilbron, "Fin-de-siècle Physics," in *Science, Technology and Society in the Time of Alfred Nobel*, ed. C. Bernhard, E. Crawford, and P. Sörbom (1982).

117. Max Planck, "On Clausius Theorem for irreversible Cycles, and on the Increase of Entropy," *Philosophical Magazine* 9 (1905). Reproduced in Planck, *PAV II*, 100–101, on 101.

118. The exchange has been translated and published in *Physical Reality: Philosophical Essays on Twentieth-Century Physics*, ed. S. Toulmin (1970).

119. Ernst Mach, *History and Root of the Principle of the Conservation of Energy* (1911); Max Planck, "Das Einheit des physikalischen Weltbildes (Lecture, 9 December 1908, at the University of Leiden)," in Planck, *PAV III*, 6–29, on 23.

120. Seth, "Allgemeine Physik? Max Planck und die Gemeinschaft der theoretischen Physik."

121. Planck, *PAV III*, 27.

122. Ibid., 29. Cf. Heilbron, *The Dilemmas of an Upright Man*, 47–60.

123. Ernst Mach, "Die Leitgedanken meiner naturwissenschaftlichen Erkenntnislehre und ihre Aufnahme durch die Zeitgenossen," *Scientia* 7 (1910): 233. Needell ("Irreversibility and the Failure of Classical Dynamics") uses the Leiden speech and Planck's Columbia lectures to explicate Planck's ongoing commitment to a non-statistical understanding of the second law.

124. Planck, *PAV III*, 8.

125. Ibid., 10.

126. Ibid., 11.

127. Ibid., 14. In contrast, it would be precisely the connection to the needs of steam engine construction that Sommerfeld would vaunt in his own approach to thermodynamics. See F. Bopp and J. Meixner, eds., *Thermodynamik und Statistik von Arnold Sommerfeld*, Vorlesungen über Theoretische Physik, V (1952), v.

128. Planck, *PAV III*, 14.

129. Ibid., 15.

130. Ibid., 16–7.

131. This argument, strictly speaking, applied to ideal processes used as a means to *define* the principles of physics. Planck continued to use such processes in his thermochemistry. See his discussions of dilute solutions in his Columbia lectures.

132. As Planck noted, Boltzmann would not have followed him in arguing for the absolute character of the second law.

133. Planck, *PAV III*, 21.

134. Ibid., 20–1.

135. Max Planck, "Selbstdarstellung," in *Max Planck: Selbstdarstellung (1942)*, ed. C. Kirsten (1982), 6–7.

136. Max Planck, *Wissenschaftliche Selbstbiographie* (1948). Reproduced in Planck, *PAV III*, 374–401, on 383.

137. Buchwald, "Electrodynamics in Context," 346. For Planck on the least action principle, see Max Planck, "Das Prinzip der Kleinsten Wirkung," in *Der Kultur der Gegenwart* (1914), reproduced in Planck, *PAV III*, 91–101.

138. Schirrmacher, "Experimenting Theory," 316–7.

Chapter 5

1. Planck, "Wissenschaftliche Selbstbiographie," 401. Sommerfeld: "Der sorgsam Urbar macht das neue Land,/Dieweil ich hier und da ein Blumensträusschen fand." Planck: "Was ich gepflückt, was Du gepflückt,/Das wollen wir verbinden,/Und da sich eins zum andern schickt,/Den schönsten Kranz draus winden."

2. The Solvay congress and Planck and Sommerfeld's papers have been discussed in detail in the literature. In the present chapter, without seeking to offer a full bibliography, I draw with gratitude, on the following: Hermann, *The Genesis of the Quantum Theory* (1971); Max Jammer, *The Conceptual Development of Quantum Mechanics* (1966); Jagdish Mehra, *The Solvay Conferences on Physics* (1975); Needell, "Irreversibility and the Failure of Classical Dynamics"; Sigeko Nisio, "The Formation of the Sommerfeld Quantum Theory of 1916," *Japanese Studies in the History of Science* 12 (1973); Jagdish Mehra and Helmut Rechenberg, *The Historical Development of Quantum Theory*, volume 1 (1982); Bruce R. Wheaton, *The Tiger and the Shark: Empirical Roots of Wave-Particle Dualism* (1983); Kuhn, *Black-Body Theory*. See also the essays in Eckert and Märker, eds., *ASWB I*.

3. The ratio of the change of energy (dE) to the change in temperature (dT), the *specific heat* of a body ($c = dE/dT$), is a measure of how much energy must be added to a substance to raise its temperature by a certain amount. It becomes important when one is working with high-

temperature filaments, which must be heated until incandescent. The lower a substance's specific heat, the more easily it can be raised to high temperatures. The problem was thus central to Walther Nernst's work on the development of carbon-filament lamps. See Diana Kormos Barkan, *Walther Nernst and the Transition to Modern Physical Science* (1999).

4. On the construction of this "transition," see Richard Staley, "On the Co-Creation of Classical and Modern Physics," *Isis* 96 (2005).

5. Mehra, *The Solvay Conferences on Physics: Aspects of the Development of Physics since 1911*, 6, 8–9.

6. Quoted in ibid., 5.

7. Kuhn, *Black-Body Theory*, 231–2; Russell McCormmach, "Henri Poincaré and the Quantum Theory," *Isis* 58 (1967).

8. Peter Debye, "Der Wahrscheinlichkeitsbegriff in der Theorie der Strahlung," *Annalen der Physik* 33 (1910); Hermann, *The Genesis of the Quantum Theory*, 107–11.

9. Peter Debye and Arnold Sommerfeld, "Theorie des lichtelektrischen Effektes vom Standpunkt des Wirkungsquantums," *Annalen der Physik* 41 (1913). Reproduced in Sauter, ed., *GS IV*, 78–135. Sommerfeld credited Wilhelm Lenz with a significant portion of the work presented in Arnold Sommerfeld, "Probleme der freien Weglänge," *Mathematische Vorlesungen an der Universität Göttingen* 6 (1914) (reproduced in Sauter, ed., *Arnold Sommerfeld: Gesammelte Schriften*, 287–327, on 287). Cf. Nisio, "The Formation of the Sommerfeld Quantum Theory of 1916," 53.

10. Sommerfeld, Report to the Faculty, 28 April 1914, Munich University Archive, OC I 40p

11. Kuhn, *Black-Body Theory*, 216.

12. Brillouin, quoted in Kuhn, *Black-Body Theory*, 252.

13. Paul Langevin, discussion following Arnold Sommerfeld, "Die Bedeutung des Wirkungsquantums für unperiodische Molekularprozesse in der Physik," in *Die Theorie der Strahlung und der Quanten*, ed. A. Eucken (1913), 303.

14. On the energetics debate, see Deltete, "The Energetics Controversy in Late Nineteenth-Century Germany"; Hiebert, "The Energetics Controversy and the New Thermodynamics"; Leegwater, "The Development of Wilhelm Ostwald's Chemical Energetics"; Deltete, "Gibbs and the Energeticists"; Hakfoort, "Wilhelm Ostwald's Energeticist Worldview and the History of Scientism"; Jungnickel and McCormmach, *Intellectual Mastery of Nature*, volume 2, 217–27.

15. Max Planck, "Gegen die neuere Energetik," *Annalen der Physik* 57 (1896).

16. Quoted in Abraham Pais, *Subtle Is the Lord: The Science and the Life of Albert Einstein* (1982), 83.

17. However, Sommerfeld would reiterate the critique in his quantum-theoretical spectroscopy: "Here that is true which may be said of every purely energetic treatment: the equating of energy... can never furnish more than *one* equation determining the course of the phenomenon. In

the case of more than one degree of freedom the energetic view must be supplemented by a deeper dynamical treatment." (*Atomic Structure and Spectral Lines*, 255)

18. Sommerfeld, "Lectures, Theorie der Strahlung."

19. Arnold Sommerfeld, "Theoretisches über die Beugung der Röntgenstrahlen," *Zeitschrift für Mathematik und Physik* 46 (1901). Reproduced in Sauter, ed., *GS IV*, 240–326, on 240.

20. Sommerfeld's association of Planck's major works with the "less satisfying" thermodynamic approach comes through clearly in his recommendation of his colleague as a corresponding member of the Bavarian academy of sciences in 1911. "Planck's scientific personality," he wrote, "is rooted in thermodynamics." Citing Planck's work in thermochemistry as a prime example of this claim, he continued by noting that, since that work in the 1890s, "Electrodynamics has, through Maxwell and Hertz, assumed the leading role in theoretical natural knowledge. Following the work of Willy Wien, Planck put forward the program of integrating electrodynamics with the principles of thermodynamics in the area of radiation theory, a program that he carried out with rare determination and complete success in the years from 1895 to 1902." (Eckert and Märker, eds., *ASWB I*, 298)

21. Elizabeth Garber, "Some Reactions to Planck's Law, 1900–1914," *Studies in History and Philosophy of Science* 7 (1976): 105. Nor did the situation change in succeeding years. "Planck's publications after 1901 on heat radiation did not help to clarify any ideas he had on the nature of the interaction between ether and matter." (Garber, "Some Reactions to Planck's Law, 1900–1914," 125) For Lorentz's emphasis on the necessity of an account of electron-theoretical mechanism in explaining the black-body problem "in the most satisfactory way," see H. A. Lorentz, "The Theory of Radiation and the Second Law of Thermodynamics," *Proc. Amsterdam Acad.* 3 (1901), in H. A. Lorentz, *Collected Papers*, volume VI (1938), 265–79, esp. 270–1; H. A. Lorentz, "On the Radiation of Heat in a System of Bodies Having a Uniform Temperature," *Proc. Amsterdam Acad.* 7 (1905), in Lorentz, *Collected Papers*, 293–317, quote on 293.

22. Eckert and Märker, eds., *ASWB I*; Sommerfeld, *Mathematical Theory of Diffraction*.

23. Arnold Sommerfeld, "Zur mathematischen Theorie der Beugungserscheinungen," *Nachrichten der K. Gesellschaft der Wissenschaften zu Göttingen (Mathematisch-physikalische Klasse)* (1894), reproduced in Sauter, ed., *GS I*, 86–90. Sommerfeld defines k as the "reciprocal wavelength" and notes that for the two-dimensional case the partial differential equation simplifies to $\partial^2 u/\partial x^2 + \partial^2 u/\partial y^2 + k^2 u = 0$. If the wavelength is taken to be vanishingly small, then k is infinitely large. In this case, however, the wave-like properties of the phenomena are lost: "[T]he light can be decomposed into individual rays, which propagate independently of one another with constant velocity.... While one has theoretically assumed the undulation theory throughout, the actual treatment of the optical problem is, in many respects, left to stand on the standpoint of the emission theory." Diffraction, however, cannot be understood from a perspective where light-rays do not interfere. "It is very remarkable," Sommerfeld concludes, "that in spite of these grave concerns the formulas reflect the observations so well." (Sauter, ed., *GS I*, 89)

24. Sauter, ed., *GS I*, 86.

25. Arnold Sommerfeld, "Mathematische Theorie der Diffraction," *Mathematische Annalen* 47 (1896). Reproduced in Sauter, ed., *GS I*, 114–171, on 170.

26. Sauter, ed., *GS I*, 170.

27. Wheaton, *The Tiger and the Shark*, esp. 29–33.

28. Indeed, Wiechert's mention of the topic—brief and ambivalent as it was—preceded Stokes's by six months. Ibid., 31.

29. Wilfried Schröder, "Arnold Sommerfeld und Emil Wiechert," *Archive for History of Exact Sciences* 32 (1985).

30. Sauter, ed., *GS IV*, 240.

31. Ibid., 240–1.

32. Wheaton, *The Tiger and the Shark*, 43.

33. Arnold Sommerfeld, "Über die Mechanik der Elektronen," *Verhandlungen des III. Internationalen Mathematiker-Kongresses* (1904). Reproduced in Sauter, ed., *Arnold Sommerfeld: Gesammelte Schriften*, 1–16, on 1.

34. Sommerfeld, "Lectures, Maxwell'sche Theorie u. Elektronenth."

35. Arnold Sommerfeld, "Über die Verteilung der Intensität bei der Emission der Röntgenstrahlen," *Physikalische Zeitschrift* 10 (1909). Reproduced in Sauter, ed., *GS IV*, 369–76.

36. Stark, in turn, was drawing on work by Wien, published in 1907. See Wheaton, *The Tiger and the Shark*, 114–6.

37. On the acrimonious debate between Sommerfeld and Stark, see Armin Hermann, "Die frühe Diskussion zwischen Stark und Sommerfeld über die Quantenhypothese," *Centaurus* 12 (1968); Wheaton, *The Tiger and the Shark*, 116–32.

38. Sauter, ed., *GS IV*, 370.

39. Arnold Sommerfeld, "Über die Struktur der γ-Strahlen," *Sitzungsberichte der Mathemat.-physikal. Klasse der Kgl.-Bayerischen Akademie der Wissenschaft zu München, München* (1911), reproduced in Sauter, ed., *GS IV*, 377–436, on 419. Sommerfeld also briefly elaborated on his previous suggestion that the unpolarized fluorescent part of X-ray radiation might be explained in terms of Planck's quantum, suggesting that the energy of the unpolarized part might be related to its frequency, given as the inverse of the duration of the *Eigenschwingungen* of the molecule: $\tau'E' = h$.

40. Sommerfeld, and most others, understood the process of λ-ray production as the inverse of that for X-rays. There, a decelerating electron produced Röntgen rays. Here, a β particle (i.e. an electron), emitted from a radioactive source, produced gamma rays during its acceleration out of the molecule. The expression Sommerfeld derived for the ratio of energies was

$$\frac{E_\beta}{E_\gamma} = \frac{6\pi c^2 l}{e(e/m_0)} \frac{\sqrt{1-\beta^2}}{\beta},$$

with c representing the speed of light, β the ratio of the electron's velocity to the speed of light (i.e. $\beta = v/c$), and e and m_0 representing the charge and the rest mass of the electron, respectively. On the analogy between X-rays and γ rays, see Wheaton, *The Tiger and the Shark*, 49–67.

41. Arnold Sommerfeld, "Das Plancksche Wirkungsquantum und seine allgemeine Bedeutung für die Molekularphysik," *Verhandlungen der Gesellschaft Deutscher Naturforscher und Ärzte* 83 (1912). Reproduced in Sauter, ed., *GS III*, 1–19, on 1.

42. On Planck's role in this modification, see below.

43. On Sommerfeld at Solvay, see Hermann, *The Genesis of the Quantum Theory*, 102–22. On the extent of Planck's "influence" on Sommerfeld, see Nisio, "The Formation of the Sommerfeld Quantum Theory of 1916," esp. 46–52. See also Jammer, *The Conceptual Development of Quantum Mechanics*, 42–5. Mehra and Rechenberg, *The Historical Development of Quantum Theory*, volume 1, 132–5.

44. Sommerfeld, "Bedeutung des Wirkungsquantums," 252.

45. Ibid., 253.

46. Sommerfeld explicitly discussed the contradiction with ballistics in notes appended to the discussion following his paper. See ibid., 304.

47. Ibid., 254.

48. Nisio, "The Formation of the Sommerfeld Quantum Theory of 1916," 51. Nisio's analysis, published before Wheaton's history of the impulse theory, does not trace the roots of the 1911 papers back to Sommerfeld's earlier work, and thus tends to overemphasize the extent to which he was "profoundly influenced by Planck." Nisio, "The Formation of the Sommerfeld Quantum Theory of 1916," 46.

49. Planck's letter to Sommerfeld of 6 April 1911, in which he suggests an integral form of Sommerfeld's action-expression, contains a notation, in Sommerfeld's hand, of the word "relativity." Sommerfeld would thenceforth use Planck's expression in his papers. Hermann, *The Genesis of the Quantum Theory*, 118.

50. Sommerfeld, "Bedeutung des Wirkungsquantums," 257.

51. Ibid., 258–9.

52. See Sommerfeld, "Über die Struktur der γ-Strahlen," reproduced in Sauter, ed., *GS IV*, 377–436. The reference to Abraham's paper is on 382.

53. The equation numbers here are mine. Except where noted, I follow Sommerfeld's notation.

54. The report published as the proceedings of the Solvay conference was not published until 1913. Sommerfeld had originally (i.e., at Solvay) used other results, those of Bassler instead of Friedrich, and those of Wien instead of his student Edna Carter in the expression 8. This leads to different numerical results, but the main argument seems unchanged. I have followed the text of the (German) Solvay report.

55. Wien's paper ("Ueber die Energie der Kathodenstrahlen im Verhältnis zur Energie der Röntgen- und Sekundär-strahlen," in *Festschrift Adolph Wüllner gewidmet zum 70. Geburtstage, 13 Juni 1905* (1905), 1–14), in which this result was reported, was included in a 1905 Festschrift for the physics professor at Aachen, Adolf Wüllner. Sommerfeld helped to edit that volume, so he would have known Wien's paper in detail.

56. Sommerfeld uses v to represent a time-variable velocity at an earlier stage in his calculation, then introduces the same symbol to represent the initial velocity of the braking electrons. For the sake of clarity, I have introduced the symbol v_i for the initial velocity.

57. Wilhelm Wien, "Über die Berechnung der Impulsbreite der Röntgenstrahlen aus ihrer Energie," *Annalen der Physik* 22 (1907); Wien, "Über eine Berechnung der Wellenlänge der Röntgenstrahlen aus dem Planckschen Energie Element," *Göttingen Nachrichten* (1907).

58. Sommerfeld, "Bedeutung des Wirkungsquantums," 267.

59. Ibid., 294.

60. Ibid., 267.

61. Ibid., 290. Cf. Hermann, *The Genesis of the Quantum Theory*, 122.

62. Sommerfeld, "Bedeutung des Wirkungsquantums," 276. Nonetheless, the photoelectric effect would remain a consistent problem for Sommerfeld's theory, which seemed to provide an overall description of the phenomena but stumbled over essential details, including the problem of the "accumulation time" (i.e. the amount of time required for a quasi-elastically bound electron to absorb a whole quantum of energy). Sommerfeld's theory seemed to give far too great a value for this quantity. See Debye and Sommerfeld, "Theorie des lichtelektrischen Effektes"; Wheaton, *The Tiger and the Shark*, 180–9.

63. Debye and Sommerfeld, "Theorie des lichtelektrischen Effektes," 79–80.

64. According to Leon Brillouin, the French theoreticians Poincaré, Langevin, and Marcel Brillouin "were all very much impressed with this Sommerfeld discussion." Leon cited this as one of the reasons he decided to go to Munich. Quoted in Hermann, *The Genesis of the Quantum Theory*, 121. On responses to Sommerfeld's theory between 1911 and 1915, see Nisio, "The Formation of the Sommerfeld Quantum Theory of 1916," 51–2.

65. Arnold Sommerfeld, *Atombau und Spektrallinien*, first edition (1919), 2; Sommerfeld, *Atomic Structure and Spectral Lines*, 2.

66. Sommerfeld, *Atombau* (first edition), 3; Sommerfeld, *Atomic Structure and Spectral Lines*, 2–3.

67. Planck to Lorentz, 7 January 1910, quoted in Kuhn, *Black-Body Theory*, 236.

68. In other words, equating the terms for entropy given in the thermodynamic expression, $dS/dE = 1/T$, with Boltzmann's probabilistic definition, $S = k \log W$.

69. Max Planck, "Die Gesetze der Wärmestrahlung und die Hypothese der elementaren Wirkungsquanten," in *Die Theorie der Strahlung und der Quanten*, ed. A. Eucken (1913), 80–1.

70. Klein, "Max Planck and the Beginnings of the Quantum Theory"; Klein, "Planck, Entropy, and Quanta."

71. See the preceding chapter.

72. Here g is the moment corresponding to the position coordinate f, and L and K are constants.

73. Planck, *Vorlesungen über die Theorie der Wärmestrahlung*, 156.

74. Needell notes that the integral definition of h was presented at Solvay as "the most general and profound way of stating the quantum hypothesis" but does not elaborate upon the substantial break with *Wärmestrahlung* that it represents. Needell, "Irreversibility and the Failure of Classical Dynamics," 219.

75. Ibid. One can begin to trace the turn to such an understanding in Planck's Columbia lectures. Max Planck, *Acht Vorlesungen über Theoretische Physik* (1910).

76. Planck to Sommerfeld, 6 April 1911; Eckert and Märker, eds., *ASWB I*, 392–5, on 393.

77. Max Planck, "Zur Dynamik bewegter System," *Annalen der Physik* (4) 26 (1908). Reproduced in Planck, *PAV II*, 198.

78. Eckert and Märker, eds., *ASWB I*, 393; Hermann, *The Genesis of the Quantum Theory*, 118.

79. This general dismissal of the light-quantum hypothesis may well have contributed to Einstein's negative judgement of the conference as a whole. At the end of December 1911, he wrote to his friend Michele Besso: "I have not made any further progress in electron theory. In Brussels, too, one lamented at the failure of the theory without finding a remedy. This Congress had an aspect similar to the wailing at the ruins of Jerusalem. Nothing positive came out of it. My treatment of fluctuations aroused great interest, but elicited no serious objection. I did not benefit much, as I did not hear anything which was not known to me already." Quoted in Mehra, *The Solvay Conferences on Physics: Aspects of the Development of Physics since 1911*, xiv.

80. Contrast this formulation to that in the Columbia lectures, which seeks to determine "the existing probability that the resonator at any fixed time possesses a given energy"—that is, where the energy is given and the probability to be calculated. Max Planck, *Eight Lectures on Theoretical Physics* (1998), 88–9.

81. Planck, "Die Gesetze der Wärmestrahlung und die Hypothese der elementaren Wirkungsquanten," 93.

82. Ibid., 85 (emphasis added).

83. These approaches were Planck's modification of Boltzmann's statistical mechanics, Einstein's quantized version of Gibb's canonical ensemble, Lorentz's technique of counting only the most probable distribution of energies, and Nernst's more visualisable, but less rigorous method in which oscillators were assumed to be in equilibrium with the molecules of an ideal gas.

84. Planck, "Die Gesetze der Wärmestrahlung und die Hypothese der elementaren Wirkungsquanten," 90.

85. Lorentz had first raised this problem in 1910. H. A. Lorentz, "Alte und neue Fragen der Physik," *Physikalische Zeitschrift* 11 (1910).

86. Although only noted briefly, the most lasting result of Planck's Solvay paper would arise in this section. Planck calculated the average energy of an oscillator, given by

$\bar{E} = E_N/N = h\nu/2(e^{h\nu/kT} + 1/e^{h\nu/kT} - 1)$.

As T approaches zero, the average energy thus approaches a non-zero value, $h\nu/2$. This zero-point energy was later experimentally verified, although Planck's derivation was abandoned very rapidly.

87. Planck, "Die Gesetze der Wärmestrahlung und die Hypothese der elementaren Wirkungsquanten," 91.

88. See Kuhn, *Black-Body Theory*, 243–4.

89. Max Planck, *The Theory of Heat Radiation*, second edition (1988), preface, ix.

90. Cf. Needell, "Irreversibility and the Failure of Classical Dynamics." 238–9. "In the context of the evolution of Planck's ideas about the physical significance of quanta...the second edition of *Wärmestrahlung* appears much more as part of an attempt to separate all dynamical assumptions about the quantum hypothesis from what Planck had come to view as the more general and trustworthy statistical approach. The extremely tentative discussion of quantum emission that appears near the end of the text and the care with which Planck pointed out the failure of the theory to actually account for the redistribution of energy between oscillators and radiation of different frequencies suggest that Planck believed the significance of the theory he presented lay elsewhere." In 1914, Planck would provide a different formulation of the dichotomy between the statistical and the dynamical. In the first speech, as Needell has shown, in which he accepted the non-absolute character of the second law of thermodynamics, he argued that *both* dynamics and statistics were to be understood in terms of thermodynamics. Dynamical laws were governed by the first law, statistical laws by the second. Max Planck, *Dynamische und statistische Gesetzmässigkeit (Rede in der Berliner Universität am 3. 8. 1914)* (1914). Reproduced in Planck, *PAV III*, 77–90.

91. Max Planck, "Die gegenwärtige Bedeutung der Quantenhypothese für kinetische Gastheorie," in *Vorträge über die kinetische Theorie der Materie und der Elektrizität*, ed. M. Planck and P. Debye (1914), 4–5.

92. Planck to Sommerfeld, 6 April 1911. In Eckert and Märker, eds., *ASWB I*, 393.

93. Sommerfeld, "Bedeutung des Wirkungsquantums," 295.

94. Arnold Sommerfeld, "Zur Theorie der Balmerschen Serie," *Sitzungsberichte der Bayerischen Akademie* (1915). The discussion below draws particularly on Eckert and Märker, eds., *ASWB I*, 431–45, Helge Kragh, "The Fine Structure of Hydrogen and the Gross Structure of the Physics Community, 1916–26," *Historical Studies in the Physical Sciences* 15 (1985), Helge Kragh, "Magic Number: A Partial History of the Fine-Structure Constant," *Archive for History of Exact Sciences* 57 (2003), Mehra and Rechenberg, *The Historical Development of Quantum Theory*, volume 1; Nisio, "The Formation of the Sommerfeld Quantum Theory of 1916."

95. Bohr, quoted in Darrigol, *From c-Numbers to q-Numbers*, 89.

96. Sommerfeld, "Zur Theorie der Balmerschen Serie," 425.

97. Bohr had already suggested the extension of his model to elliptical orbits and had discussed the Stark and Zeeman effects in 1914. In 1915 he introduced a relativistic correction, but found a relatively poor agreement from available experimental data. Niels Bohr, "On the Effect of Electric and Magnetic Fields on Spectral Lines," *Philosophical Magazine* 27 (1914); Bohr, "On the Series Spectrum of Hydrogen and the Structure of the Atom," *Philosophical Magazine* 29 (1915).

98. H. A. Lorentz, "Zur Theorie des Zeemaneffektes," *Physikalische Zeitschrift* 1 (1899); Lorentz, "Über den Einfluss magnetischer Kräfte auf die Emission des Lichtes," *Annalen der Physik* 63 (1897).

99. Sommerfeld, *Atomic Structure and Spectral Lines*, 385.

100. Sommerfeld to Carl Runge, 10 October 1907; Eckert and Märker, eds., *ASWB I*, 314–6, on 316. Sommerfeld's use of the term *klassischen* here presumably refers to the meaning of the term as "definitive," rather than that opposed to "modern" physics and the quantum and relativity theories. See Staley, "On the Co-Creation of Classical and Modern Physics."

101. Sommerfeld to Runge, 12 June 1904; Eckert and Märker, eds., *ASWB I*, 231–2, on 232; Sommerfeld, "Zur Elektronentheorie 2. Grundlagen für eine allgemeine Dynamik des Elektrons," *Nachrichten der K. Gesellschaft der Wissenschaften zu Göttingen (Mathematisch-physikalische Klasse)* (1904), reproduced in Sauter, ed., *Arnold Sommerfeld: Gesammelte Schriften*, 71 147—see esp. 139–44.

102. Note the earlier correspondence with Paschen in 1904 and 1905.

103. Arnold Sommerfeld, "Der Zeemaneffekt eines anisotrop gebundenen Elektrons und die Beobachtungen von Paschen-Back," *Annalen der Physik* 40 (1913), reproduced in Sauter, ed., *GS III*, 20–46, on 20.

104. Sommerfeld to Voigt, 24 March 1913. Eckert and Märker, eds., *ASWB I*, 471–4, on 471.

105. Voigt to Sommerfeld, 26 January 1913, DM, HS 1977–28/A, 347. Cf. ibid., 432–3.

106. Arnold Sommerfeld, "Zur Voigtschen Theorie des Zeeman-Effektes," *Nachrichten der K. Gesellschaft der Wissenschaften zu Göttingen (Mathematisch-physikalische Klasse)* (1914), reproduced in Sauter, ed., *GS III*, 47–69.

107. Sommerfeld to Bohr, 4 September 1913. Eckert and Märker, eds., ASWB I.

108. Sommerfeld to Langevin, Jun 1 1914. Ibid., 484–5, on 484.

109. Sommerfeld to Wien, 3 May 1915. Ibid., 493–4.

110. A letter from Paschen that declared Bohr's theory to be "exactly right" in its explanation of the origin of certain Helium lines must have helped. See Paschen to Sommerfeld, 24 November 1915. Ibid., 499–500, on 500. See also 437.

111. Sommerfeld to Bohr, 4 September 1913 in ibid., 477.

112. Arnold Sommerfeld, "Die allgemeine Dispersionsformel nach dem Bohrschen Modell," *Arbeiten aus den Gebieten der Physik, Mathematik, Chemie* (1915). Reproduced in Sauter, ed., *GS III*, 136–71, on 167.

113. Sauter, ed., *GS III*, 136.

114. Arnold Sommerfeld, "Die Feinstruktur der wasserstoff- und wasserstoffähnlichen Linien," *Sitzungsberichte der Bayerischen Akademie* (1915): 459.

115. Sommerfeld, "Zur Theorie der Balmerschen Serie," 427.

116. Oddly, having made this point, Sommerfeld did not draw the ellipses either.

117. Sommerfeld, "Zur Theorie der Balmerschen Serie," 428.

118. Ibid., 458.

119. Sommerfeld, "Das Plancksche Wirkungsquantum und seine allgemeine Bedeutung für die Molekularphysik," reproduced in Sauter, ed., *GS III*, 1–19, on 4–5.

120. "Die physikalische Struktur des Phasenraumes" [The Physical Structure of Phase Space] was the title of a summary of these two papers published in *Annalen der Physik* in 1916. Planck, *PAV II*, 386–419.

121. Max Planck, "Bemerkung über die Emission von Spektrallinien," *Sitzungsberichte der Preussischen Akademie der Wissenschaften* (1915). Reproduced in Planck, *PAV II*, 376–80. Cf. Mehra and Rechenberg, *The Historical Development of Quantum Theory*, volume 1, 209; Nisio, "The Formation of the Sommerfeld Quantum Theory of 1916," 71.

122. Sommerfeld to Schwarzschild, 19 February 1916. Eckert and Märker, eds., *ASWB I*, 529. Cf. Michael Eckert, "Planck vorsichtig u. abstrakt, ich etwas draufgängerisch," *Kultur und Technik* 4 (1997); Eckert and Märker, eds., *ASWB I*, 445.

123. Sommerfeld, "Zur Theorie der Balmerschen Serie," 439; Mehra and Rechenberg, *The Historical Development of Quantum Theory*, volume 1, 217.

124. Sommerfeld, "Zur Theorie der Balmerschen Serie."

125. Bohr to Sommerfeld, 19 March 1916, document 245 in Eckert and Märker, eds., *ASWB I*, 540.

126. Sommerfeld, "Die Feinstruktur der wasserstoff- und wasserstoffähnlichen Linien," 484.

127. Arnold Sommerfeld, "Zur Quantentheorie der Spektrallinien. Intensitätsfragen," *Sitzungsberichte der Bayerischen Akademie* (1917), reproduced in Sauter, ed., *GS III*, 432–58, on 434.

128. Kragh, "Fine Structure," 76.

129. Sommerfeld's Report to the Faculty, 15 Feb 1917, Munich University Archive OC I 43p.

130. Sommerfeld, *Atomic Structure and Spectral Lines*, 531.

131. Niels Bohr, "On the Program of the Newer Atomic Physics (2 December 1919)," in *Niels Bohr Collected Works*, ed. J. Rud Nielsen (1976), 222.

132. Both men provided theoretical explanations of the Stark effect. Schwarzschild had put forward a classical solution in 1914, modeling it after a special case of the celestial three-body problem, one in which a body (the electron) moved under the Newtonian attraction of two fixed centers, with the added condition that one center was positioned infinitely far away. (K. Schwarzschild, "Bemerkungen zur Aufspaltung der Spektrallinien im elektrischen Feld," *Verhandlungen der deutschen physikalischen Gesellschaft* 16 (1914)) His 1916 paper quantized the motion by elaborating on Planck and Sommerfeld's statistical divisions of phase space. "The quantum hypothesis was extended recently by Planck and Sommerfeld," he wrote, "in the case of mechanical systems of several degrees of freedom. The problem consists of indicating principles for the organization of phase space in elementary areas. I would like to show here, in a clear way, an organization which can occur for an important group of mechanical problems, by making use of canonical variables of a certain type." K. Schwarzschild, "Zur Quantenhypothese," *Berliner Berichte* (1916): 548. The canonical variables to which Schwarzschild referred—the so-called "action-angle variables"—were valuable tools for mathematical astronomers at the time, since working with them provided immediate access to the periods of celestial motions. Their introduction to the quantum theory was a master stroke, for it avoided the need to solve a mechanical problem in its entirety in order to apply the Bohr Sommerfeld quantization conditions. Schwarzschild's paper, in effect, was an application of his rule for dividing phase space to a number of exemplary problems, including the Stark effect. Epstein's paper, in contrast, was a much more straightforward solution to the problem of line splitting in an electric field, one much closer to Sommerfeld's own methods. Paul Sophus Epstein, "Zur Quantentheorie," *Annalen der Physik* 51 (1916); Epstein, "Zur Theorie des Starkeffektes," *Annalen der Physik* 50 (1916). On Debye's contributions, especially those in the study of X-ray spectra and structure, see chapter 7 of the present volume.

133. Sommerfeld, *Atomic Structure and Spectral Lines*, 610. That said, the immediate application of these methods to the Zeeman effect was not a great success, as papers by both Debye and Sommerfeld showed. Arnold Sommerfeld, "Zur Theorie des Zeeman-Effekts der Wasserstofflinien mit einem Anhang über den Stark-Effekt," *Physikalische Zeitschrift* 17 (1916). Reproduced in Sauter, ed., *GS III*, 309–25. Peter Debye, "Quantenhypothese und Zeeman-Effekt," *Physikalische Zeitschrift* 17 (1916). Neither solution was able to do much more, it seemed, than reproduce the classical result of Lorentz and Larmor. Sommerfeld's calculation of the Zeeman Effect in the absence of relativistic corrections produced a result, he claimed, that "does not coincide with the result of the classical theory, but is very close." The splitting induced by the magnetic field in Sommerfeld's treatment was equal to a whole number times the classical value obtained through Lorentz's electron theory. Sauter, ed., *GS III*, 312. The relativistic Zeeman Effect, to Sommerfeld's considerable surprise, was no different to the non-relativistic. Sauter, ed., *GS III*, 320. A good general discussion of the history of the Zeeman effect and its importance to the older quantum theory may be found in Jammer, *The Conceptual Development of Quantum Mechanics*, 118–33. Also Mehra and Rechenberg, *The Historical Development of Quantum Theory*, volume 1. Good discussions of the Hamilton-Jacobi method in the context of the quantum theory are given in Jammer, *The Conceptual Development of Quantum Mechanics*, 89–109; Darrigol, *From c-Numbers to q-Numbers*, 102–16.

134. Hermann, *The Genesis of the Quantum Theory*, 120–1.

135. Sommerfeld to Runge, 10 December 1907. Eckert and Märker, eds., *ASWB I*, 314–6, on 315. On Sommerfeld's emphasis on including all calculation within a paper, see chapter 2 of the present volume.

136. Sauter, ed., *GS IV*, 267, Sommerfeld, "Theoretisches über die Beugung der Röntgenstrahlen."

137. Sommerfeld, "Bedeutung des Wirkungsquantums," 295.

138. Arnold Sommerfeld, "Zur Quantentheorie der Spektrallinien," *Annalen der Physik* 51 (1915). Reproduced in Sauter, ed., *GS III*, 172–308, on 205 and 206.

139. Bopp and Meixner, eds., *Thermodynamik und Statistik von Arnold Sommerfeld*, Vorwort, v.

140. Nisio, "The Formation of the Sommerfeld Quantum Theory of 1916," 77.

Chapter 6

1. Max Planck to Albert Einstein, 26 October 1918, document 640 in Einstein, *Collected Papers*, volume 8 (1998), 683–4.

2. Detlef Peukert, *The Weimar Republic* (1993).

3. Sommerfeld to Einstein, 3 December 1918, document 295 in Eckert and Märker, eds., *ASWB I*, 612–3.

4. Einstein to Besso, 4 December 1918, document 663 in Einstein, *The Collected Papers of Albert Einstein: English Translation*, 703–4.

5. Einstein to Sommerfeld, 6 December 1918, document 665 in ibid., 705–6.

6. Peter Gay, *Weimar Culture: The Outsider as Insider* (1968), xiv.

7. Gerald D. Feldman, *The Great Disorder: Politics, Economics, and Society in the German Inflation, 1914–1924* (1993), 546. Much of Feldman's data is derived from Anon., "Das Einkommen der Hochschullehrer: Denkschrift des Verbandes der Deutschen Hochschulen," *Mitteilungen des Verbandes der deutschen Hochschulen* 2 (1922): 247–59.

8. Feldman, *The Great Disorder*, 546–7.

9. Wilhelm Wien, "Ziele und Methoden der theoretischen Physik," *Jahrbuch der Radioaktivität und Elektronik* 12 (1915).

10. Jungnickel and McCormmach, *Intellectual Mastery of Nature*, volume 2, 355–6.

11. On this exchange, see Lewis Elton, "Einstein, General Relativity, and the German Press, 1919–1920," *Isis* 77 (1986): 100–101. A translation of Einstein's "My Reply. On the Anti-Relativity Theoretical Co., Ltd." may be found in Klaus Hentschel and Ann M. Hentschel, eds., *Physics and National Socialism: An Anthology of Primary Sources* (1996), 1–5.

12. Elton ("Einstein, General Relativity, and the German Press, 1919–1920," 101) notes that, in spite of later testimony to the contrary, contemporary accounts of the debate in Nauheim reveal

it to have been an event kept in good order by the chair, Max Planck. For details of the discussions, see "Vorträge und Diskussionen von der 86. Naturforscherversammlung in Nauheim vom 19.-25. September 1920," *Physikalische Zeitschrift* 21 (1920); H[ermann] Weyl, "Die Relativitätstheorie auf der Naturforscherversammlung in Bad Nauheim," *Jahresbericht der Deutschen Mathematiker-Vereinigung* 31 (1922).

13. Johannes Stark, *Die gegenwärtige Krisis in der deutschen Physik* (1922). "Vorwort." Sommerfeld was singled out by Stark as particularly guilty of dogmatism about the quantum theory.

14. This review is translated in Hentschel and Hentschel, eds., *Physics and National Socialism*, 6–7.

15. Arnold Sommerfeld, "The Differences Between American and German Universities (undated, untitled manuscript, presumably written 1922/3)." See, similarly, Sommerfeld to the Carlsberg-Fonds, 25 October 1919. Eckert and Märker, eds., *ASWB 2*, 62–3. Heisenberg made a similar argument for (part of) the cause of strife between theoreticians and experimentalists in interviews with Thomas S. Kuhn in the 1960s: "I could imagine that this difference, or this gap, between experimental and theoretical physics had a political component already at that time. Experimental physics after the first world war, was in a very difficult situation in Germany, just for economic reasons. They had no equipment. So they had rather little success, while the theoretical physicists like Sommerfeld had very big success. Also Einstein was now very famous for what he had done and the experimental physicists were not so well known. So it was this feeling of frustration from economic reasons and other reasons which made them uneasy and in some way they started hating the theoreticians who had a so much easier life." Kuhn, "Interview with Werner Heisenberg, 02/07/1963," 18.

16. Sommerfeld, "The Differences Between American and German Universities."

17. The principal published source for this story appears to be George Gamow, *Thirty Years That Shook Physics: The Story of Quantum Theory* (1966), 64. Pauli began at the ETH in Zurich in 1928, and Franck left his position in Göttingen in 1933, with the National Socialist rise to power, thus giving a five-year window of possibility for the events of this story.

18. Traweek, *Beamtimes and Lifetimes: The World of High Energy Physicists*, esp. 74–105.

19. Quoted in Eckert, "Mathematics, Experiments, and Theoretical Physics," 242.

20. Kuhn, "Interview with Hans Bethe, 01/17/1964," 15.

21. Thomas S. Kuhn, "Interview with Werner Heisenberg, 02/11/1963" (1963), 3. The fact that Heisenberg is relating here not a lack of skill with experimental equipment, but a lack of theoretical knowledge about it, makes this story less prone to the particular exaggerations of the discourse of the Pauli effect. The explanation for Heisenberg's poor performance can only (it would seem) be his lack of interest and study, rather than an innate and essential lack of ability.

22. Born, quoted in Skuli Sigurdsson, "Hermann Weyl, Mathematics and Physics, 1900–1927" (1991), 80.

23. Ibid.

24. Wolfgang Pauli, *Exclusion Principle and Quantum Mechanics: Lecture Given in Stockholm After the Award of the Nobel Prize of Physics 1945* (Neuchatel, 1945), 8. Not until page 18 is Planck's "discovery of the quantum of action" mentioned. Pauli's failure to mention Göttingen and Max Born in this context is peculiar. The six months he spent as Born's assistant is not mentioned in the history he provides. One suspects a remnant of his long-lasting hostility toward what he referred to in the 1920s as "Göttingen formalism."

25. Born, "Sommerfeld als Begründer einer Schule," 1036.

26. Holton, *Thematic Origins of Scientific Thought*; Martin J. Klein, "Thermodynamics in Einstein's Thought," *Science* 157 (1967); John Stachel, *Einstein from "B" to "Z"* (2002).

27. On Eddington, see Matthew Stanley, *Practical Mystic: Religion, Science, and A. S. Eddington* (2007).

28. Ehrenfest to Einstein, 16 September 1925. My thanks to Jeroen van Dongen for drawing my attention to this correspondence.

29. Einstein to Ehrenfest, 18 September 1925. *Prinzipienfuchser* is near untranslatable, and is not listed in any dictionary I could find. Native speakers, however, seem to have no problem gleaning its meaning. The suffix "*-fuchser*" [*Fuchs* meaning "fox"] is not entirely uncommon, and seems to imply a certain level of obsession, thus *Pfennigfuchser* (penny-pincher) and *Pfederfuchser* (pedant).

30. Paul Ehrenfest, "Adiabatische Transformationen in der Quantentheorie und ihre Behandlung durch Niels Bohr," *Naturwissenschaften* 11 (1923): 550. Cf. Benz, *Arnold Sommerfeld*, 121. The "Bohr-Festival" was the affectionate name given to a two-week-long set of lectures that Bohr delivered in June 1922, lectures that have been credited with inspiring an enormous amount of significant research in the years that followed. Cassidy, *Uncertainty: The Life and Science of Werner Heisenberg*, 127–9.

31. Kuhn, "Interview with G. Uhlenbeck, 03/30/1962," 1.

32. Ibid.

33. See Seth, "Allgemeine Physik? Max Planck und die Gemeinschaft der theoretischen Physik."

34. On Einstein's political elitism, see Britta Scheideler, "The Scientist as Moral Authority: Albert Einstein between Elitism and Democracy, 1914–1933," *Historical Studies in the Physical and Biological Sciences* 32 (2002).

35. Albert Einstein, "Principles of Research," in *Essays in Science* (1934), 1.

36. Ibid., 1–2.

37. Ibid., 2.

38. Ibid., 4.

39. On Mach and Einstein, see Gerald Holton, "Mach, Einstein, and the Search for Reality," in *Thematic Origins of Scientific Thought* (1973).

40. Einstein, "Principles of Research," 4. In Einstein's later writings a meta-principle (of simplicity or beauty) would serve to select between equally suitable theories, allowing one to realize which was "best adapted" to the environment of the phenomenological world.

41. Einstein, cited in Holton, "Mach, Einstein, and the Search for Reality," 244. Similarly, in 1911–12, Einstein wrote that "I can't quite understand how Planck has so little understanding for your efforts," and in 1913: "Next year at the solar eclipse it will turn out whether the light rays are bent by the sun, in other words whether the basic and fundamental assumption of the equivalence of the acceleration of the reference frame and of the gravitational field really holds. If so, then your inspired investigations into the foundations of mechanics—despite Planck's unjust criticism—will receive a splendid confirmation." Holton, "Mach, Einstein, and the Search for Reality," 246.

42. For an introduction, see Holton, "Mach, Einstein, and the Search for Reality," Stachel, *Einstein from "B" to "Z"*, and the first two volumes of *Science and Society: The History of Modern Physical Science in the Twentieth Century*, ed. P. Galison, M. Gordin, and D. Kaiser (2001). Einstein's first explicit enunciation of his famous principle theory/constructive theory distinction (in 1919) is to be found in Albert Einstein, "What Is the Theory of Relativity?" in *The Collected Papers of Albert Einstein: English Translation* (2002).

43. Albert Einstein, "Bemerkungen zu der Notiz von Hrn. Paul Ehrenfest: "Die Translation deformierbarer Elektronen und der Flächensatz," *Annalen der Physik* 23 (1907). Translated and reprinted as Doc. 44 in Albert Einstein, *The Collected Papers of Albert Einstein: English Translation*, volume 2: *The Swiss Years: Writings, 1900–1909* (1989), 236–7 (emphasis added). I am indebted to Richard Staley for drawing my attention to this paper, and for several discussions on this point.

44. Einstein, *The Collected Papers of Albert Einstein: English Translation*, 236.

45. Pais, *Subtle Is the Lord*, 309.

46. Born to Sommerfeld, 5 March 1920, Eckert and Märker, eds., *ASWB 2*, 74.

47. Sommerfeld to Einstein, 4 July 1921. Ibid., 100.

48. Sommerfeld, "The Differences Between American and German Universities (undated, untitled manuscript, presumably written 1922/3)."

49. Einstein, "What Is the Theory of Relativity?" in *The Collected Papers of Albert Einstein: English Translation* (2002): 100–105.

50. Ibid., 101.

51. Ibid.

52. Alistair Sponsel, "Constructing a 'Revolution in Science': The Campaign to Promote a Favorable Reception for the 1919 Solar Eclipse Experiments," *British Journal for the History of Science* 35 (2002): 441.

53. Ibid., 465.

54. Warwick, *Masters of Theory*, 395.

55. Ibid., 443–500.

56. Stanley, *Practical Mystic*, esp. 153–93.

57. A. S. Eddington, *Report on the Relativity Theory of Gravitation* (1918). Quoted in Matthew Stanley, "Practical Mystic: Religion and Science in the Life of A. S. Eddington" (2004).

58. A. S. Eddington, *Space, Time, and Gravitation: An Outline of the General Relativity Theory* (1920), 180.

59. J. H. Jeans et al., "Discussion on the Theory of Relativity," *Proceedings of the Royal Society of London* 97 (1 March 1920): 78.

60. Jeans's comments are on pages 66–72 of ibid.

61. Ibid., 66. Jeans remained unconvinced of the validity of Einstein's idea of a "warped" space-time. "If we do not insist on Einstein's interpretation of *ds* [the line-element], the theory becomes less rich and less inclusive," Jeans noted in his concluding lines, "but at the same time less beset by difficulties. The reality of the four-dimensional continuum is, I think, beyond dispute, but the reality of the twists and kinks in it do not appear to be." Jeans et al., "Discussion on the Theory of Relativity," 72.

62. P. A. M. Dirac, *The Principles of Quantum Mechanics*, fourth edition (1981), vii.

63. Warwick is less illuminating on the question of why Eddington himself should have felt the appeal of Einstein's axiomatic presentation, why he should have called the deductivism of Einstein's approach its "most attractive aspect." See Stanley, "Practical Mystic: Religion and Science in the Life of A. S. Eddington."

64. Niels Bohr to Arnold Sommerfeld, 30 April 1922. Document 55 in Eckert and Märker, eds., *ASWB 2*, 116–7.

65. Ibid., 117.

66. Niels Bohr, "On the Application of the Quantum Theory to Periodic Systems (Unpublished Paper, Intended for Publication in the Phil. Mag., April 1916)," in *Niels Bohr Collected Works*, volume 2, ed. U. Hoyer (1981), 258.

67. Paul Ehrenfest, "A Mechanical Theorem of Boltzmann and Its Relation to the Theory of Energy Quanta," *Proc. Amsterdam Acad.* 16 (1913). In *Paul Ehrenfest: Collected Scientific Papers* (1959), ed. M. Klein.

68. Niels Bohr, "On the Quantum Theory of Line Spectra, Parts I–III [1918–1922]," in *Niels Bohr Collected Works*, ed. J. Rud Nielsen (1976), 75.

69. On Ehrenfest and the adiabatic principle see Martin J. Klein, *Paul Ehrenfest: The Making of a Theoretical Physicist*, volume 1 (1970), esp. 264–92.

70. Bohr, "On the Quantum Theory of Line Spectra, Parts I–III [1918–1922]," 91.

71. For systems of more than one degree of freedom, the ratio of the kinetic energy to the frequency is no longer an adiabatic invariant. In 1917, Ehrenfest's student J. M. Burgers extended the analysis to arbitrarily many degrees of freedom by proving the adiabatic invariance of Karl Schwarzschild's "action-angle variables." Burgers's result ultimately opened the way for Bohr to increase the scope of his model to include the very broad class of conditionally periodic motions.

72. M. Norton Wise, "How Do Sums Count? On the Cultural Origins of Statistical Causality," in *The Probabilistic Revolution*, ed. L. Krüger, L. Daston, and M. Heidelberger (1989). The argument for a connection between Bohr's thinking and that of Høffding has a long history in the literature. As early as 1966, Jammer (*The Conceptual Development of Quantum Mechanics*) claimed that Bohr's repeated emphases on the pragmatic nature of truth, and on the impossibility of completely divorcing the observer from the subject of his or her observation, could be traced to the writing of the Danish philosopher Søren Kierkegaard, whose ideas would have been made known to Bohr through its exposition in Høffding's works. It was also through Høffding, Jammer suggests, that Bohr came across the American pragmatist William James, whose work he cited often and with approval. Holton (*Thematic Origins of Scientific Thought*, 99–145; "The Roots of Complementarity," *Daedalus* 99 (1970)) canvassed a similar list of thinkers in an attempt, rather more specifically, to outline the philosophical background to Bohr's complementarity principle. The most detailed account of Bohr's intellectual antecedents to date has been that of Jan Faye ("The Influence of Harald Høffding's Philosophy on Niels Bohr's Interpretation of Quantum Mechanics," *Danish Yearbook of Philosophy* 16 (1979)), who argued that attempts to trace Bohr's ideas back to writings by Kierkegaard through Høffding suffered from the mistaken notion that Høffding himself had drawn key elements of his own ontology and theory of knowledge from Kierkegaard's work. Faye contends convincingly that one need look no further than Høffding's own distinctive writings to find ideas that Bohr would later employ in dramatically different contexts. "Høffding's philosophical influence on Bohr," Faye writes, "was direct and exceedingly significant." Of particular importance was Høffding's "conception of reality, his analysis of the relation between subject and object, and the complementary conditions for description in connection with psychical experience." ("The Influence of Harald Høffding's Philosophy on Niels Bohr's Interpretation of Quantum Mechanics," 38–9) In short: "Bohr developed his interpretation of quantum mechanics in analogy with Høffding's treatment of the problems concerning psychical experience in psychology, as he considered the problems in both domains to be identical in epistemological aspects." (Faye, "The Influence of Harald Høffding's Philosophy on Niels Bohr's Interpretation of Quantum Mechanics," note 10, 39–40; see also Faye, *Niels Bohr: His Heritage and Legacy* (1991)) Wise makes clear his indebtedness to Faye's earlier analysis. Most recently, Darrigol (*From c-Numbers to q-Numbers*) has used Wise's account to explain Bohr's ideas on language. The claim that Bohr's ideas might be traced to an "external" source nonetheless has been and remains contested. Leon Rosenfeld, for example, has claimed that "Bohr was a completely independent thinker; from early youth, he developed his epistemological ideas single-handed and with no more philosophical preparation than Høffding's elementary course of lectures." This passage, from a review of Jammer's book, is cited with approval in David Favrholdt, *Niels Bohr's Philosophical Background* (1992), 40. This last book provides more details, but substantially

the same argument as David Favrholdt, "Niels Bohr and Danish Philosophy," *Danish Yearbook of Philosophy* 13 (1976). Faye, in "The Influence of Harald Høffding's Philosophy on Niels Bohr's Interpretation of Quantum Mechanics," explicitly disagrees with Favrholdt's 1976 position in *Niels Bohr's Philosophical Background*. On Bohr's philosophy and that of quantum mechanics generally, see David Kaiser, "More Roots of Complementarity: Kantian Aspects and Influences," *Studies in History and Philosophy of Science* 23 (1992); Catherine Chevalley, "Niels Bohr's Words and the Atlantis of Kantianism," in *Niels Bohr and Contemporary Philosophy*, ed. J. Faye and H. Folse (1994); Catherine Chevalley, "Philosophy and the Birth of Quantum Theory," in *Physics, Philosophy, and the Scientific Community: Essays in the Philosophy and History of the Natural Sciences and Mathematics. In Honor of Robert S. Cohen*, ed. K. Gavroglu et al. (1995).

73. Harald Höffding, *The Problems of Philosophy* (1905), 8.

74. Ibid., 25.

75. Wise, "How Do Sums Count?" 416.

76. Höffding, *The Problems of Philosophy*, 62.

77. "The Correspondence Principle," Bohr wrote in 1923, "must be regarded as a law of the quantum theory which can in no way diminish the contrast between the postulates [of the quantum theory] and electrodynamic theory." Quoted in Darrigol, *From c-Numbers to q-Numbers*.

78. Ehrenfest, quoted in ibid., 142.

79. Kuhn, "Interview with Werner Heisenberg, 02/11/1963," 13.

80. Höffding, *The Problems of Philosophy*, 73–4.

81. Ibid., 81–2.

82. Ibid., 78.

83. Ibid., 77.

84. Ibid., 71.

85. Ibid., 171.

86. Ibid., 79–80 (emphasis added).

87. Ibid., 84–5.

Chapter 7

1. Sommerfeld to Einstein, 11 January 1922, document 50 in Eckert and Märker, eds., *ASWB 2*, 110–1.

2. Arnold Sommerfeld, "Die Bedeutung der Röntgenstrahlung für die heutige Physik," *Verhandlungen der Bayerischen Akademie der Wissenschaften, München* (1925). Reprinted in Sauter, ed., *GS IV*, 564–79.

3. See e.g. Cassidy, *Uncertainty: The Life and Science of Werner Heisenberg*, 116, Nisio, "The Formation of the Sommerfeld Quantum Theory of 1916," 77. That said, at least two scholars have identified Sommerfeld as either the "engineer" of the quantum (Daniel Serwer, "*Unmechanischer Zwang*: Pauli, Heisenberg, and the Rejection of the Mechanical Atom, 1923–25," Historical Studies in the Physical Sciences 8 (1977): 193.) or its "technologist" (Cassidy, *Uncertainty: The Life and Science of Werner Heisenberg*, 129).

4. *Oxford English Dictionary online*: "technique"

5. Ibid., 1884 Grove Dict. Mus. IV, 66

6. See introduction.

7. Arnold Sommerfeld, "Ein Zahlenmysterium in der Theorie des Zeeman-Effektes," *Die Naturwissenschaften* 8 (1920). For another attempt to reconcile Sommerfeld's "pragmatism" with this talk of mysticism, see Benz, *Arnold Sommerfeld*, 120–23.

8. Sommerfeld, *Atomic Structure and Spectral Lines*, viii. This first English edition is a translation of Arnold Sommerfeld, *Atombau und Spektrallinien*, third edition (1922). Where the German is identical between the third German edition and earlier editions, I have used Brose's translation.

9. Arnold Sommerfeld, "Über kosmische Strahlung," *Suddeutsche Monatshefte* 24 (1927). Reprinted in Sauter, ed., *GS IV*, 580–3. Translation above taken from Paul Forman, "Weimar Culture, Causality, and Quantum Theory, 1918–1927: Adaptation by German Physicists and Mathematicians to a Hostile Intellectual Environment," *Historical Studies in the Physical Sciences* 3 (1971): 13. Sommerfeld distanced himself from the implications of his own previous evocations of a Keplerian "music of the spheres" by noting that "around 1600 the attitude to religious and worldview questions must have been necessarily very different to the attitude around 1900 and that the mature Kepler in his own researches himself overhauled his youthful ideas, the *Mysterium Cosmographicum*." Sauter, ed., *GS IV*, 580.

10. Werner Heisenberg discussed the antagonism between Wien's and Sommerfeld's institutes in Kuhn, "Interview with Werner Heisenberg, 02/07/1963," 16.

11. The response to this aspect of Wien's talk was not positive. As Heisenberg phrased it: "I do remember that Willy Wien gave, when he was rector of the University, a speech about atomic physics and never mentioned the name of Sommerfeld. Everyone felt that this is a thing that one can't do because Sommerfeld after all was a very famous and certainly very good physicist. So there was lots of trouble between the two...." Kuhn, "Interview with Werner Heisenberg, 11/30/1962," 10.

12. Wilhelm Wien, *Vergangenheit, Gegenwart und Zukunft der Physik: Rede Gehalten beim Stiftungsfest der Universität München am 19. Juni 1926* (1926), 15.

13. Born, "Sommerfeld als Begründer einer Schule," 1035–6. Reprinted in Akademie der Wissenschaften Göttingen, ed., *Max Born: Ausgewählte Abhandlungen*, volume 2 (1963), 604–6. Born may have had in mind something like the specific situation he would later describe involving

Fritz London, who arrived at Göttingen in the late 1920s: "He wanted to work on philosophy for his thesis—the philosophy of the new quantum mechanics. And I said, "No, my fellow. You must do real work—calculations. Work out a special problem. Before that I wouldn't give you any such question, even if I knew one." And he was quite intolerable and insisted on having such a fundamental problem—the other didn't interest him. I tried to persuade him and Franck tried and although he seemed like a nice kind of fellow, we couldn't do anything with him. And since I knew no way, I wrote to Sommerfeld to see whether he would like to take him, for I knew that Sommerfeld had much more interest in young people than I had. And so I sent him to Sommerfeld, and Sommerfeld put him right. He persuaded him by the force of his personality to do a very simple and straightforward calculation. I don't know what it was, but he got his thesis and he never became a philosopher again." Thomas S. Kuhn and F. Hund, "Interview with Max Born, 10/17/1962" (1962), 27. Cf Sigurdsson, "Hermann Weyl, Mathematics and Physics, 1900–1927." 217–18.

14. Pauli, *Exclusion Principle and Quantum Mechanics: Lecture Given in Stockholm After the Award of the Nobel Prize of Physics 1945*, 8.

15. Paul Forman suggests that the origin of the characterization of the results as a "Number-mystery" came from a letter written to Sommerfeld by A. Rubinowicz. Paul Forman, "Alfred Landé and the Anomalous Zeeman Effect, 1919–1921," *Historical Studies in the Physical Sciences* 2 (1970): 188, note 85.

16. Sommerfeld, "Ein Zahlenmysterium in der Theorie des Zeeman-Effektes," 64.

17. In treating symmetrically Sommerfeld's talk of "Mystik" and "Technik" I thus seek to problematize Paul Forman's contention that one should understand the language of number-mysticism as merely "the readiness of even a Sommerfeld to flirt with the very antiscientific tendencies he deplored," and see in it rather a description of the scientific methodology that Sommerfeld increasingly advocated after 1919. This is not, however, to disagree with Forman's characterization of the Weimar period as one in which talk of mysticism was common. It is rather to suggest that Sommerfeld drew upon and modified this language as a means of expressing his own ideas on the importance of an aesthetic sense in his—on the surface at least—phenomenological spectroscopy. Forman, "Weimar Culture, Causality, and Quantum Theory," quotation on 55.

18. Hilbert to Sommerfeld, 21 January 1920. Eckert and Märker, eds., *ASWB 2*, 73. Beggerow to Sommerfeld, 21 February 1920. DM NL 089 (022). Zeeman to Sommerfeld, 16 January 1920. Eckert and Märker, eds., *ASWB 2*, 72.

19. Born to Sommerfeld, 5 March 1920. Eckert and Märker, eds., *ASWB 2*, 74–5, on 75. For further criticisms, see Eckert and Märker, eds., *ASWB 2*, 42–3.

20. Arnold Sommerfeld, "Schwebende Fragen der Atomphysik," *Physikalische Zeitschrift* 21 (1920). Reproduced in Sauter, ed., *GS III*, 496–7. The text derived from a lecture delivered at the Naturforscher-Versammlung in Bad-Nauheim, 19–25 September 1920.

21. Arnold Sommerfeld, "Über die Feinstruktur der K-Beta Linie," *Sitzungsberichte der Bayerischen Akademie* (1918). Reproduced in Sauter, ed., *GS III*, 470–5.

22. Sommerfeld had first introduced this result in Arnold Sommerfeld, "Zur Quantentheorie der Spektrallinien, Ergänzungen und Erweiterungen," *Sitzungsberichte der Bayerischen Akademie* (1916). Reproduced in Sauter, ed., *GS III*, 326–77, esp. 347–61.

23. Arnold Sommerfeld and W. Kossel, "Auswahlprinzip und Verschiebungssatz bei Serienspektren," *Verhandlungen der deutschen physikalischen Gesellschaft* 21 (1919). Reproduced in Sauter, ed., *GS III*, 476–95, on 480.

24. Sauter, ed., *GS III*, 496.

25. Ibid., 497.

26. Ibid.

27. Sommerfeld, "Zur Theorie des Zeeman-Effekts der Wasserstofflinien mit einem Anhang über den Stark-Effekt." Reproduced in Sauter, ed., *GS III*, 309–25, quote on 312. Debye, "Quantenhypothese und Zeeman-Effekt."

28. Sauter, ed., *GS III*, 320.

29. Sommerfeld to Carl Runge, 16 August 1919. Eckert and Märker, eds., *ASWB 2*, 53–5.

30. Sommerfeld mentions lecturing on the "Zahlenmysterium" in Sweden in a letter to Zeeman, 29 January 1920. Ibid., 73.

31. Sommerfeld, "Ein Zahlenmysterium in der Theorie des Zeeman-Effektes." Reproduced in Sauter, ed., *GS III*, 511–4.

32. Sauter, ed., *GS III*, 513.

33. Ibid., 514.

34. Sommerfeld to Runge, 16 August 1919. Eckert and Märker, eds., *ASWB 2*, 55.

35. Arnold Sommerfeld, "Allgemeine spektroskopische Gesetze, insbesondere ein magnetooptischer Zerlegungssatz," *Annalen der Physik* 63 (1920). Reprinted in Sauter, ed., *GS III*, 523–65, on 523. A translation of the introduction and section I and II is given in W. R. Hindmarsh, *Atomic Spectra* (1967), 145–59. Forman discusses both the background to, and the more detailed content of the paper in Forman, "Alfred Landé and the Anomalous Zeeman Effect, 1919–1921," 179–95.

36. Sommerfeld, "Allgemeine spektroskopische Gesetze," 523.

37. Ibid., 524.

38. Ibid.

39. Ibid., 560.

40. Ibid., 532. Quoted in Forman, "Alfred Landé and the Anomalous Zeeman Effect, 1919–1921," 191.

41. Sauter, ed., *GS III*, 536.

42. Sommerfeld and Kossel, "Auswahlprinzip und Verschiebungssatz bei Serienspektren." Reproduced in Sauter, ed., *GS III*, 476–95.

43. Sauter, ed., *GS III*, 481. Where the description of the *modellmässig* sense of the displacement law was reproduced essentially verbatim in the 1920 paper, this additional sentence on the limitations of the model-understanding was not. The line does recur, however, in the third edition of *Atombau*, indicating that Sommerfeld had not changed his position on this point up to 1922/23. Sommerfeld, *Atomic Structure and Spectral Lines*, 373.

44. The terminology of "arc" and "spark" spectra originally referred to the lines' mode of production. Sommerfeld defined the term theoretically to associate spark spectra with ionized atoms, and arc spectra with neutral atoms. See Sommerfeld, *Atomic Structure and Spectral Lines*, 372.

45. Sauter, ed., *GS III*, 481. Cf. Forman, "Alfred Landé and the Anomalous Zeeman Effect, 1919–1921," 192.

46. Forman, "Alfred Landé and the Anomalous Zeeman Effect, 1919–1921," 192.

47. Woldemar Voigt, "Weiteres zum Ausbau der Koppelungstheorie der Zeemaneffekte," *Annalen der Physik* 41 (1913): 440. See also Woldemar Voigt, "Die anomalen Zeemaneffekte der Spektrallinien vom D-Typus," *Annalen der Physik* 42 (1913).

48. Sommerfeld's expression for Voigt's equations was

$$\left(\frac{d^2}{dt^2} + ih\frac{d}{dt} + n_x^2\right)\zeta_x = \pm\frac{ih}{3}\frac{d}{dt}(\zeta_1 + \zeta_2 + \zeta_3), \text{ where } h = \frac{e}{m}H$$

and $x = 1, 2, 3$. The positive sign denoted oscillations parallel to the magnetic field, H, the negative sign those perpendicular to it. One could re-write this, however, in terms of another variable, $\xi_x = c_x\zeta_x$. For the parallel case, all three values of c are identical and equal to 1. For the perpendicular oscillation, however, $c_1 = 1$, $c_2 = e^{2i\pi/3}$, and $c_3 = e^{4i\pi/3}$. Sommerfeld, "Zur Voigtschen Theorie des Zeeman-Effektes." Reproduced in Sauter, ed., *GS III*, 47–69.

49. Sauter, ed., *GS III*, 65–6.

50. Ibid., 69.

51. Arnold Sommerfeld, "Quantentheoretische Umdeutung der Voigtschen Theorie des anomalen Zeemaneffektes vom D-Linientypus," *Zeitschrift für Physik* 8 (1922). Eing. 12 December 1921. Reproduced in Sauter, ed., *GS III*, 69.

52. Sommerfeld, *Atomic Structure and Spectral Lines*. v.

53. Ibid.

54. The discussion in this section is indebted to John L. Heilbron, "The Kossel-Sommerfeld Theory and the Ring Atom," *Isis* 58 (1967). See also Sigeko Nisio, "X-Rays and Atomic Structure at the Early Stages of the Old Atomic Theory," *Japanese Studies in the History of Science* 8 (1969).

55. Arnold Sommerfeld, "Zur Theorie der Multipletts und ihrer Zeeman-Effekt," *Annalen der Physik* 73 (1924). Reproduced in Sauter, ed., *GS III*, 713–31, on 716.

56. Sommerfeld to Bohr, 5 February 1919. Eckert and Märker, eds., *ASWB 2*, 47.

57. Sommerfeld to Siegbahn, 27 July 1919. Ibid., 49. Siegbahn was awarded the Nobel Prize for his work on precision X-ray spectroscopy in 1924.

58. Sommerfeld, "Die Feinstruktur der wasserstoff- und wasserstoffähnlichen Linien," 490–98.

59. Heilbron, "The Kossel-Sommerfeld Theory and the Ring Atom," esp. 463–70.

60. Terms increase as $\alpha^{2n} \cdot (Z - 1)^{2n+2}$ for $n = 1, 2, 3, \ldots$, where α is the fine-structure constant $\approx$ 1/137. Clearly, when Z is of the order of $1/\alpha$ these terms cannot be omitted.

61. Sommerfeld, "Zur Quantentheorie der Spektrallinien." Reproduced in Sauter, ed., *GS III*, 172–308, on 289.

62. An important complication here concerns the fact that the "L-orbital" is, in fact, a doublet. The quantity Sommerfeld considers is the energy difference between the elliptical orbit ($n + n'$ = 1 + 1) and the N or M orbital, i.e., $\alpha' = L - N$, $\lambda = L - O$.

63. Sommerfeld, "Zur Quantentheorie der Spektrallinien." Sauter, ed., *GS III*, 302.

64. Sommerfeld, "Zur Quantentheorie der Spektrallinien, Ergänzungen und Erweiterungen." Reproduced in Sauter, ed., *GS III*, 326–77.

65. Sauter, ed., *GS III*, 375.

66. Ibid., 375 and 347. Cf. Heilbron, "The Kossel-Sommerfeld Theory and the Ring Atom," 472.

67. Sauter, ed., *GS III*, 375.

68. Arnold Sommerfeld, "Atombau und Röntgenspektren," *Physikalische Zeitschrift* 19 (1918). Reproduced in Sauter, ed., *GS III*, 459–69, on 460.

69. Nisio, "X-Rays and Atomic Structure at the Early Stages of the Old Atomic Theory," 66–71.

70. This was not because external rings did not affect inner ones, but because their effect tended to cancel when one subtracted the energy of one ring from another.

71. Sauter, ed., *GS III*, 462.

72. Sommerfeld to Bohr, 18 May 1918. Eckert and Märker, eds., *ASWB I*, 594–5, on 595. Cf. Heilbron, "The Kossel-Sommerfeld Theory and the Ring Atom," 478, translation mine.

73. A critique in the first edition concerned Burgers's demonstration that the elliptical orbits of the *Ellipsenverein* would have to intersect the circular orbit of the K-shell. Were this so, of course, one could no longer assume that the shielded Coulomb force on each electron was a constant throughout its motion. Sommerfeld declared the objection "fatal." Sommerfeld, *Atombau* (first edition), 368. The text is the same for the second edition, Arnold Sommerfeld, *Atombau und Spektrallinien*, second edition (1921), 365.

74. See Heilbron, "The Kossel-Sommerfeld Theory and the Ring Atom," esp. 476–9.

75. Forman, "Alfred Landé and the Anomalous Zeeman Effect, 1919–1921."

76. Sommerfeld, "Schwebende Fragen," 620.

77. Bohr to Sommerfeld, 19 November 1919. Eckert and Märker, eds., *ASWB 2*, 68–70, on 68.

78. Three ellipses and a circle, for $n + n' = 4 = 4 + 0 = 3 + 1 = 2 + 2 = 1 + 3$.

79. $L_\beta - L_\alpha = (M_2 - L_2) - (M_1 - L_1)$, but $L_\beta - L_{\alpha'} = (M_2 - L_2) - (M_2 - L_1) = L_1 - L_2$.

80. Arnold Sommerfeld, "Bemerkungen zur Feinstruktur der Röntgenspektren I," *Zeitschrift für Physik* 1 (1920). Reproduced in Sauter, ed., *GS III*, 566–77, data on 566. See also Sommerfeld to Bohr, 26 October 1919. Eckert and Märker, eds., *ASWB 2*, 63–66, esp. 65. In spite of this success, the combination-defect, "the greatest difficulty in the theory of Röntgen-spectra" remained. The size of Δ in the expression $K_\alpha + L_\alpha - K_\beta = \Delta$ depends upon the meaning ascribed to K_β. If K_β is given by $M_2 \to K$ (as Sommerfeld maintained), then $\Delta = g_2 - g_1$. If, on the other hand, as Kossel suspected, K_β is given by $M_3 \to K$, then $\Delta = g_3 - g_1$. This latter expression matched the data better than Sommerfeld's did, but still only accounted for roughly half of the measured defect. Sauter, ed., *GS III*, 575. On 18 February 1921 he wrote to Bohr: "I'm racking my brains over the combination-defect of Röntgen-Rad.[iation]. Nothing works." Eckert and Märker, eds., *ASWB 2*, 94.

81. Arnold Sommerfeld, "Bemerkungen zur Feinstruktur der Röntgenspektren II," *Zeitschrift für Physik* 5 (1921). Reproduced in Sauter, ed., *GS III*, 578–93, on 581 and 593. The first argument given above was Sommerfeld's reiteration of the logic of his 1918 paper (included in the first two volumes of *Atombau*). In 1921, however, M was deemed to be a triplet and two transitions to the K-orbital were now possible. Both β' and β" began at M_1, while β, in the new notation, began at M_2.

82. Sauter, ed., *GS III*, 593.

83. Niels Bohr, "Atomic Structure," *Nature* 107 (1921), Bohr, "Atomic Structure."

84. Sommerfeld to Landé, 3 March 1921. Eckert and Märker, eds., *ASWB 2*, 97.

85. Sommerfeld to Bohr, 7 March 1921. Ibid.

86. In the intervening lines, Sommerfeld reiterated his claim (now in the main body of the text rather than a note) that the peculiar structure of the *Ellipsenverein* should not be taken as an argument against it: "One need not, it seems to us, take exception to the artful interlocking of the q elliptic orbits as being something unnatural; one can see therein much more a sign of the high harmony of motion that must rule within the atom." Sommerfeld, *Atombau* (third edition), 613. I include the German here, first for the third edition of *Atombau* and then for Sommerfeld, "Atombau und Röntgenspektren." "Man braucht sich zwar, wie uns scheint, an der kunstreiche Aneinanderpassung der q Ellipsenbahnen nichts als etwas Widernatürlichem zu stoßen, kann vielmehr hierin ein Anzeichen sehen für die hohe Bewegungsharmonie, die im Innern des Atoms herrscht." Cf. "Für mein Gefühl hat die kunstreiche Aneinanderpassung der n Elektronenbahnen in unserem "Ellipsenverein" nichts Unnatürliches, vielmehr sehe ich darin ein Anzeichen für die hohe Bewegungsharmonie, die im Atominnern herrschen muß."

87. Sommerfeld, *Atomic Structure and Spectral Lines*, 143–5.

88. Sommerfeld, *Atombau* (third edition), 613–4. Translation from Sommerfeld, *Atomic Structure and Spectral Lines*, 503–4.

89. Sommerfeld, *Atomic Structure and Spectral Lines*, 505. Note that Burgers's objection to the *Ellipsenverein*, earlier pronounced "fatal," is now deemed essential to any atomic model. This followed from Bohr's claims about the necessary "coupling" between different orbitals. "Another essential feature of the constitution described lies in the configuration of the orbits of the electrons in the relevant groups relative to each other. Thus for each group the electrons within certain sub-groups will penetrate during their revolution into regions which are closer to the nucleus than the mean distances of the electrons belonging to groups of fewer-quanta orbits.... This circumstance... is intimately connected with the fact that the electrons in the various sub-groups, although they may be said to play equivalent parts in the harmony of the interatomic motions, are not at every moment arranged in configurations of simple axial or polyhedral symmetry as in Sommerfeld's or Landé's work, but that their motions are, on the contrary, linked to each other in such a way that it is possible to remove any one of the electrons from the group by a process whereby the orbits of the remaining electrons are altered in a continuous manner." Bohr, "Atomic Structure," 105–6.

90. Sommerfeld, *Atomic Structure and Spectral Lines*, 505.

91. Ibid., vi. If Bohr had provided the motive for Sommerfeld's sharp change in thinking about Röntgen spectra, then one of Sommerfeld's students, Gregor Wentzel, had provided the means. Sommerfeld's *Votum Informativum* on Wentzel's dissertation (translated in full here) makes clear both the nature and the significance of the results: "Herr G. Wentzel has brought the systematics of Röntgen-spectra a good amount further through his work. It was a lucky coincidence that, as Herr W. began his investigations, new, trustworthy material about lines became known, partly from the American, partly from the Swedish side, and that already several fine line-combinations were explained through theoretical preliminary works by Smekal and Coster. Further, the observations of G. Hertz (Autumn 1920) about the L_3 edge and Kossel's idea (beginning of 1920), that the combination-defect found its cause in the difference between respective initial shells, were essential. The many lines of the *L*-series—still puzzling only a short time ago—are now completely classified into definite initial- and final levels, with such security that Herr W. could predict many of the results from Coster's concurrently running measurements. The distinction between regular and irregular doublets, which was adumbrated by me in a lecture during Winter 1920/21 proved very important. Through this distinction it becomes possible to formulate a radical selection principle that provides the necessary and sufficient condition for the occurrence of almost all lines. The few exceptions to this principle (L_ψ, $L\beta_2$) do not speak against it. The quantitative character of the combination-defect is completely explained through the general *Gesetzmässigkeit* of the irregular doublets. All the results of this work are of an essentially empirical nature. The time is not yet ripe for a *modellmässige* derivation of them. The security and soundness of Wentzel's work resides in this more empirical than calculational character. The numerical material of the work, recorded in 33 tables is also first-rate. I regard the work as a promising, significant scientific effort and recommend its acceptance as a dissertation." Arnold

Sommerfeld, "Report to the Faculty on the Dissertation of Gregor Wentzel" (1921). See also Arnold Sommerfeld and Gregor Wentzel, "Über reguläre und irreguläre Dubletts," *Zeitschrift für Physik* 7 (1921), Gregor Wentzel, "Zur Systematik der Röntgenspektren," *Zeitschrift für Physik* 6 (1921).

92. Sommerfeld, *Atomic Structure and Spectral Lines*. vii.

93. Born to Sommerfeld, 5 January 1923. Document 63 in Eckert and Märker, eds., *ASWB 2*, 138.

94. Pauli to Sommerfeld, 6 December 1924, document 83 in ibid., 176–9.

95. Schrödinger to Sommerfeld, 21 July 1925, document 89 in ibid., 192–5.

96. John L. Heilbron, "Interview with Otto Klein, 25/02/1963" (1963), 3. Forman cites the story, and argues that the lecture Sommerfeld gave was probably the text of his Naturwissenschaften article on "Ein Zahlenmysterium in der Theorie des Zeemaneffektes." Forman, "Alfred Landé and the Anomalous Zeeman Effect, 1919–1921," 188.

97. Kuhn and Hund, "Interview with Max Born, 10/17/1962," 13.

98. Oseen, together with Allvar Gullstrand and Manne Siegbahn wielded a great deal of power on the Prize committee, a fact that led to their nickname: the "small popes of Uppsala." Robert Marc Friedman, *The Politics of Excellence: Behind the Nobel Prize in Science* (2001), 122. On the discussion concerning Sommerfeld's nomination in 1925 for the 1924 prize, Friedman writes: "Sommerfeld was chanceless: Oseen was dead-set opposed to his nomination. He saw little reason to reward a partial solution, even if it was fruitful for stimulating further work. He also opposed Sommerfeld's style of theorizing. Beginning with the "answer," Sommerfeld then worked backward mathematically to create a model that would account for experimental data. The mathematics might work, but what did it mean physically? Oseen opposed such so-called mathematical formalisms in theoretical physics; he insisted that physical interpretation and visual comprehensibility must remain central to the study of physics." Friedman, *The Politics of Excellence: Behind the Nobel Prize in Science*, 153–4 and note on 324.

99. Sommerfeld, *Atomic Structure and Spectral Lines*, 317–8.

100. Forman, "Alfred Landé and the Anomalous Zeeman Effect, 1919–1921," 186. "This new approach, which Sommerfeld and his students began applying to X-ray and optical spectra in 1919, 1920, and 1921, we may call *a posteriori*. It began with the observed spectra lines, and worked back to the energy levels. These levels were then characterized by quantum numbers and selection rules—invented *ad hoc* if necessary."

101. Sommerfeld, "Die Bedeutung der Röntgenstrahlung für die heutige Physik." Reprinted in Sauter, ed., *GS IV*, 564–578. As before, however, Sommerfeld claimed that his was an unconventional mysticism, having nothing to do with its "astrological, metaphysical and spiritistic" connotation. "I speak only," he said, "of the laws of nature and the way to fathom them, not of human things." Sauter, ed., *GS IV*, 575.

102. Sauter, ed., *GS IV*, 576.

103. The quotation is from Bohr in 1923, cited in Darrigol, *From c-Numbers to q-Numbers*, xvi–xvii.

104. Ibid., xvi.

105. Niels Bohr, "On the Series Spectra of the Elements (Lecture before the German Physical Society in Berlin, 27 April 1920)," in *Niels Bohr: Collected Works*, ed. J. Rud Nielsen (1920), 241–82, here 245–6; Darrigol, *From c-Numbers to q-Numbers*, 137–8.

106. Darrigol, *From c-Numbers to q-Numbers*, 138.

107. Ibid., 139, 144–5.

108. Ibid., xx.

109. Ibid., 144.

110. Darrigol is not alone in understanding Sommerfeld's reaction to the correspondence principle in this way. Max Jammer writes: "While its versatility and fertility made it attractive to the more synthetically minded physicists, such as Bohr, its flexibility and lack of rigidity made it repugnant to the more analytically oriented theoreticians, such as Sommerfeld." Jammer, *The Conceptual Development of Quantum Mechanics*, 199.

111. Part of the reason for Darrigol's reading surely springs from the fact that he does not describe a position for Sommerfeld that parallels Bohr's own. Thus comparatively little importance is ascribed to anything other than Sommerfeld's analysis of multiperiodic systems. Theoreticians in Munich and Göttingen, Darrigol claims, concentrated most of their attention on the procedures that developed from the methods of Bohr-Sommerfeld quantization: "they tended to neglect all aspects of quantum phenomena that did not fit into this well defined mathematical framework (for instance, the intensities of spectral lines)." Again, this serves to define Bohr's success as others' failure, and to ascribe a mechanical rigidity of approach to those who did not, apparently, emphasize "the still provisional and incomplete character of the newly extended quantum theory." Darrigol, *From c-Numbers to q-Numbers*, 118.

112. Arnold Sommerfeld, "Grundlagen der Quantentheorie und des Bohr'schen Atommodelles," *Die Naturwissenschaften* 12 (1924).

113. Sommerfeld, *Atombau* (first edition). "Vorwort," v.

114. Sommerfeld to Einstein, 3 December 1918, document 295 in Eckert and Märker, eds., *ASWB I*, 612–3.

115. The second edition differed only to a small extent from the first in the main text, but the appendices were thoroughly reworked, making it more useful for a more expert audience. Sommerfeld, *Atombau* (second edition). "Vorwort zur zweiten Auflage," viii.

116. The foreword to the book notes that "the molecular models which are briefly described in the second chapter appear there perhaps more certain than they deserve, and are critiqued at the book's conclusion." Sommerfeld, *Atombau* (first edition), vii.

117. Ibid., viii.

118. Ibid., vii.

119. Ibid., 380. Translation from Sommerfeld, *Atomic Structure and Spectral Lines*, 256. The spherical wave here is not assumed to be "ideal." It is only approximately monochromatic, and hence possesses a finite coherence length; spherical symmetry only applies to surfaces of equal phase, not to surfaces of equal intensity, a fact that allows the definition of a unique directional axis. Sommerfeld has in mind a wave produced by an oscillating or rotating electron. For the former case, "the direction of vibration of the electron is at the same time a *unique axis of the distribution of intensity*." For the latter case, this axis is perpendicular to the plane of rotation and is at the same time the unique axis of polarization. Sommerfeld, *Atomic Structure and Spectral Lines*, 256.

120. Sommerfeld, *Atombau* (first edition), 380. Translation from Sommerfeld, *Atomic Structure and Spectral Lines*, 256.

121. Sommerfeld, *Atombau* (first edition), 380. Translation from Sommerfeld, *Atomic Structure and Spectral Lines*, 256–7.

122. Sommerfeld, *Atombau* (first edition), 389–90. Translation from Sommerfeld, *Atomic Structure and Spectral Lines*, 263.

123. Sommerfeld, *Atombau* (first edition), 389.

124. Adalbert Rubinowicz, "Bohrsche Frequenzbedingung und Erhaltung des Impulsmomentes I & II," *Physikalische Zeitschrift* 19 (1918).

125. Sommerfeld, *Atombau* (first edition), 393.

126. Kuhn, "Interview with Werner Heisenberg, 02/11/1963," 11.

127. Bohr, "On the Quantum Theory of Line Spectra, Parts I–III [1918–1922]," 81.

128. On the importance to Bohr of Einstein's 1916 papers, which derived Planck's radiation law by assuming that the probabilities of transition could be determined by analogy with the theory of radioactivity, see Jammer, *The Conceptual Development of Quantum Mechanics*, 112–4; Darrigol, *From c-Numbers to q-Numbers*, 118–28.

129. Bohr, "On the Quantum Theory of Line Spectra, Parts I–III [1918–1922]," 82.

130. Both Bohr and Sommerfeld noted the asymmetry that such an approach required. As Sommerfeld put it: "The coincidence of the vibration numbers calculated on the quantum and on the classical theory, respectively, for high quantum numbers and relatively small quantum leaps is perfect from the formal aspect. Nevertheless there remains a considerable difference of view in the matter itself. From the classical point of view all vibrations, overtones and combination tones are emitted *simultaneously* when the orbit is being traversed. The whole vibration spectrum owes its origin to one uniform event. From the quantum point of view, however, each line of the spectrum corresponds to a different single event and a different kind of quantum leap. The individual events do not necessarily occur simultaneously, but rather *independently of each other*." Sommerfeld, *Atomic Structure and Spectral Lines*, 580. For Bohr's acknowledgement of the same point, see Bohr, "On the Quantum Theory of Line Spectra, Parts I–III

[1918–1922]," 81. Darrigol explains Bohr's utilization of the method, nonetheless, by noting that Bohr was probably thinking, again following Einstein, in terms of a statistical ensemble of atoms. Darrigol, *From c-Numbers to q-Numbers*, 126.

131. Sommerfeld, *Atombau* (first edition), 402.

132. Ibid., 402–3.

133. Sommerfeld to Bohr, 11 November 1920, in J. Rud Nielsen, ed., *Niels Bohr Collected Works: The Correspondence Principle (1918–1923)*, volume 3 (1976), 690–1.

134. Sommerfeld, *Atombau* (second edition), 527.

135. Sommerfeld, *Atomic Structure and Spectral Lines*, 584.

136. Ibid., vi.

137. Sommerfeld was less circumspect in a letter to Einstein, written around the same time as the foreword to the third edition. "Inwardly," he confessed, "I also no longer believe in the spherical wave." Sommerfeld to Einstein, 11 January 1922. Eckert and Märker, eds., *ASWB 2*, 110–1.

138. Sommerfeld, *Atomic Structure and Spectral Lines*. vi.

139. Ibid., 276.

140. This aspect of the Bohr-Sommerfeld controversy is discussed in an interview between Thomas Kuhn and Werner Heisenberg. Kuhn was, I think, the first to note Sommerfeld's antipathy to models, and to connect this with his unenthusiastic response to the correspondence principle. I quote the relevant exchange here between Heisenberg and Kuhn from Kuhn, "Interview with Werner Heisenberg, 02/11/1963," 13. TSK: [G]ranting that the Correspondence Principle tells you that you cannot take the model altogether seriously, there is nevertheless a sense in which the Correspondence Principle is a much more *modellmässig* approach than Sommerfeld's approach. And when Sommerfeld discusses the Correspondence principle in *Atombau* at a time before he is yet giving it a central place, but still as late as 1923, he said, "After all, Bohr's success does indicate that we may have to take models seriously." So there is the other sense in which the Correspondence Principle is the place that is allowing one to use models in fact. You might say that Bohr is for models in one sense in which Sommerfeld is against them. H: Yes, perhaps one can put it that way. When you speak about the model, you mean something which can only be described by means of classical physics. As soon as you go away from classical physics, then, in a strict sense, you don't even know what a model could possibly mean because then the words haven't got any meaning any more. Now this was a dilemma. Because if one said that one used a model to calculate the levels and then at the same time one goes away from classical physics by introducing integral pdq, then one does something which already is inconsistent. Now, having this inconsistency, Sommerfeld wanted to go into rigorous mathematical laws which in some way went away from the model, while Bohr tried to keep to the picture while at the same time omitting classical mechanics. He tried to keep the words and the pictures without keeping the meaning of the words or the pictures. Both things are possible in such a situation because

your words don't really tackle the situation any more. You can't get hold of the things by means of your words and so what shall you do? But Sommerfeld's escape would always be into the rigorous mathematical scheme, while Bohr's escape would be into the philosophy of things. Perhaps this is the fundamental difference between the two people. Sommerfeld was never a philosopher...."

141. Sommerfeld, *Atomic Structure and Spectral Lines*, 276.

142. Sommerfeld, "Grundlagen der Quantentheorie und des Bohr'schen Atommodelles," 1048.

143. Roger H. Stuewer, *The Compton Effect: Turning Point in Physics* (1975).

144. Stuewer credits Sommerfeld, traveling the United States at the time, with spreading the word of Compton's results to other *American* cities as well. Ibid., 240–42.

145. Quoted in ibid., 241.

146. Arnold Sommerfeld, *Atombau und Spektrallinien*, fourth edition (1924), vii–viii.

147. Ibid., 338.

148. Helge Kragh, "Niels Bohr's Second Atomic Theory," *Historical Studies in the Physical Sciences* 10 (1979).

149. Sommerfeld, *Atomic Structure and Spectral Lines*, 69 (emphasis added).

150. Ibid., 109 (emphasis added). See also Kragh, "Niels Bohr's Second Atomic Theory," 161.

151. Gregor Wentzel, "Fortschritte der Atom- und Spektraltheorie," *Ergebnisse der Exakten Naturwissenschaften* 1 (1922).

152. Kragh, "Niels Bohr's Second Atomic Theory," 163.

153. Wentzel, "Fortschritte der Atom- und Spektraltheorie," 299.

154. Sommerfeld, *Atombau* (fourth edition), vi.

155. Ibid., vii.

156. Sommerfeld, "Grundlagen der Quantentheorie und des Bohr'schen Atommodelles," 1049.

157. Sommerfeld had first emphasized this "demand" in the context of a discussion of an oft-mentioned peculiarity of the quantum theory, namely that according to Bohr's condition governing radiative emission, an electron would seem to "know" in advance the state to which it would finally jump, before it could emit the appropriate amount of radiation. This aspect seemed to introduce a greater element of teleology into the quantum theory than classical theory. Essentially as a response to this peculiarity, Sommerfeld wrote "what we have to demand, at any rate, so long as there should be a natural science, is the unique [*eindeutige*] determination of observable events, the mathematical security of natural laws. How this uniqueness [*Eindeutigkeit*] is achieved, whether it is given merely by the initial state or through the initial and final state together, we cannot know *a priori*, but must learn from Nature." Ibid.

158. Niels Bohr, "Speech of Appreciation to Sommerfeld (22 September 1919)," in *Niels Bohr Collected Works*, ed. J. Rud Nielsen (1976).

159. Sommerfeld, "Grundlagen der Quantentheorie und des Bohr'schen Atommodelles," 1048. Cf. Benz, *Arnold Sommerfeld*, 120. Sommerfeld's rejection of any claim that quantum theory was not yet ripe for a deductive (read: mathematical) treatment may well have been a pointed reference to Bohr. It has been suggested that Bohr's avoidance of mathematical calculations was due to "his conviction that, at this stage of his work, many ideas could be best expressed in carefully chosen words"; that mathematics expressed more certainty than was warranted at the time. Kragh, "Niels Bohr's Second Atomic Theory," 161, fn. 118.

160. Sommerfeld, "Grundlagen der Quantentheorie und des Bohr'schen Atommodelles," 1048.

161. Ibid. (emphasis added).

162. Kuhn, "Interview with Werner Heisenberg, 02/07/1963," 5.

163. Werner Heisenberg, "Über Stabilität und Turbulenz von Flüssigkeitsströmen," *Annalen der Physik* 74 (1924). Reproduced in W. Blum, H.-P. Dürr, and H. Rechenberg, *Werner Heisenberg: Collected Works*, volume A: *Original Scientific Papers* (1985), 31–81. The "odd man out" was R. von Mises. Sommerfeld suggested the turbulence problem after Heisenberg successfully dealt with an earlier question concerning vortex motion, published as Werner Heisenberg, "Die absoluten Dimensionen der Karmanschen Wirbelbewegung," *Physikalische Zeitschrift* 23 (1922). Reproduced in Blum, Dürr, and Rechenberg, *Werner Heisenberg: Collected Works*, 27–30. Heisenberg described the project and its later history in Kuhn, "Interview with Werner Heisenberg, 02/07/1963," 8.

164. Karl Schlayer, "Über die Stabilität der Karman'schen Wirbelstrasse gegenüber beliebigen Störungen in drei Dimensionen" (Munich University, 1927), Arnold Sommerfeld, "Report to the Faculty on the Dissertation of Karl Schlayer" (1927).

165. Arnold Sommerfeld, "Report to the Faculty on the Dissertation of Hans Pfrang" (1925).

166. Arnold Sommerfeld, "Report to the Faculty on the Dissertation of Ernst Guillemin" (1926). Sommerfeld noted that Guillemin's father had repeatedly provided money for the physical institute and for the "Studentenhaus." Victor Guillemin, Ernst's brother, also completed a dissertation under Sommerfeld in 1926.

167. Arnold Sommerfeld, "Report to the Faculty on the Dissertation of Karl Bechert" (1925).

168. Kuhn, "Interview with Werner Heisenberg, 02/07/1963," 3.

169. Kuhn, "Interview with Werner Heisenberg, 02/11/1963," 16 (emphasis added).

170. Sommerfeld's changing attitude toward Heisenberg's model is nicely illustrative of his overall position with regard to model-based accounts in the early 1920s. Initially enthusiastic about the work of his "gifted" student—who had managed to explain the laws governing the anomalous Zeeman effect in a *modellmässig* form "based on simple assumptions"—his feelings had soured by the end of March 1922. At that time, he wrote to Bohr, noting that "Heisenberg's representation is not suitable and should have been more straightened out.... I couldn't bridle

his eagerness to publish any longer and find his results so important that I assented to their publication in spite of the fact that the form of the derivation may not yet be the definitive one." Sommerfeld to Pieter Zeeman, 2 October 1921; Sommerfeld to Bohr, March 25 1922. Eckert and Märker, eds., *ASWB 2*, 105 and 114. When Sommerfeld and Heisenberg published together on "The Intensity of Multiple Lines and their Zeeman Components" in 1922, they limited themselves to a purely *kinematic* representation, "requiring no more detailed [*näheren*] conception about atomic structure, whereby the security of our conclusions is strengthened." Arnold Sommerfeld and Werner Heisenberg, "Die Intensität der Mehrfachlinien und ihrer Zeemankomponenten," *Zeitschrift für Physik* 11 (1922). Reproduced in Sauter, ed., *GS III*, 625–48, on 648. On Heisenberg's model more generally, see David C. Cassidy, "Heisenberg's First Core Model of the Atom: The Formation of a Professional Style," *Historical Studies in the Physical Sciences* 10 (1979).

171. Arnold Sommerfeld, "Report to the Faculty on the Dissertation of Josef Krönert" (1920).

172. Sommerfeld, "Report to the Faculty on the Dissertation of Gregor Wentzel." Wentzel was Sommerfeld's assistant in these early years.

173. Wolfgang Pauli, "Über das Modell des Wasserstoffmolekülions," *Annalen der Physik* 68 (1922). Reproduced in R. Kronig and V. F. Weisskopf, eds., *Wolfgang Pauli: Collected Scientific Papers*, volume 2 (1964), 70–133, here 72. The article is an improved and expanded version of Pauli's Munich dissertation.

174. Arnold Sommerfeld, "Report to the Faculty on the Dissertation of Wolfgang Pauli" (1921). Pauli noted in 1926 that the model "must be considered outmoded in the light of present-day physical knowledge." Kronig and Weisskopf, eds., *Wolfgang Pauli: Collected Scientific Papers*. x.

Conclusion

1. Quoted in Eckert and Märker, eds., *ASWB 2*, 23.

2. Pauli to Kronig, 21 May 1925. A. Hermann et al. (eds.), *Wolfgang Pauli: Wissenschaftlicher Briefwechsel mit Bohr, Einstein, Heisenberg u. a.* (2005) (hereafter cited as *PWB*), 214–16, on 216.

3. Pauli to Kramers, 27 July 1925. Ibid., 232–34, on 234. "So fühle ich mich denn jetzt weniger einsam als etwa vor einem halben Jahr, wo ich mich (geistig wie raumlich) zwischen der Scylla der zahlenmystischen Münchener Schule und der Charybdis des von Ihnen mit zelotischen Exzessen propagierten reaktionären Kopenhager Putsches ziemlich allein befand!"

4. Wolfgang Pauli, "Über den Zusammenhang des Abschlusses der Elektronengruppen im Atom mit der Komplexstruktur der Spektren," *Zeitschrift der Physik* 31 (1925). Reproduced in R. Kronig and V. F. Weisskopf, eds., *Wolfgang Pauli: Collected Scientific Papers*, volume II (1964), 214–32, on 225.

5. Pauli, *Exclusion Principle and Quantum Mechanics: Lecture Given in Stockholm After the Award of the Nobel Prize of Physics 1945*.

6. See, e.g., John L. Heilbron, "The Origins of the Exclusion Principle," *Historical Studies in the Physical Sciences* 13 (1983); John Hendry, *The Creation of Quantum Mechanics and the Bohr-Pauli Dialogue* (1984); Serwer, "*Unmechanischer Zwang*."

7. B. L. van der Waerden, ed., *Sources of Quantum Mechanics* (1968).

8. Serwer, "*Unmechanischer Zwang*," 193.

9. Ibid.

10. Michel Janssen and Anthony Duncan, "On the Verge of *Umdeutung*: John van Vleck and the Correspondence Principle, Parts I & II," *Archive for History of the Exact Sciences* 61 (2007).

11. Heilbron, "Exclusion Principle"; B. L. van der Waerden, "Exclusion Principle and Spin," in *Theoretical Physics in the Twentieth Century: A Memorial Volume to Wolfgang Pauli*, ed. M. Fierz and V. Weisskopf (1960).

12. Pauli to Sommerfeld, 6 June 1923. Hermann et al. (eds.), *PWB*, 97. Quotation from Heilbron, "Exclusion Principle," 292.

13. Forman, "Alfred Landé and the Anomalous Zeeman Effect, 1919–1921."

14. Not only was the idea incomprehensible in classical terms, it contradicted a basic principle of the quantum theory, Bohr's so-called *Aufbauprinzip*. This "construction principle" held that quantum numbers were conserved in building up atoms. Assuming, as was standard, that a rare gas atom (where all shells were filled with the maximum possible number of electrons) possessed zero net angular momentum, then the core of an alkali atom which immediately followed it in the periodic table of elements should also have zero net momentum. In Heisenberg's model, in contrast, it possessed a non-zero value, namely $\frac{1}{2}$.

15. A. Landé, "Termstruktur und Zeemaneffekt der Multipletts," *Zeitschrift der Physik* 15 (1923). The capitalized variables bear simple relations to their uncapitalised counterparts. Landé's version was one of many renormalizations, with $K = k - \frac{1}{2}$, $J = j$ for even multiplets, $J = j - \frac{1}{2}$ for odd; and $R = r/2$.

16. Pauli to Lande, 23 May 1923. Hermann et al. (eds.), *PWB*, 87–90, on 87. Cf. Heilbron, "Exclusion Principle," 296.

17. Pauli suggested in the same letter that his publication of the offending article only occurred under pressure from Bohr. Pauli to Sommerfeld, 6 June 1923. Hermann et al. (eds.), *PWB*, 94–99, on 97. The paper appeared as Wolfgang Pauli, "Über die Gesetzmässigkeiten des anomalen Zeemaneffektes," *Zeitschrift für Physik* 16 (1923). Reproduced in Kronig and Weisskopf, eds., *PCSP II*, 151–60.

18. The term "rein phänomenologische" appears in Pauli's letter to Lande, 23 May 1923. Hermann et al. (eds.), *PWB*, 87–90, on 87. Pauli's approach in this paper was strongly reminiscent of Woldemar Voigt's (and subsequently Sommerfeld's) similarly phenomenological treatment of D-line splitting.

19. For more detailed accounts of Pauli's paper, see Heilbron, "Exclusion Principle" and van der Waerden, "Exclusion Principle and Spin."

20. Pauli to Kronig, 21 May 1925. Quoted in Darrigol, *From c-Numbers to q-Numbers*, 249.

21. Pauli, "Über den Zusammenhang des Abschlusses der Elektronengruppen im Atom mit der Komplexstruktur der Spektren."

22. Kragh, "Niels Bohr's Second Atomic Theory."

23. Sommerfeld to Landé, 3 March 1921. Cf. Kragh, "Niels Bohr's Second Atomic Theory."

24. Niels Bohr, "Essay III: The Structure of the Atom and the Physical and Chemical Properties of the Elements (Based on a Danish address, delivered in October, 1921)," in *The Theory of Spectra and Atomic Constitution: Three Essays by Niels Bohr* (1922). Text of second edition (1924) reproduced in *Niels Bohr Collected Works*, volume 4, ed. N. Bohr and J. Rud Nielsen (1977).

25. Unlike Sommerfeld's *Ellipsenverein*, where electrons in each group were always at the same distance from the atomic core (on "pulsing" circular orbits), the electrons in Bohr's model were finely tuned so that only one penetrated the lower electronic shells at a time. A structure like Sommerfeld's, Bohr wrote, "may be described as one where the motions of the electrons within the groups are coupled together in a manner which is largely independent of the interaction between the various groups. On the contrary, the characteristic feature of a structure like that I have suggested is the *intimate coupling between the motions of the electrons in the various groups* characterized by different quantum numbers, as well as the *greater independence in the mode of binding within one and the same group of electrons* the orbits of which are characterized by the same quantum number." Bohr and Nielsen (eds.), *BCW 4*, 294.

26. "With increasing nuclear charge and the consequent decrease in the difference between the fields of force inside and outside the region of the orbits of the first 18 bound electrons, the dimensions of those parts of the 4_1 orbit which fall outside will approach more and more to the dimensions of a 4-quantum orbit calculated on the assumption that the interaction between the electrons in the atom may be neglected, *With increasing atomic number a point will therefore be reached where a 3_3 orbit will correspond to a firmer binding of the 19th electron than a 4_1 orbit*, and this occurs as early as the beginning of the fourth period." Bohr does not include this calculation. Bohr and Nielsen (eds.), *BCW 4*, 303.

27. Thus, for example, Bohr writes, "with the binding of the 4th, 5th, and 6th electrons in 2_1 orbits, the spatial symmetry of the regular configuration of the orbits must be regarded as steadily increasing, until with the binding of the 6th electron the orbits of the four last bound electrons may be expected to form an exceptionally symmetrical configuration.... It cannot be expected that the 7th electron will be bound in a 2_1 orbit equivalent to the orbits of the four preceding electrons. The occurrence of five such orbits would so definitely destroy the symmetry in the interaction of these electrons that it is inconceivable that a process resulting in the accession of a fifth electron to this group would be in agreement with the correspondence principle." Ibid. See also Heilbron, "Exclusion Principle"; Kragh, "Niels Bohr's Second Atomic Theory."

28. Kragh, "Niels Bohr's Second Atomic Theory."

29. Sommerfeld, *Atombau* (fourth edition). vi.

30. Arnold Sommerfeld, "Zur Theorie des periodischen Systems," *Physikalische Zeitschrift* 26 (1925). in Sauter, ed., *GS III*, 757–61, on 757.

31. As we shall see below, Bohr also used a third quantum number as a means of classifying and ordering the multitude of x-ray lines. However, he did not use the third number to further subdivide orbits.

32. E.g., for the 2_{11} and 2_{21} subgroups (numerically equating the conceptually distinct k_2 and Stoner's j), $m_1 = \pm\frac{1}{2}$. For the 2_{22} subgroup, $m_1 = \pm\frac{1}{2}, \pm\frac{3}{2}$.

33. Pauli to Bohr, 12 December 1924. In Hermann et al. (eds.), *PWB*, 189.

34. PCW 225.

35. Pauli to Sommerfeld, 6 December 1924. Hermann et al. (eds.), *PWB*, 182–84, on 183.

36. Sommerfeld to Pauli, 18 June 1925. Ibid., 216–7, on 217.

37. Bohr to Pauli, 22 December 1924. Ibid., 193–96, on 195, Danish on 193.

38. Pauli to Sommerfeld, 6 December 1924. Ibid., 182–84, on 183.

39. The term emerged from Bohr and Pauli's work on the Zeeman effect. Cassidy, *Uncertainty: The Life and Science of Werner Heisenberg*, 167. "In his new scheme, Bohr allowed only integral k but hypothesized the presence of a vague Zwang. He first reduced the usual $2k + 1$ magnetic states by eliminating the orientation that is perpendicular to the magnetic field. The Zwang did the rest: mysteriously, it reduced the allowed states even further to $2k - 1$, then forced a Zweideutigkeit (two-valuedness, or ambiguity) on the core such that it took up two positions under the action of the Zwang, rather than the two positions that would arise if the core shared half a unit of momentum with the electron, as in Heisenberg's model. The proper number of states, $2(2k - 1)$, resulted!" See also, and more recently, Charles P. Enz, *No Time to Be Brief: A Scientific Biography of Wolfgang Pauli* (2001), 107. "The notion of *Komplexstruktur* and *Zweideutigkeit* may be traced back to an important review article by Bohr in the previous year. In this article Bohr, after observing that the necessity of introducing half-integer quantum numbers has rendered the use of the theory of multiply periodic systems questionable, states in his well-known cautious style: 'we are rather led to the view that, because of the mechanically indescribable stability properties of the carriers of a force (*Zwang*) which is not analogous to the action of external fields of force, and which implies that the atomic core, instead of having one single possibility of alignment in a constant external field, is forced upon two different alignments while, on the other hand, because of the same force (*Zwang*) the outer electron disposes in the atomic compound of only $2k - 1$ possibilities of alignment, instead of the $2k$ possibilities of alignment in an external field.'" The internal quotation is from Niels Bohr, "Linienspektren und Atombau," *Annalen der Physik* 71 (1923): 276. It is to be emphasized, however, that Bohr does *not* use the term *Zweideutigkeit* in this paper.

40. Pauli to Landé, 23 September 1923. In Hermann et al. (eds.), *PWB*, 119–123, on 122–3. Cf. Michela Massimi, *Pauli's Exclusion Principle: The Origin and Validation of a Scientific Principle* (2005).

41. He is followed in this by Jammer (*The Conceptual Development of Quantum Mechanics*, 141).

42. Pauli to Bohr, 12 December 1924. In Hermann et al. (eds.), *PWB*, 187.

43. Niels Bohr and Dirk Coster, "Röntgenspektren und periodisches System der Elemente," *Zeitschrift für Physik* 12 (1923). Translated as "X-Ray Spectra and the Periodic System of the Elements," in Bohr and Nielsen (eds.), *BCW 4*, 519–48, on 540.

44. Sommerfeld, *Atomic Structure and Spectral Lines*, 520.

45. Paul Forman, "The Doublet Riddle and Atomic Physics circa 1924," *Isis* 59 (1968): 170.

46. Proportional to the fourth power of the nucleus's effective charge, the width of "relativistic" doublets in the Röntgen spectra of uranium ($Z = 92$) was roughly 70 million times as large as that for hydrogen.

47. Forman, "Doublet Riddle," 171; Kragh, "Fine Structure."

48. Forman, "Doublet Riddle," 171.

49. Sommerfeld, *Atombau* (fourth edition), 312.

50. Ibid., 453–4.

51. The German word for "two-valuedness" should be *Zweiwertigkeit*. *Zweideutigkeit* implies a double meaning or signification.

52. Pauli to Bohr, 12 December 1924. In Hermann et al. (eds.), *PWB*, 189.

53. Werner Heisenberg, "Über quantentheoretische Umdeutung kinematischer und mechanischer Beziehungen," *Zeitschrift für Physik* 33 (1925). Translated and reproduced in van der Waerden, ed., *Sources of Quantum Mechanics*, 261–76, on 262.

54. Werner Heisenberg, *Encounters with Einstein and Other Essays on People, Places, and Particles* (1989), 113–4.

55. Cassidy, *Uncertainty: The Life and Science of Werner Heisenberg*, 198.

56. On Pauli's operationalism in this period, see Hendry, *The Creation of Quantum Mechanics and the Bohr-Pauli Dialogue*, esp. 6–23.

57. Sommerfeld, "Quantentheoretische Umdeutung der Voigtschen Theorie des anomalen Zeemaneffektes vom D-Linientypus," 624.

58. Ibid., 609.

59. Jammer, *The Conceptual Development of Quantum Mechanics*, 196.

60. Kuhn, "Interview with Hans Bethe, 01/17/1964," 9.

61. Arnold Sommerfeld, "Zur Frage nach der Bedeutung der Atommodelle," *Zeitschrift für Elektrochemie und angewandte physikalische Chemie* 34 (1928). Reproduced in Sauter, ed., *GS III*, 845–9, here 847.

62. Kuhn, "Interview with Hans Bethe, 01/17/1964," 9.

63. Suman Seth, "Crisis and the Construction of Modern Theoretical Physics," *British Journal for the History of Science* 40 (2007).

64. Forman, "Weimar Culture, Causality, and Quantum Theory."

65. Darrigol, *From c-Numbers to q-Numbers*, 177.

66. Pauli, quoted in Abraham Pais, *Niels Bohr's Times: In Physics, Philosophy, and Polity* (1991), 199. Pauli also provides Thomas S. Kuhn (*The Structure of Scientific Revolutions*, enlarged second edition, 1970, 83–4) with an example of the crisis before the advent of matrix mechanics.

67. Forman, "Doublet Riddle"; Forman, "Alfred Landé and the Anomalous Zeeman Effect, 1919–1921"; Darrigol, *From c-Numbers to q-Numbers*; Serwer, "*Unmechanischer Zwang*." As Helge Kragh has noted, however, the hydrogen atom—the subject of Sommerfeld's famous work on fine structure—was not implicated in the growing dissatisfaction with standard explanations: "In the process that eventually transformed the old quantum theory into quantum mechanics, experimental anomalies contributed strongly, but there was no feeling of crisis at all as far as the hydrogen atom was concerned." Kragh, "Fine Structure," 84.

68. Thomas S. Kuhn, "Interview with P. A. M. Dirac, 04/01/1962" (1962), 14. Dirac seems to have had a similar response to Bohr's worries, in 1929, about the need for a "drastic alteration" to quantum mechanics as it then stood. "I cannot see any reason for thinking that quantum mechanics has already reached the limit of its development. I think it will undergo a number of small changes, namely with regard to its method of application, and by those means most of the difficulties now confronting the theory will be removed. If any of the concepts now used (e.g. potentials at a point) are found to be incapable of having an exact meaning, one will have to replace them by something a little more general, rather than make some drastic alteration in the whole theory." Quoted in Silvan S. Schweber, *QED and the Men Who Made It: Dyson, Feynman, Schwinger, and Tomonaga* (1994), 65.

69. Kuhn, "Interview with Werner Heisenberg, 02/11/1963," 7.

70. Certainly Jammer's descripion, cited above, does not indicate the requisite level of foundational stability or coherence necessary for a Kuhnian paradigm. I suspect that, as with the question of the existence or not of a "crisis," the question of the existence of a paradigm may well have a different answer depending upon which group of practitioners the historian studies. I am indebted to Jed Buchwald for bringing this point to my attention.

71. In the interview, Pauling suggests that Sommerfeld's comment dates from 1924 or 1925.

72. Heilbron, "Interview with Linus Pauling, 03/27/1964," 17.

73. Arnold Sommerfeld, *Atombau und Spektrallinien: Wellenmechanischer Ergänzungsband*, first edition (1929).

74. Kuhn, "Interview with Hans Bethe, 01/17/1964," 12.

75. Ibid. Bethe would note that "looking back on this period from five years later I had the impression that I had really missed the development." A much more recent interview, however, would suggest that his approach to physics was very close indeed to that of his former teacher. "I'm not searching that much," he said, "In fact, I've never tried to go into the really deep fundamentals of physics, into the philosophical part of physics. I'm much more interested in phenomena that you can observe. Now, I am searching for solutions for some equations also, but that's not the kind of thing that Bohr was doing." Judith Goodstein, "A Conversation with Hans Bethe," *Phys. Persp.* 1 (1999): 279.

76. Heilbron, "Interview with Linus Pauling, 03/27/1964," 11.

77. Kuhn, "Interview with Hans Bethe, 01/17/1964," 12.

78. This is too little emphasized by those who would point to the importance of the helium and Zeeman crises as *causes* for the search for new bases for the quantum theory.

79. Forman, "Weimar Culture, Causality, and Quantum Theory."

80. Kuhn, *The Structure of Scientific Revolutions*, 103.

81. S. S. Schweber has described a similar situation with regard to the development of quantum electrodynamics. Where "leaders of the discipline" like Bohr, Pauli, Heisenberg and Dirac were insistent on the need for a "revolution," the actual solution of the divergence difficulties was provided by those who looked rather for incremental, technical solutions to particular problems, like Kramers, Schwinger, Tomonaga, Feynman, Dyson and Bethe. Schweber, *QED*, xxv–xxvi. Bethe's significant contribution provides a certain continuity in what could be described as another round in a debate about principles/revolutions versus problems/incremental technical development.

82. Kuhn, *The Structure of Scientific Revolutions*, 24.

Bibliography

Albisetti, James C. *Secondary School Reform in Imperial Germany*. Princeton University Press, 1983.

Anonymous. "Das Einkommen der Hochschullehrer: Denkschrift des Verbandes der Deutschen Hochschulen." *Mitteilungen des Verbandes der deutschen Hochschulen* 2, no. 19 (1922): 247–259.

Arrhenius, Svante. *Theories of Solutions*. Yale University Press, 1912.

Arrhenius, Svante. "Über die Gültigkeit des Beweises von Herrn Planck für das Van 't Hoffsche Gesetz." *Zeitschrift für physikalische Chemie* 9 (1892): 330–334.

Barkan, Diana Kormos. *Walther Nernst and the Transition to Modern Physical Science*. Cambridge University Press, 1999.

Beauchamp, Ken. *History of Telegraphy*. Institution of Electrical Engineers, 2001.

Benz, Ulrich. *Arnold Sommerfeld: Lehrer und Forscher an der Schwelle zum Atomzeitalter, 1868–1951*, ed. H. Degen. Wissenschaftliche Verlagsgesellschaft, 1975.

Bernstein, Jeremy. *Hans Bethe, Prophet of Energy*. Basic Books, 1980.

Blum, W., H.-P. Dürr, and H. Rechenberg. *Werner Heisenberg: Collected Works*, volume A. Springer-Verlag, 1985.

Bohr, Niels. "Atomic Structure." *Nature* 107 (1921): 104–107.

Bohr, Niels. "Atomic Structure." *Nature* 108 (1921): 208–209.

Bohr, Niels. "Essay III: The Structure of the Atom and the Physical and Chemical Properties of the Elements (Based on a Danish address, delivered in October, 1921)." In *The Theory of Spectra and Atomic Constitution: Three Essays by Niels Bohr*. Cambridge University Press, 1922.

Bohr, Niels. "Linienspektren und Atombau." *Annalen der Physik* 71 (1923): 228–288.

Bohr, Niels. "On the Application of the Quantum Theory to Periodic Systems (Unpublished Paper, Intended for Publication in the *Phil. Mag.*, April 1916)." In *Niels Bohr Collected Works*, volume 2, ed. U. Hoyer. North-Holland, 1981.

Bohr, Niels. "On the Effect of Electric and Magnetic Fields on Spectral Lines." *Philosophical Magazine* 27 (1914): 506–524.

Bohr, Niels. "On the Program of the Newer Atomic Physics (2 December 1919)." In *Niels Bohr: Collected Works*, ed. J. Rud Nielsen. North-Holland, 1976.

Bohr, Niels. "On the Quantum Theory of Line Spectra, Parts I–III." In *Niels Bohr: Collected Works*, ed. J. Rud Nielsen. North-Holland, 1976.

Bohr, Niels. "On the Series Spectra of the Elements (Lecture before the German Physical Society in Berlin, 27 April 1920)." In *Niels Bohr: Collected Works*, ed. J. Rud Nielsen. North-Holland, 1976.

Bohr, Niels. "On the Series Spectrum of Hydrogen and the Structure of the Atom." *Philosophical Magazine* 29 (1915): 332–335.

Bohr, Niels. "Speech of Appreciation to Sommerfeld (22 September 1919)." In *Niels Bohr Collected Works*, ed. J. Rud Nielsen. North-Holland, 1976.

Bohr, Niels. "The Structure of the Atom (Nobel Lecture, delivered December 11, 1922)." *Nature* (1923): 1–16.

Bohr, Niels, and Dirk Coster. "Röntgenspektren und periodisches System der Elemente." *Zeitschrift für Physik* 12 (1923): 342–374.

Bohr, Niels, and J. Rud Nielsen, eds. *Niels Bohr Collected Works*, volume 4. North-Holland, 1977.

Boltzmann, Ludwig. *Lectures on Gas Theory*. Dover, 1995.

Boltzmann, Ludwig. "Über irreversible Strahlungsvorgänge I." *Berliner Berichte* (1897): 660–662.

Boltzmann, Ludwig. "Über irreversible Strahlungsvorgänge II." *Berliner Berichte* (1897): 1016–1018.

Bopp, F., and J. Meixner, eds. *Thermodynamik und Statistik von Arnold Sommerfeld, Vorlesungen über Theoretische Physik, V*. Dieterich'sche Verlagsbuchhandlung, 1952.

Born, Max. "Arnold Johannes Wilhelm Sommerfeld." *Obituary Notices of the Fellows of the Royal Society* 8, November (1952): 275–296.

Born, Max. *Ausgewählte Abhandlungen*, volume 2. Vandenhoeck & Ruprecht, 1963.

Born, Max. *My Life: Recollections of a Nobel Laureate*. Taylor and Francis, 1978.

Born, Max. "Sommerfeld als Begründer einer Schule." *Naturwissenschaften* 16 (1928): 1035–1036.

Broelmann, Jobst. *Intuition und Wissenschaft in der Kreiseltechnik, 1750 bis 1930*. Deutsches Museum, 2002.

Buchwald, Jed Z. *The Creation of Scientific Effects: Heinrich Hertz and Electric Waves*. University of Chicago Press, 1994.

Buchwald, Jed Z. "Electrodynamics in Context: Object States, Laboratory Practice, and Anti-Romanticism." In *Hermann von Helmholtz and the Foundations of Nineteenth-Century Science*, ed. D. Cahan. University of California Press, 1993.

Buchwald, Jed Z. "Reflections on Hertz and the Hertzian Dipole." In *Heinrich Hertz: Classical Physicist, Modern Philosopher*, ed. D. Baird, R. Hughes, and A. Nordmann. Kluwer, 1998.

Burgess, George K. "Applications of Science to Warfare in France." *The Scientific Monthly* 5, no. 4 (1917): 289–297.

Cahan, David. *An Institute for an Empire: the Physikalisch-Technische Reichsanstalt, 1871–1918*. Cambridge University Press, 1989.

Cahan, David. "The Institutional Revolution in German Physics, 1865–1914." *Historical Studies in the Physical Sciences* 15 (1985): 1–65.

Cantor, Mathias. "Über die Grundlage der Lösungstheorie." *Annalen der Physik* 10 (1903): 205–213.

Cassidy, David C. "Heisenberg's First Core Model of the Atom: The Formation of a Professional Style." *Historical Studies in the Physical Sciences* 10 (1979): 187–224.

Cassidy, David C. *Uncertainty: The Life and Science of Werner Heisenberg*. Freeman, 1992.

Chevalley, Catherine. "Niels Bohr's Words and the Atlantis of Kantianism." In *Niels Bohr and Contemporary Philosophy*, ed. J. Faye and H. Folse. Kluwer, 1994.

Chevalley, Catherine. "Philosophy and the Birth of Quantum Theory." In *Physics, Philosophy, and the Scientific Community: Essays in the Philosophy and History of the Natural Sciences and Mathematics. In Honor of Robert S. Cohen*, ed. K. Gavroglu, J. Stachel, and M. Wartofsky. Kluwer, 1995.

Chickering, Roger. *Imperial Germany and the Great War, 1914–1918*. Cambridge University Press, 1998.

Collins, Harry M. *Changing Order: Replication and Induction in Scientific Practice*. University of Chicago Press, 1992.

Crawford, Elisabeth. *Arrhenius: From Ionic Theory to the Greenhouse Effect:* Science History Publications, 1996.

Cronin, James W., ed. *Fermi Remembered*. University of Chicago Press, 2004.

Cushing, James T. *Quantum Mechanics: Historical Contingency and the Copenhagen Hegemony*. University of Chicago Press, 1994.

Darrigol, Olivier. "The Electrodynamic Origins of Relativity Theory." *Historical Studies in the Physical and Biological Sciences* 26 (1996): 241–312.

Darrigol, Olivier. *Electrodynamics from Ampere to Einstein*. Oxford University Press, 2000.

Darrigol, Olivier. *From c-Numbers to q-Numbers: The Classical Analogy in the History of Quantum Theory*. University of California Press, 1992.

Darrigol, Olivier. "Henri Poincaré's Criticism of Fin-de-siècle Electrodynamics." *Studies in History and Philosophy of Modern Physics* 26, no. 1 (1995): 1–44.

Darrigol, Olivier. "The Historians' Disagreement over the Meaning of Planck's Quantum." *Centaurus* 43 (2001): 219–239.

Darrigol, Olivier. "Turbulence in 19th-Century Hydrodynamics." *Historical Studies in the Physical and Biological Sciences* 32, no. 2 (2002): 207–262.

Debye, Peter. "Der Wahrscheinlichkeitsbegriff in der Theorie der Strahlung." *Annalen der Physik* 33 (1910): 1427–1434.

Debye, Peter. "Quantenhypothese und Zeeman-Effekt." *Physikalische Zeitschrift* 17 (1916): 507–512.

Debye, Peter, Otto Blumenthal, S. Bochner, and Eberhard Buchwald, eds. *Probleme der modernen Physik: Arnold Sommerfeld zum 60. Geburtstage gewidmet von seinen Schülern*. Hirzel, 1928.

Debye, Peter, and Arnold Sommerfeld. "Theorie des lichtelektrischen Effektes vom Standpunkt des Wirkungsquantums." *Annalen der Physik* 41 (1913): 873–930.

Deltete, Robert J. The Energetics Controversy in Late Nineteenth-Century Germany: Helm, Ostwald, and Their Critics. Dissertation, Yale University, 1983.

Deltete, Robert J. "Gibbs and the Energeticists." In *No Truth Except in the Details*, ed. A. Kox and D. Siegel. Kluwer, 1995.

Dirac, P. A. M. *The Principles of Quantum Mechanics*, fourth edition. Clarendon Press, 1981.

Eckert, Michael. "Arnold Sommerfeld: Theoretische Physik und Technik." Deutscher Wissenschaftshistorikertag, 1996.

Eckert, Michael. *Die Atomphysiker: Eine Geschichte der theoretischen Physik am Beispiel der Sommerfeldschule*. Vieweg, 1993.

Eckert, Michael. "Mathematics, Experiments, and Theoretical Physics: The Early Days of the Sommerfeld School." *Physics in Perspective* 1 (1999): 238–252.

Eckert, Michael. "Planck vorsichtig u. abstrakt, ich etwas draufgängerisch." *Kultur und Technik* 4 (1997): 38–45.

Eckert, Michael. "Sommerfeld in World War I: Military Research and Political Attitude." Conference talk delivered at EASST-4S-Meeting, University of Bielefeld, 1996.

Eckert, Michael, and Walther Kaiser. "An der Nahstelle von Theorie und Praxis: Arnold Sommerfeld und der Streit um die Wellenausbreitung in der drahtlosen Telegraphie." In *Chemie-Kultur-Geschichte*, ed. A. Schurmann and B. Weiss, GNT-Verlag, 2002.

Eckert, Michael, and Karl Märker, eds. *Arnold Sommerfeld: Wissenschaftlicher Briefwechsel*, volume 1. Deutsches Museum, Verlag für Geschichte der Naturwissenschaften und der Technik, 2000.

Eckert, Michael, and Karl Märker, eds. *Arnold Sommerfeld: Wissenschaftlicher Briefwechsel*, volume 2. Deutsches Museum, Verlag für Geschichte der Naturwissenschaften und der Technik, 2004.

Eckert, Michael, and Willibald Pricha. "Boltzmann, Sommerfeld und die Berufungen auf die Lehrstühle für theoretische Physik in München und Wien, 1890–1914." *Mitteilungen der Österreichischen Gesellschaft für Geschichte der Naturwissenschaften* 4 (1984): 101–119.

Eddington, A. S. *Report on the Relativity Theory of Gravitation*. Fleetway, 1918.

Eddington, A. S. *Space, Time, and Gravitation: An Outline of the General Relativity Theory*. Cambridge University Press, 1920.

Editor. "Zur Entstehung der Technischen Berichte der Flugzeugmeisterei." *Technische Berichte: Herausgegeben von der Flugzeugmeisterei der Inspektion der Fliegertruppen Charlottenburg* I (1917).

Ehrenfest, Paul. "Adiabatische Transformationen in der Quantentheorie und ihre Behandlung durch Niels Bohr." *Naturwissenschaften* 11 (1923): 543–550.

Ehrenfest, Paul. "A Mechanical Theorem of Boltzmann and its Relation to the Theory of Energy Quanta." *Proceedings of the Amsterdam Academy* 16 (1913): 591–597.

Ehrenfest, Paul. "Zur Planckschen Strahlungstheorie." *Physikalische Zeitschrift* 7 (1906): 528–532.

Ehrenfest, Paul. "Über die physikalischen Voraussetzungen der Planck'schen Theorie der irreversiblen Strahlungsvorgänge." *Wiener Berichte* 114 (1905): 1301–1314.

Einstein, Albert. "Bemerkungen zu der Notiz von Hrn. Paul Ehrenfest: Die Translation deformierbarer Elektronen und der Flächensatz." *Annalen der Physik* 23 (1907): 206–208.

Einstein, Albert. *The Collected Papers of Albert Einstein: English Translation*, volume 2. Princeton University Press, 1989.

Einstein, Albert. *The Collected Papers of Albert Einstein: English Translation*, volume 8. Princeton University Press, 1998.

Einstein, Albert. "Principles of Research." In *Essays in Science*. Philosophical Library, 1934.

Einstein, Albert. "Wahlvorschlag für A. Sommerfeld und P. Debye zur Aufnahme als korrespondierende Mitglieder in die Akademie d. Wiss." In *Albert Einstein in Berlin, 1913–1933*, ed. C. Kirsten and H. Treder. Akademie Verlag, 1979.

Einstein, Albert. "What Is the Theory of Relativity?" In *The Collected Papers of Albert Einstein: English Translation*. Princeton University Press, 2002.

Elton, Lewis. "Einstein, General Relativity, and the German Press, 1919–1920." *Isis* 77 (1986): 95–103.

Enz, Charles P. *No Time to Be Brief: A Scientific Biography of Wolfgang Pauli*. Oxford University Press, 2001.

Epstein, Paul Sophus. "Zur Quantentheorie." *Annalen der Physik* 51 (1916): 168–188.

Epstein, Paul Sophus. "Zur Theorie des Starkeffektes." *Annalen der Physik* 50 (1916): 489–520.

Ewald, P. P. "Arnold Sommerfeld als Mensch, Lehrer und Freund: Rede, gehalten zur Feier der 100sten Wiederkehr seiner Geburt." In *Physics of the One- and Two-Electron Atoms: Proceedings of the Arnold Sommerfeld Centennial Meeting and of the International Symposium on the Physics of the One- and Two-Electron Atoms, Munich, 10–14 September 1968*, ed. F. Bopp and H. Kleinpoppen. Wiley Interscience, 1969.

Ewald, P. P. "Erinnerungen an die Anfänge des Münchener Physikalischen Kolloquiums." *Physikalische Blätter* 24 (1968): 538–542.

Ewald, P. P. "The Setting for the Discovery of X-Ray Diffraction by Crystals. Speech Given at the First General Assembly of the International Union of Crystallography at Harvard University, 2 August 1948." DM: NL 089 (026), 1948.

Favrholdt, David. "Niels Bohr and Danish Philosophy." *Danish Yearbook of Philosophy* 13 (1976): 206–220.

Favrholdt, David. *Niels Bohr's Philosophical Background.* Copenhagen: Royal Danish Academy of Sciences and Letters, 1992.

Faye, Jan. "The Influence of Harald Høffding's Philosophy on Niels Bohr's Interpretation of Quantum Mechanics." *Danish Yearbook of Philosophy* 16 (1979): 37–72.

Faye, Jan. *Niels Bohr: His Heritage and Legacy.* Kluwer, 1991.

Feldman, Gerald D. *Army, Industry and Labor in Germany, 1914–1918.* Princeton University Press, 1966.

Feldman, Gerald D. *The Great Disorder: Politics, Economics, and Society in the German Inflation, 1914–1924.* Oxford University Press, 1993.

Fleming, John Ambrose. *An Elementary Manual of Radiotelegraphy and Radiotelephony for Students and Operators*, first edition. Longmans, Green, 1908.

Fleming, John Ambrose. *An Elementary Manual of Radiotelegraphy and Radiotelephony for Students and Operators*, third edition. Longmans, Green, 1916.

Forman, Paul. "Alfred Landé and the Anomalous Zeeman Effect, 1919–1921." *Historical Studies in the Physical Sciences* 2 (1970): 153–261.

Forman, Paul. "The Doublet Riddle and Atomic Physics circa 1924." *Isis* 59 (1968): 156–174.

Forman, Paul. "Review: Intellectual Mastery of Nature." *Philosophy of Science* 58 (1991): 129–132.

Forman, Paul. "Weimar Culture, Causality, and Quantum Theory, 1918–1927: Adaptation by German Physicists and Mathematicians to a Hostile Intellectual Environment." *Historical Studies in the Physical Sciences* 3 (1971): 1–115.

Forman, Paul, John L. Heilbron, and Spencer Weart. "Physics circa 1900: Personnel, Funding, and Productivity of the Academic Establishments." *Historical Studies in the Physical Sciences* 5 (1975): 1–175.

Foucault, Michel. *Discipline and Punish: The Birth of the Prison*, second edition. Vintage, 1995.

Foucault, Michel. *The History of Sexuality, Volume 1: An Introduction.* Allen Lane, 1979.

Friedman, Robert Marc. *The Politics of Excellence: Behind the Nobel Prize in Science.* Freeman, 2001.

Fruton, Joseph. *Contrasts in Scientific Style: Research Groups in the Chemical and Biochemical Sciences.* American Philosophical Society, 1990.

Fuchs, R., and L. Hopf. "Die allgemeine Längsbewegung des Flugzeuges." *Technische Berichte: Herausgegeben von der Flugzeugmeisterei der Inspektion der Fliegertruppen Charlottenburg* 3 (1918): 317–330.

Fuchs, Richard, and Ludwig Hopf. *Aerodynamik*. Carl Schmidt, 1922.

Galison, Peter. "Einstein's Clocks: The Place of Time." *Critical Inquiry* 26 (2000): 355–389.

Galison, Peter. "Minkowski's Space-Time: From Visual Thinking to the Absolute World." *Historical Studies in the Physical Sciences* 10 (1979): 85–121.

Galison, Peter, Michael Gordin, and David Kaiser, eds. *Science and Society: The History of Modern Physical Science in the Twentieth Century.* Routledge, 2001.

Galison, Peter Louis. *Image and Logic: A Material Culture of Microphysics.* University of Chicago Press, 1997.

Galison, Peter, and Andrew Warwick. "Introduction: Cultures of Theory." *Studies in History and Philosophy of Modern Physics* 29, no. 3 (1998): 287–294.

Gamow, George. *Thirty Years That Shook Physics: The Story of Quantum Theory.* Doubleday, 1966.

Garber, Elizabeth. *Language of Physics: The Calculus and the Development of Theoretical Physics in Europe, 1750–1914.* Birkhäuser, 1999.

Garber, Elizabeth. "Some Reactions to Planck's Law, 1900–1914." *Studies in History and Philosophy of Science* 7 (1976): 89–126.

Gay, Peter. *Weimar Culture: The Outsider as Insider.* Secker & Warburg, 1968.

Gearhart, Clayton A. "Planck, the Quantum, and the Historians." *Physics in Perspective* 4 (2002): 170–215.

Geison, Gerald. *Michael Foster and the Cambridge School of Physiology: The Scientific Enterprise in Late Victorian Society.* Princeton University Press, 1978.

Geison, Gerald. "Scientific Change, Emerging Specialties, and Research Schools." *History of Science* 19 (1981): 20–40.

Geison, Gerald, and Frederic L. Holmes, eds. *Research Schools: Historical Reappraisals*. University of Chicago Press, 1993.

Giedymin, Jerzy. "The Physics of the Principles and its Philosophy: Hamilton, Poincaré and Ramsey." In *Science and Convention: Essays on Henri Poincaré's Philosophy of Science and The Conventionalist Tradition*. Pergamon, 1982.

Gispen, Kees. *New Profession, Old Order: Engineers and German Society, 1815–1914*. Cambridge University Press, 1989.

Goldberg, Stanley. "Max Planck's Philosophy of Nature and His Elaboration of the Special Theory of Relativity." *Historical Studies in the Physical Sciences* 7 (1976): 125–160.

Goldstein, Jan. "Foucault Among the Sociologists: The 'Disciplines' and the History of the Professions." *History and Theory* 23, no. 2 (1984): 170–192.

Goodstein, Judith. "A Conversation with Hans Bethe." *Physics in Perspective* 1 (1999): 253–281.

Günther, Siegmund. "Die Meteorologie im Kriege." In *Deutsche Naturwissenschaft, Technik und Erfindung im Weltkriege*, ed. B. Schmid. Otto Nemnich, 1919.

Hakfoort, C. "Wilhelm Ostwald's Energeticist World-View and the History of Scientism." *Annals of Science* 49 (1992): 525–544.

Hartcup, G. *The War of Invention: Scientific Developments, 1914–1918*. Brassey's Defence Publishers, 1988.

Hasenöhrl, Fritz, ed. *Wissenschaftliche Abhandlungen von Ludwig Boltzmann*, volume 3, 1882–1905. Johann Ambrosius Barth, 1909.

Heilbron, John L. *The Dilemmas of an Upright Man: Max Planck as Spokesman for German Science*. University of California Press, 1986.

Heilbron, John L. "Fin-de-siècle Physics." In *Science, Technology and Society in the Time of Alfred Nobel*, ed. C. Bernhard, E. Crawford, and P. Sörbom. Pergamon, 1982.

Heilbron, John L. "Interview with Linus Pauling, 03/27/1964." AHQP, 1964.

Heilbron, John L. "Interview with Otto Klein, 25/02/1963." AHQP, 1963.

Heilbron, John L. "The Kossel-Sommerfeld Theory and the Ring Atom." *Isis* 58 (1967): 450–485.

Heilbron, John L. "The Origins of the Exclusion Principle." *Historical Studies in the Physical Sciences* 13 (1983): 261–310.

Heisenberg, Werner. "Die absoluten Dimensionen der Karmanschen Wirbelbewegung." *Physikalische Zeitschrift* 23 (1922): 363–366.

Heisenberg, Werner. *Encounters with Einstein and Other Essays on People, Places, and Particles*. Princeton University Press, 1989.

Heisenberg, Werner. "Über quantentheoretische Umdeutung kinematischer und mechanischer Beziehungen." *Zeitschrift für Physik* 33 (1925): 879–893.

Heisenberg, Werner. "Über Stabilität und Turbulenz von Flüssigkeitströmen." *Annalen der Physik* 74 (1924): 577–627.

Hendry, John. *The Creation of Quantum Mechanics and the Bohr-Pauli Dialogue.* Reidel, 1984.

Hentschel, Klaus, and Ann M. Hentschel, eds. *Physics and National Socialism: An Anthology of Primary Sources.* Birkhäuser, 1996.

Hermann, Armin. "Der Brückenschlag zwischen Mathematik und Technik." *Physikalische Blätter* 23 (1967): 442–449.

Hermann, Armin. "Die frühe Diskussion zwischen Stark und Sommerfeld über die Quantenhypothese." *Centaurus* 12 (1968): 38–59.

Hermann, Armin. *The Genesis of the Quantum Theory, 1899–1913.* MIT Press, 1971.

Hermann, Armin, Karl von Mayenn, and V. F. Weisskopf, eds. *Wolfgang Pauli: Wissenschaftlicher Briefwechsel mit Bohr, Einstein, Heisenberg, u. a.* Springer, 1979.

Hertz, Heinrich. "The Forces of Electric Oscillations, Treated According to Maxwell's Theory," in *Electric Waves: Being Researches on the Propagation of Electric Action with Finite Velocity Through Space.* Macmillan, 1893.

Hertz, Heinrich. *Untersuchungen über die Ausbreitung der elektrischen Kraft.* J. A. Barth, 1891.

Hiebert, Erwin N. "The Conception of Thermodynamics in the Scientific Thought of Mach and Planck." In *Wissenschaftlicher Bericht Nr. 5/68*: Ernst Mach Institut, 1968.

Hiebert, Erwin N. "The Energetics Controversy and the New Thermodynamics." In *Perspectives in the History of Science and Technology*, ed. D. Roller. University of Oklahoma Press, 1971.

Hindmarsh, W. R. *Atomic Spectra.* Pergamon, 1967.

Holton, Gerald. "Einstein and the Cultural Roots of Modern Science." *Daedalus* 127 (1998): 1–44.

Holton, Gerald. "Mach, Einstein, and the Search for Reality." In *Thematic Origins of Scientific Thought.* Harvard University Press, 1973.

Holton, Gerald. "The Roots of Complementarity." *Daedalus* 99 (1970): 1015–1055.

Holton, Gerald. *Thematic Origins of Scientific Thought.* Harvard University Press, 1973.

Hopf, Ludwig. *Hydrodynamische Untersuchungen: Turbulenz bei einem Flusse. Über Schiffswellen. Dissertation, München 1909.* Johann Ambrosius Barth, 1910.

Hull, Andrew. "War of Words: The Public Science of the British Scientific Community and the Origins of the Department of Scientific and Industrial Research, 1914–16." *British Journal for the History of Science* 32 (1999): 461–481.

Höffding, Harald. *The Problems of Philosophy.* Macmillan, 1905.

Jammer, Max. *The Conceptual Development of Quantum Mechanics.* McGraw-Hill, 1966.

Janssen, Michel, and Anthony Duncan. "On the Verge of Umdeutung: John van Vleck and the Correspondence Principle, Parts I & II." *Archive for History of the Exact Sciences* 61 (2007): 553–671.

Jeans, J. H. "Non-Newtonian Mechanical Systems, and Planck's Theory of Radiation." *Philosophical Magazine* 20 (1910): 943–954.

Jeans, J. H., A. S. Eddington, F. W. Dyson, A. Fowler, E. Cunningham, H. F. Newall, F. A. Lindemann, and L. Silberstein. "Discussion on the Theory of Relativity." *Proceedings of the Royal Society of London* 97 (1920): 66–79.

Johnson, Jeffrey A. *The Kaiser's Chemists: Science and Modernization in Imperial Germany.* University of North Carolina Press, 1990.

Jungnickel, Christa, and Russell McCormmach. *Intellectual Mastery of Nature: Theoretical Physics from Ohm to Einstein*, volume 1. University of Chicago Press, 1990.

Jungnickel, Christa, and Russell McCormmach. *Intellectual Mastery of Nature: Theoretical Physics from Ohm to Einstein*, volume 2. University of Chicago Press, 1990.

Jurkowitz, Edward. "Helmholtz and the Liberal Unification of Science." *Historical Studies in the Physical Sciences* 32 (2002): 291–317.

Jünger, Ernst. *The Storm of Steel: From the Diary of a German Storm-Troop Officer on the Western Front.* Howard Fertig, 1996.

Kaiser, David. *Drawing Theories Apart: The Dispersion of Feynman Diagrams in Postwar Physics.* University of Chicago Press, 2005.

Kaiser, David. "More Roots of Complementarity: Kantian Aspects and Influences." *Studies in History and Philosophy of Science* 23 (1992): 213–239.

Kaiser, David. "The Postwar Suburbanization of American Physics." *American Quarterly* 56, no. 4 (2004): 851–888.

Kangro, Hans. *Early History of Planck's Radiation Law.* Taylor and Francis, 1976.

Kangro, Hans. *Vorgeschichte des Planckschen Strahlungsgesetzes.* Steiner, 1970.

Katzir, Shaul. "On 'The Electromagnetic World-View': A Comment on an Article by Suman Seth." *Historical Studies in the Physical and Biological Sciences* 36 (2005): 189–192.

Kellermann, H. *Der Krieg der Geister.* Dresden, 1915.

Kevles, Daniel J. "George Ellery Hale, the First World War, and the Advancement of Science in America." *Isis* 59 (1968): 427–437.

Kevles, Daniel J. "Into Hostile Political Camps: The Reorganization of International Science after World War I." *Isis* 62 (1971): 47–60.

Kevles, Daniel J. *The Physicists: The History of a Scientific Community in Modern America.* Harvard University Press, 1995.

Kirkpatrick, Paul. "Address of Recommendation to Arnold Sommerfeld upon the award of the 1948 Oersted Medal for Notable Contributions to the Teaching of Physics." *American Journal of Physics* 12 (1949): 312–314.

Kirsten, Christa, and Hans-Günther Körber, eds. *Physiker über Physiker: Wahlvorschläge zur Aufnahme von Physikern in die Berliner Akademie 1870 bis 1929, von Hermann v. Helmholtz bis Erwin Schrödinger.* Akademie Verlag, 1975.

Klein, Felix. *Ausgewählte Kapitel der Zahlentheorie.* B. G. Teubner, 1896.

Klein, Felix. "On the Stability of the Sleeping Top." In *Felix Klein: Gesammelte Mathematische Abhandlungen*, ed. R. Fricke and H. Vermeil. Julius Springer, 1922.

Klein, Felix. "Über die Gründung eines physikalisch-technischen Universitätsinstituts in Göttingen." *Zeitschrift des Vereines deutscher Ingenieure* 40 (1896): 75–78.

Klein, Felix, and Arnold Sommerfeld. *Über die Theorie des Kreisels.* Johnson Reprint, 1965.

Klein, Felix, and Arnold Sommerfeld. *Über die Theorie des Kreisels, volume 1: Einführung in Kinematik und Kinetik des Kreisels (1897); Vol. 2: Durchführung der Theorie im Falle des Schweren symmetrischen Kreisels (1898); Vol. 3: Die störende Einflüsse, Astronomische und geophysikalische Anwendungen (1903); Vol. 4: Die technische Anwendungen der Kreiseltheorie (1910).* B. G. Teubner, 1897–1910.

Klein, Martin J. "Max Planck and the Beginnings of the Quantum Theory." *Archive for History of Exact Sciences* 1 (1962): 459–479.

Klein, Martin J. *Paul Ehrenfest: The Making of a Theoretical Physicist*, volume 1. North-Holland, 1970.

Klein, Martin J. "Planck, Entropy, and Quanta." *The Natural Philosopher* 1 (1963): 83–108.

Klein, Martin J. "Thermodynamics in Einstein's Thought." *Science* 157 (1967): 509–516.

Klein, Martin J., ed. *Paul Ehrenfest: Collected Scientific Papers.* North-Holland, 1959.

Koelzer, Dr. "Der Militärische Wetterdienst." In *Die Technik im Weltkriege*, ed. M. Schwarte. Mittler, 1920.

Koenigsberger, Leo. *Hermann von Helmholtz.* Clarendon, 1906.

Kragh, Helge. "The Fine Structure of Hydrogen and the Gross Structure of the Physics Community, 1916–26." *Historical Studies in the Physical Sciences* 15, no. 2 (1985): 67–125.

Kragh, Helge. "Magic Number: A Partial History of the Fine-Structure Constant." *Archive for History of Exact Sciences* 57 (2003): 395–431.

Kragh, Helge. "Niels Bohr's Second Atomic Theory." *Historical Studies in the Physical Sciences* 10 (1979): 123–186.

Kronig, R., and V. F. Weisskopf, eds. *Wolfgang Pauli: Collected Scientific Papers.* 2 vols, volume 2. Interscience, 1964.

Krüger, Lorenz, Lorraine J. Daston, and Michael Heidelberger, eds. *The Probabilistic Revolution*, volume 1: *Ideas in History*. MIT Press, 1987.

Kuhn, K. "Erinnerungen an die Vorlesungen von W. C. Röntgen und L. Grätz." *Physikalische Blätter* 18 (1962): 314–316.

Kuhn, Thomas S. *Black-Body Theory and the Quantum Discontinuity, 1894–1912*, second edition. University of Chicago Press, 1987.

Kuhn, Thomas S. *Black-Body Theory and the Quantum Discontinuity, 1894–1912*. Oxford University Press, 1978.

Kuhn, Thomas S. "The Essential Tension: Tradition and Innovation in Scientific Research." In *The Essential Tension: Selected Studies in Scientific Tradition and Change*. University of Chicago Press, 1977.

Kuhn, Thomas S. "Interview with G. Uhlenbeck, 03/30/1962." AHQP, 1962.

Kuhn, Thomas S. "Interview with Hans Bethe, 01/17/1964." AHQP, 1964.

Kuhn, Thomas S. "Interview with P. A. M. Dirac, 04/01/1962." AHQP, 1962.

Kuhn, Thomas S. "Interview with Paul Epstein, 05/25/1962." AHQP, 1962.

Kuhn, Thomas S. "Interview with Walther Friedrich, 05/15/1963." AHQP, 1963.

Kuhn, Thomas S. "Interview with Werner Heisenberg, 02/07/1963." AHQP, 1963.

Kuhn, Thomas S. "Interview with Werner Heisenberg, 02/11/1963." AHQP, 1963.

Kuhn, Thomas S. "Interview with Werner Heisenberg, 02/15/1963." AHQP, 1963.

Kuhn, Thomas S. "Interview with Werner Heisenberg, 02/19/1963." AHQP, 1963.

Kuhn, Thomas S. "Interview with Werner Heisenberg, 11/30/1962." AHQP, 1962.

Kuhn, Thomas S. *The Structure of Scientific Revolutions*, enlarged second edition. University of Chicago Press, 1970.

Kuhn, Thomas S., and F. Hund. "Interview with Max Born, 10/17/1962." AHQP, 1962.

Kuhn, Thomas S., and G. Uhlenbeck. "Interview with Peter Debye, 05/03/1962." AHQP, 1962.

Landé, A. "Termstruktur und Zeemaneffekt der Multipletts." *Zeitschrift der Physik* 15 (1923): 189–205.

Leegwater, Arie. "The Development of Wilhelm Ostwald's Chemical Energetics." *Centaurus* 29 (1986): 314–337.

Lenoir, Timothy. "The Discipline of Nature and the Nature of Disciplines." In *Instituting Science: The Cultural Production of Scientific Disciplines*. Stanford University Press, 1997.

Lorentz, H. A. "Alte und neue Fragen der Physik." *Physikalische Zeitschrift* 11 (1910): 1234–1257.

Lorentz, H. A. *Collected Papers*, volume 6. Martinus Nijhoff, 1938.

Lorentz, H. A. *Collected Papers*, volume 7. The Hague: Martinus Nijhoff, 1934.

Lorentz, H. A. "On the Radiation of Heat in a System of Bodies Having a Uniform Temperature." *Proceedings of the Amsterdam Academy* 7 (1905): 401.

Lorentz, H. A. "The Theory of Radiation and the Second Law of Thermodynamics." *Proceedings of the Amsterdam Academy* 3 (1901): 436.

Lorentz, H. A. "Zur Strahlungstheorie." *Physikalische Zeitschrift* 9 (1908): 562–563.

Lorentz, H. A. "Zur Theorie des Zeemaneffektes." *Physikalische Zeitschrift* 1 (1899): 39–41.

Lorentz, H. A. "Über den Einfluss magnetischer Kräfte auf die Emission des Lichtes." *Annalen der Physik* 63 (1897): 279–284.

Mach, Ernst. "Die Leitgedanken meiner naturwissenschaftlichen Erkenntnislehre und ihre Aufnahme durch die Zeitgenossen." *Scientia* 7 (1910): 225–240.

Mach, Ernst. *History and Root of the Principle of the Conservation of Energy*. Open Court, 1911.

Macleod, R., and E. K. Andrews. "Scientific Advice for the War at Sea, 1915–1917: The Board of Invention and Research." *Journal of Contemporary History* 6 (1971): 3–40.

Macleod, R., and K. Macleod. "War and Economic Development: Government and the Optical Industry in Britain, 1914–1918." In *War and Economic Development*, ed. J. Winter. Cambridge University Press, 1975.

Macleod, Roy. "The Chemists Go to War: The Mobilization of Civilian Chemists and the British War Effort, 1914–1918." *Annals of Science* 50 (1993): 455–481.

Macleod, Roy, and Jeffrey Allan Johnson, eds. *Frontline and Factory: Comparative Perspectives on the Chemical Industry at War, 1914–1924*. Springer, 2006.

Manegold, Karl-Heinz. *Universität, Technische Hochschule, und Industrie: Ein Beitrag zur Emanzipation der Technik im 19. Jahrhundert unter besonderer Berücksichtigung der Bestrebungen Felix Kleins*. Duncker und Humblot, 1970.

Mann, Thomas. "Gladius Dei." In *Tonio Kröger and Other Stories*. Bantam Books, 1970.

Massimi, Michela. *Pauli's Exclusion Principle: The Origin and Validation of a Scientific Principle*. Cambridge University Press, 2005.

McClelland, Charles E. *State, Society, and University in Germany, 1700–1914*. Cambridge University Press, 1980.

McCormmach, Russell. "H. A. Lorentz and the Electromagnetic View of Nature." *Isis* 61 (1970): 459–497.

McCormmach, Russell. "Henri Poincaré and the Quantum Theory." *Isis* 58 (1967): 37–55.

Mehra, Jagdish. *The Solvay Conferences on Physics: Aspects of the Development of Physics since 1911.* Reidel, 1975.

Mehra, Jagdish, and Helmut Rechenberg. *The Historical Development of Quantum Theory*, volume 1. Springer, 1982.

Moore, Walter. *Schrödinger: Life and Thought.* Cambridge University Press, 1992.

Morrell, J. B. "The Chemist Breeders: The Research Schools of Liebig and Thomas Thomson." *Ambix* 19 (1972): 1–46.

Morrow, John Howard. *German Air Power in World War I.* University of Nebraska Press, 1982.

Needell, Allan. Irreversibility and the Failure of Classical Dynamics: Max Planck's work on the Quantum Theory, 1900–1915. Dissertation, Yale University, 1980.

Nielsen, J. Rud, ed. *Niels Bohr Collected Works: The Correspondence Principle (1918–1923)*, ed. L. Rosenfeld, volume 3. North-Holland, 1976.

Nisio, Sigeko. "The Formation of the Sommerfeld Quantum Theory of 1916." *Japanese Studies in the History of Science* 12 (1973): 39–78.

Nisio, Sigeko. "X-Rays and Atomic Structure at the Early Stages of the Old Atomic Theory." *Japanese Studies in the History of Science* 8 (1969): 55–75.

Noether, Fritz. "Über analytische Berechnung der Geschosspendelungen." *Artilleristische Monatsheft* 149/50 (1919).

Noether, Fritz. "Über analytische Berechnung der Geschosspendelungen." *Nachrichten der K. Gesellschaft der Wissenschaften zu Göttingen (Mathematisch-physikalische Klasse)* (1919): 1–19.

Olesko, Kathryn Mary. The Emergence of Theoretical Physics in Germany: Franz Neumann and the Königsberg School of Physics, 1830–1890. Dissertation, Cornell University, 1980.

Olesko, Kathryn Mary. *Physics as a Calling: Discipline and Practice in the Königsberg Seminar for Physics, Cornell History of Science Series.* Cornell University Press, 1991.

Pais, Abraham. *Niels Bohr's Times: In Physics, Philosophy, and Polity.* Clarendon, 1991.

Pais, Abraham. *Subtle Is the Lord.: The Science and the Life of Albert Einstein.* Oxford University Press, 1982.

Pang, Alex Soojun-Kim. *Empire and the Sun: Victorian Solar Eclipse Expeditions.* Stanford University Press, 2002.

Pattison, Michael. "Scientists, Inventors and the Military in Britain, 1915–19: The Munitions Inventions Department." *Social Studies of Science* 13, no. 4 (1983): 521–568.

Pauli, Wolfgang. *Exclusion Principle and Quantum Mechanics: Lecture Given in Stockholm after the award of the Nobel Prize of Physics 1945.* Griffon, 1945.

Pauli, Wolfgang. "Über das Modell des Wasserstoffmolekülions." *Annalen der Physik* 68 (1922): 177–240.

Pauli, Wolfgang. "Über den Zusammenhang des Abschlusses der Elektronengruppen im Atom mit der Komplexstruktur der Spektren." *Zeitschrift der Physik* 31 (1925): 765–783.

Pauli, Wolfgang. "Über die Gesetzmässigkeiten des anomalen Zeemaneffektes." *Zeitschrift für Physik* 16 (1923): 155–164.

Peukert, Detlef. *The Weimar Republic.* Hill and Wang, 1993.

Pickering, Andrew, ed. *Science as Practice and Culture.* University of Chicago Press, 1992.

Planck, Max. "Absorption und Emission elektrischer Wellen durch Resonanz." *Wiedemanns Annalen* 57 (1896): 1–14.

Planck, Max. *Acht Vorlesungen über Theoretische Physik.* S. Hirzel, 1910.

Planck, Max. "Allgemeines zur neuren Entwicklung der Wärmetheorie." *Zeitschrift für physikalische Chemie* 8 (1891): 647–656.

Planck, Max. "Antrittsrede zur Aufnahme in die Akademie vom 28. Juni 1894." In *Physikalische Abhandlungen und Vorträge*, 1–5. Vieweg, 1958.

Planck, Max. "Bemerkung über die Emission von Spektrallinien." *Sitzungsberichte der Preussischen Akademie der Wissenschaften* (1915): 909–913.

Planck, Max. "Das chemische Gleichgewicht in verdünnten Lösungen." *Wiedemanns Annalen* 34 (1888): 139–154.

Planck, Max. "Das Einheit des physikalischen Weltbildes (Lecture, 9 December 1908, at the University of Leiden)." *Physikalische Zeitschrift* 10 (1909): 62–75.

Planck, Max. "Das Prinzip der kleinsten Wirkung." In *Der Kultur der Gegenwart.* J. A. Barth, 1914.

Planck, Max. "Die gegenwärtige Bedeutung der Quantenhypothese für kinetische Gastheorie." In *Vorträge über die kinetsiche Theorie der Materie und der Elektrizität*, ed. M. Planck and P. Debye. B. G. Teubner, 1914.

Planck, Max. "Die Gesetze der Wärmestrahlung und die Hypothese der elementaren Wirkungsquanten." In *Die Theorie der Strahlung und der Quanten: Verhandlungen auf einer von E. Solvay einberufenen Zusammenkunft (30 Oktober bis 3. November 1911)*, ed. A. Eucken. Verlag Chemie GmbH, 1913.

Planck, Max. "Die Kaufmannschen Messungen der Ablenkbarkeit der γ-Strahlen für die Dynamik der Elektronen." *Physikalische Zeitschrift* 7 (1906): 753–759.

Planck, Max. "Die Stellung der neueren Physik zur mechanischen Naturanschauung." *Physikalische Zeitschrift* 11 (1910): 922–932.

Planck, Max. *Dynamische und statistische Gesetzmässigkeit (Rede in der Berliner Universität am 3. 8. 1914)*. J. A. Barth, 1914.

Planck, Max. *Eight Lectures on Theoretical Physics*. Dover, 1998.

Planck, Max. "Erwiderung auf einen von Herrn Arrhenius erhobenen Einwand." *Zeitschrift für physikalische Chemie* 9 (1892): 636–637.

Planck, Max. "Gegen die neuere Energetik." *Annalen der Physik* 57 (1896): 72–78.

Planck, Max. *Gleichgewichtszustände isotroper Körper in verschiedenen Temperaturen*. Th. Ackerman, 1880.

Planck, Max. *Grundriss der allgemeinen Thermochemie*. Eduard Trewendt, 1893.

Planck, Max. "On Clausius Theorem for irreversible Cycles, and on the Increase of Entropy." *Philosophical Magazine* 9 (1905): 167–168.

Planck, Max. *Physikalische Abhandlungen und Vorträge*, volume 1. Vieweg, 1958.

Planck, Max. *Physikalische Abhandlungen und Vorträge*, volume 2. Vieweg, 1958.

Planck, Max. *Physikalische Abhandlungen und Vorträge*, volume 3. Vieweg, 1958.

Planck, Max. "Selbstdarstellung." In *Max Planck: Selbstdarstellung (1942)*, ed. Christa Kirsten. Akademie-Verlag, 1982.

Planck, Max. *The Theory of Heat Radiation*, second edition. Tomash and American Institute of Physics, 1988.

Planck, Max. *Treatise on Thermodynamics*, third edition. Dover, 1945.

Planck, Max. "Verdampfen, Schmelzen, und Sublimiren." *Wiedemanns Annalen* 13 (1882): 446–475.

Planck, Max. *Vorlesungen über die Theorie der Wärmestrahlung*. Johann Ambrosius Barth, 1906.

Planck, Max. *Vorlesungen über Thermodynamik*. Veit, 1897.

Planck, Max. "Wissenschaftliche Selbstbiographie." In *Max Planck: Physikalische Abhandlungen und Vorträge III*. Vieweg, 1958.

Planck, Max. *Wissenschaftliche Selbstbiographie*. J. A. Barth, 1948.

Planck, Max. "Zur Dynamik bewegter System." *Annalen der Physik* (4) 26 (1908): 1–34.

Planck, Max. "Zur Geschichte der Auffindung des physikalischen Wirkungsquantums." *Naturwissenschaften* 31 (1943): 153–159.

Planck, Max. "Zur Theorie des Gesetzes der Energieverteilung im Normalspektrum." *Verhandlungen der deutschen physikalischen Gesellschaft* 2 (1900): 237–245.

Planck, Max. "Zur Thermodynamik und Dissoziationstheorie binärer Elektrolyte." *Zeitschrift für physikalische Chemie* 41 (1902): 212–223.

Planck, Max. "Über das Gesetz der Energieverteilung im Normalspektrum." *Annalen der Physik* 4 (1901): 553–563.

Planck, Max. "Über das Prinzip der Vermehrung der Entropie. 3. Abhandlung." *Wiedemanns Annalen* 32 (1887): 462–503.

Planck, Max. "Über das Prinzip der Vermehrung der Entropie: 1. Abhandlung." *Wiedemanns Annalen* 30 (1887): 562–582.

Planck, Max. *Über den zweiten Hauptsatz der mechanischen Wärmetheorie.* Th. Ackermann, 1879.

Planck, Max. "Über die Grundlage der Lösungstheorie, eine Erwiderung." *Annalen der Physik* 10 (1903): 436–445.

Planck, Max. "Über die Verteilung der Energie zwischen Äther und Materie." *Annalen der Physik* 9 (1902): 629–641.

Planck, Max. "Über elektrische Schwingungen, welche durch Resonanz erregt und durch Strahlung gedämpft werden." *Wiedemanns Annalen* 60 (1897): 577–599.

Planck, Max. "Über irreversible Strahlungsvorgänge." *Annalen der Physik* 1 (1900): 69–122.

Planck, Max. "Über irreversible Strahlungsvorgänge. 1. Mitteilung." *Sitzungsberichte der Preussischen Akademie der Wissenschaften* (1897): 57–68.

Planck, Max. "Über irreversible Strahlungsvorgänge. 2. Mitteilung." *Sitzungsberichte der Preussischen Akademie der Wissenschaften* (1897): 715–717.

Planck, Max. "Über irreversible Strahlungsvorgänge. 3. Mitteilung." *Sitzungsberichte der Preussischen Akademie der Wissenschaften* (1897): 1122–1145.

Planck, Max. "Über irreversible Strahlungsvorgänge. 4. Mitteilung." *Sitzungsberichte der Preussischen Akademie der Wissenschaften* (1898): 449–476.

Planck, Max. "Über irreversible Strahlungsvorgänge. 5. Mitteilung." *Sitzungsberichte der Preussischen Akademie der Wissenschaften* (1899): 440–480.

Planck, Max, ed. *Abhandlungen zur Thermodynamik, von H. Helmholtz,* ed. W. Ostwald. W. Engelmann, 1902.

Pyenson, Lewis. *Neohumanism and the Persistence of Pure Mathematics in Wilhelmian Germany.* American Philosophical Society, 1983.

Pyenson, Lewis. "Physics in the Shadow of Mathematics: The Göttingen Electron-Theory Seminar of 1905." *Archive for History of Exact Sciences* 21 (1979): 55–89.

Pyenson, Lewis, and Douglas Skopp. "Educating Physicists in Germany circa 1900." *Social Studies of Science* 7 (1977): 329–366.

Quinn, Susan. *Marie Curie: A Life*: Perseus, 1996.

Rasch, Manfred. "Wissenschaft und Militär: Die Kaiser Wilhelm Stiftung für kriegstechnische Wissenschaft." *Militärgeschichtliche Mitteilungen* 49 (1991): 73–87.

Redaktion. "Vom Kriegsschauplatze Verbandsmitglieder im Felde." *Zeitschrift des Verbandes deutscher Diplom-Ingenieure* 5 (1914): 375.

Redaktion. "Übersicht über die Kriegsbeteiligung der Deutschen Physiker." *Physikalische Zeitschrift* 16 (1915): 142–145.

Riedler, Alois. *Die technsiche Hochschule und die Wissenschaftliche Forschung, Rektoratsrede*. Berlin, 1899.

Riedler, Alois. "Die Ziele der technischen Hochschule." *Zeitschrift des Vereines deutscher Ingenieure* 40 (1896): 301–09, 337–46, 374–5.

Riedler, Alois. *Unsere Hochschulen und die Anforderungen des 20 Jh*. A. Seydel, 1898.

Riedler, Alois. *Zur Frage der Ingenieur-Erziehung*. Leonhard Simion, 1895.

Ringer, Fritz K. *The Decline of the German Mandarins: The German Academic Community, 1890–1933*. Harvard University Press, 1969.

Root-Bernstein, Robert Scott. The Ionists: Founding Physical Chemistry, 1872–1890. Dissertation, Princeton University, 1980.

Routh, Edward John. *Die Dynamik der Systeme starrer Körper*. B. G. Teubner, 1898.

Rowe, David E. "Klein, Hilbert, and the Göttingen Mathematical Tradition." *Osiris* 5 (1989): 186–213.

Rubinowicz, Adalbert. "Bohrsche Frequenzbedingung und Erhaltung des Impulsmomentes I & II." *Physikalische Zeitschrift* 19 (1918): 441–445, 465–474.

Sauter, F., ed. *Arnold Sommerfeld: Gesammelte Schriften*, volume 1. Vieweg, 1968.

Sauter, F., ed. *Arnold Sommerfeld: Gesammelte Schriften*, volume 2. Vieweg, 1968.

Sauter, F., ed. *Arnold Sommerfeld: Gesammelte Schriften*, volume 3. Vieweg, 1968.

Sauter, F., ed. *Arnold Sommerfeld: Gesammelte Schriften*, volume 4. Vieweg, 1968.

Scheideler, Britta. "The Scientist as Moral Authority: Albert Einstein between Elitism and Democracy, 1914–1933." *Historical Studies in the Physical and Biological Sciences* 32 (2002): 319–46.

Schirrmacher, Arne. "Experimenting Theory: The Proofs of Kirchhoff's Radiation Law Before and After Planck." *Historical Studies in the Physical and Biological Sciences* 33, no. 2 (2003): 299–335.

Schlayer, Karl. "Über die Stabilität der Karman'schen Wirbelstrasse gegenüber beliebigen Störungen in drei Dimensionen." Munich University, 1927.

Schröder, Wilfried. "Arnold Sommerfeld und Emil Wiechert." *Archive for History of Exact Sciences* 32 (1985): 77–93.

Schubring, Gert. "Pure and Applied Mathematics in Divergent Institutional Settings in Germany: The Role and Impact of Felix Klein." *The History of Modern Mathematics* 1 (1989): 171–220.

Schwarzschild, K. "Bemerkungen zur Aufspaltung der Spektrallinien im elektrischen Feld." *Verhandlungen der deutschen physikalischen Gesellschaft* 16 (1914): 20–24.

Schwarzschild, K. "Zur Quantenhypothese." *Berliner Berichte* (1916): 548–568.

Schweber, S. S. "The Empiricist Temper Regnant: Theoretical Physics in the United States, 1920–1950." *Historical Studies in the Physical Sciences* 17, no. 1 (1986): 55–98.

Schweber, Silvan S. *QED and the Men Who Made It: Dyson, Feynman, Schwinger, and Tomonaga.* Princeton University Press, 1994.

Serwer, Daniel. "*Unmechanischer Zwang*: Pauli, Heisenberg, and the Rejection of the Mechanical Atom, 1923–25." *Historical Studies in the Physical Sciences* 8 (1977): 189–256.

Seth, Suman. "Allgemeine Physik? Max Planck und die Gemeinschaft der theoretischen Physik." In *Der Hochsitz des Wissens: Das Allgemeine als wissenschaftlicher Wert*, ed. M. Hagner and M. Laublichler. Diaphenes, 2006.

Seth, Suman. "Crisis and the Construction of Modern Theoretical Physics." *British Journal for the History of Science* 40 (2007): 25–51.

Seth, Suman. "Quantum Theory and the Electromagnetic World-View." *Historical Studies in the Physical and Biological Sciences* 35, no. 1 (2004): 67–93.

Seth, Suman. "Response to Shaul Katzir: 'On the Electromagnetic World-View.'" *Historical Studies in the Physical and Biological Sciences* 36 (2005): 193–196.

Sigurdsson, Skuli. Hermann Weyl, Mathematics and Physics, 1900–1927. Dissertation, Harvard University, 1991.

Sommerfeld, Arnold. "Allgemeine spektroskopische Gesetze, insbesondere ein magnetooptischer Zerlegungssatz." *Annalen der Physik* 63 (1920): 221–263.

Sommerfeld, Arnold. "Atombau und Röntgenspektren." *Physikalische Zeitschrift* 19 (1918): 367–372.

Sommerfeld, Arnold. *Atombau und Spektrallinien*, first edition. Vieweg, 1919.

Sommerfeld, Arnold. *Atombau und Spektrallinien*, second edition. Vieweg, 1921.

Sommerfeld, Arnold. *Atombau und Spektrallinien*, third edition. Vieweg, 1922.

Sommerfeld, Arnold. *Atombau und Spektrallinien*, fourth edition. Vieweg, 1924.

Sommerfeld, Arnold. *Atombau und Spektrallinien: Wellenmechanischer Ergänzungsband*, first edition. Vieweg, 1929.

Sommerfeld, Arnold. *Atomic Structure and Spectral Lines.* Methuen, 1923.

Sommerfeld, Arnold. "Autobiographische Skizze." In *Gesammelte Schriften*, ed. F. Sauter. Vieweg, 1968.

Sommerfeld, Arnold. "Bemerkungen zur Feinstruktur der Röntgenspektren I." *Zeitschrift für Physik* 1 (1920): 135–146.

Sommerfeld, Arnold. "Bemerkungen zur Feinstruktur der Röntgenspektren II." *Zeitschrift für Physik* 5 (1921): 1–16.

Sommerfeld, Arnold. "Das Institut für Theoretische Physik." In *Die Wissenschaftlichen Anstalten der Ludwig-Maximilians-Universität zu München: Chronik zur Jahrhundertfeier im Auftrag des Akademischen Senats*, ed. K. von Müller. R. Oldenbourg und Dr. C. Wolf & Sohn, 1926.

Sommerfeld, Arnold. "Das Pendeln parallel geschalteter Wechselstrommaschinen (1904)." In *Arnold Sommerfeld: Gesammelte Schriften*, ed. F. Sauter. Vieweg, 1968.

Sommerfeld, Arnold. "Das Plancksche Wirkungsquantum und seine allgemeine Bedeutung für die Molekularphysik." *Verhandlungen der Gesellschaft Deutscher Naturforscher und Ärzte* 83 (1912): 31–49.

Sommerfeld, Arnold. "Der Zeemaneffekt eines anisotrop gebundenen Elektrons und die Beobachtungen von Paschen-Back." *Annalen der Physik* 40 (1913): 748–774.

Sommerfeld, Arnold. "Die allgemeine Dispersionsformel nach dem Bohrschen Modell." *Arbeiten aus den Gebieten der Physik, Mathematik, Chemie* (1915): 549–584.

Sommerfeld, Arnold. "Die Bedeutung der Röntgenstrahlung für die heutige Physik." *Verhandlungen der Bayerischen Akademie der Wissenschaften, München* (1925): 1–17.

Sommerfeld, Arnold. "Die Bedeutung des Wirkungsquantums für unperiodische Molekularprozesse in der Physik." In *Die Theorie der Strahlung und der Quanten: Verhandlungen auf einer von E. Solvay einberufenen Zusammenkunft (30 Oktober bis 3. November 1911)*, ed. A. Eucken. Verlag Chemie, 1913.

Sommerfeld, Arnold. "Die Feinstruktur der wasserstoff- und wasserstoffähnlichen Linien." *Sitzungsberichte der Bayerischen Akademie* (1915): 459–500.

Sommerfeld, Arnold. "Die naturwissenschaftlichen Ergebnissen und die Ziele der modernen technischen Mechanik." *Physikalische Zeitschrift* 4 (1903): 773–782.

Sommerfeld, Arnold. "The Differences Between American and German Universities" (undated, untitled manuscript, presumably written 1922/3). DM: NL 089 (026).

Sommerfeld, Arnold. "Ein Beitrag zur hydrodynamischen Erklärung der turbulenten Flüssigkeitsbewegung." In *Arnold Sommerfeld: Gesammelte Schriften*, ed. F. Sauter. Vieweg, 1968.

Sommerfeld, Arnold. "Ein Zahlenmysterium in der Theorie des Zeeman-Effektes." *Die Naturwissenschaften* 8 (1920): 61–64.

Sommerfeld, Arnold. "Grundlagen der Quantentheorie und des Bohr'schen Atommodelles." *Die Naturwissenschaften* 12 (1924): 1047–1049.

Sommerfeld, Arnold. "Lectures, Maxwell'sche Theorie u. Elektronenth." DM: NL 089 (028), Winter 1906/07 & Winter 1908/09.

Sommerfeld, Arnold. "Lectures, Theorie der Strahlung." DM: NL 089 (026), Sommer 1907.

Sommerfeld, Arnold. "Lectures, Wärmeleitung, Diffusion u. Elektrizitätsleitung nebst ihren molekular- und elektronentheoret. Zusammenhangen." DM: NL (089) 028, Sommer 1908.

Sommerfeld, Arnold. 2004. *Mathematical Theory of Diffraction. Progress in Mathematical Physics*, volume 35. Birkhäuser.

Sommerfeld, Arnold. "Mathematische Theorie der Diffraction." *Mathematische Annalen* 47 (1896): 317–374.

Sommerfeld, Arnold. 1952. *Mechanics*, volume 1. Academic.

Sommerfeld, Arnold. 1950. *Mechanics of Deformable Bodies*, volume 2. Academic.

Sommerfeld, Arnold. *Mechanik*. Leipzig, 1943.

Sommerfeld, Arnold. "Probleme der freien Weglänge." *Mathematische Vorlesungen an der Universität Göttingen* 6 (1914): 125–165.

Sommerfeld, Arnold. "Quantentheoretische Umdeutung der Voigtschen Theorie des anomalen Zeemaneffektes vom D-Linientypus." *Zeitschrift für Physik* 8 (1922): 257–272.

Sommerfeld, Arnold. "Report to the Faculty on the Dissertation of Ernst Guillemin." Munich: University Archive, OC-I-52p, 1926.

Sommerfeld, Arnold. "Report to the Faculty on the Dissertation of Gregor Wentzel." Munich: OC-I-47p, 1921.

Sommerfeld, Arnold. "Report to the Faculty on the Dissertation of Hans Pfrang." Munich: University Archive OC-I-51p, 1925.

Sommerfeld, Arnold. "Report to the Faculty on the Dissertation of Hermann von Hoerschelmann." Munich: University Archive OC I 37p, January 1911.

Sommerfeld, Arnold. "Report to the Faculty on the Dissertation of Hermann W. March." Munich: University Archive, OC I 37p, June 13 1911.

Sommerfeld, Arnold. "Report to the Faculty on the Dissertation of Josef Krönert." Munich: University Archive, OC-I-46p, 1920.

Sommerfeld, Arnold. "Report to the Faculty on the Dissertation of Karl Bechert." Munich: University Archive, OC-I-51p, 1925.

Sommerfeld, Arnold. "Report to the Faculty on the Dissertation of Karl Schlayer." Munich: University Archive, OC-Np-WS 1926/27, 1927.

Sommerfeld, Arnold. "Report to the Faculty on the Dissertation of Ludwig Hopf." Munich: University Archive, OC I, July 5 1909.

Sommerfeld, Arnold. "Report to the Faculty on the Dissertation of Wolfgang Pauli." Munich: University Archive, OC-I-47p, 1921.

Sommerfeld, Arnold. "Schwebende Fragen der Atomphysik." *Physikalische Zeitschrift* 21 (1920): 619–620.

Sommerfeld, Arnold. "The Scientific Results and Aims of Modern Applied Mechanics." *Mathematical Gazette* (1903).

Sommerfeld, Arnold. "Some Reminiscences of my Teaching Career." *American Journal of Physics* 12 (1949): 315–316.

Sommerfeld, Arnold. "Theoretisches über die Beugung der Röntgenstrahlen." *Zeitschrift für Mathematik und Physik* 46 (1901): 11–97.

Sommerfeld, Arnold. "undated, untitled manuscript [presumably written in Königsberg, 1890]." 32 sheets, 63 pages, pages 5 and 6 missing.: DM: NL 089 (026).

Sommerfeld, Arnold. "Untitled, undated paper [presumably written in Königsberg in 1890]." DM: NL 089 (026).

Sommerfeld, Arnold. "Zu Röntgens siebzigstem Geburtstag." *Deutsche Revue*, April (1915): 85–92.

Sommerfeld, Arnold. "Zur Elektronentheorie 2. Grundlagen für eine allgemeine Dynamik des Elektrons." *Nachrichten der K. Gesellschaft der Wissenschaften zu Göttingen (Mathematisch-physikalische Klasse)* (1904): 363–439.

Sommerfeld, Arnold. "Zur Frage nach der Bedeutung der Atommodelle." *Zeitschrift für Elektrochemie und angewandte physikalische Chemie* 34 (1928): 426–430.

Sommerfeld, Arnold. "Zur hydrodynamischen Theorie der Schmiermittelreibung (1904)." In *Arnold Sommerfeld: Gesammelte Schriften*, ed. F. Sauter. Vieweg, 1968.

Sommerfeld, Arnold. "Zur mathematischen Theorie der Beugungserscheinungen." *Nachrichten der K. Gesellschaft der Wissenschaften zu Göttingen (Mathematisch-physikalische Klasse)* (1894): 338–342.

Sommerfeld, Arnold. "Zur Quantentheorie der Spektrallinien." *Annalen der Physik* 51 (1915): 1–94, 125–167.

Sommerfeld, Arnold. "Zur Quantentheorie der Spektrallinien, Ergänzungen und Erweiterungen." *Sitzungsberichte der Bayerischen Akademie* (1916): 131–182.

Sommerfeld, Arnold. "Zur Quantentheorie der Spektrallinien. Intensitätsfragen." *Sitzungsberichte der Bayerischen Akademie* (1917): 83–109.

Sommerfeld, Arnold. "Zur Theorie der Balmerschen Serie." *Sitzungsberichte der Bayerischen Akademie* (1915): 425–458.

Sommerfeld, Arnold. "Zur Theorie der Eisenbahnbremsen (1902)." In *Arnold Sommerfeld: Gesammelte Schriften*, ed. F. Sauter. Vieweg, 1968.

Sommerfeld, Arnold. "Zur Theorie der Multipletts und ihrer Zeeman-Effekt." *Annalen der Physik* 73 (1924): 209–227.

Sommerfeld, Arnold. "Zur Theorie des periodischen Systems." *Physikalische Zeitschrift* 26 (1925): 70–74.

Sommerfeld, Arnold. "Zur Theorie des Zeeman-Effekts der Wasserstofflinien mit einem Anhang über den Stark-Effekt." *Physikalische Zeitschrift* 17 (1916): 491–507.

Sommerfeld, Arnold. "Zur Voigtschen Theorie des Zeeman-Effektes." *Nachrichten der K. Gesellschaft der Wissenschaften zu Göttingen (Mathematisch-physikalische Klasse)* (1914): 207–229.

Sommerfeld, Arnold. "Über die Ausbreitung elektrischer Wellen in der drahtlosen Telegraphie." In *Arnold Sommerfeld: Gesammelte Schriften*, ed. F. Sauter. Vieweg, 1968.

Sommerfeld, Arnold. "Über die Feinstruktur der K-Beta Linie." *Sitzungsberichte der Bayerischen Akademie* (1918): 367–272.

Sommerfeld, Arnold. "Über die Mechanik der Elektronen." *Verhandlungen des III. Internationalen Mathematiker-Kongresses* (1904): 417–432.

Sommerfeld, Arnold. "Über die Struktur der γ-Strahlen." *Sitzungsberichte der Mathemat.-physikal. Klasse der Kgl.-Bayerischen Akademie der Wissenschaft zu München, München* (1911): 1–60.

Sommerfeld, Arnold. "Über die Verteilung der Intensität bei der Emission der Röntgenstrahlen." *Physikalische Zeitschrift* 10 (1909): 969–976.

Sommerfeld, Arnold. "Über kosmische Strahlung." *Süddeutsche Monatshefte* 24 (1927): 195–198.

Sommerfeld, Arnold, ed. *Die Enzyklopädie der mathematischen Wissenschaften mit Einschluss ihrer Anwendungen*. Teubner, 1904–1926.

Sommerfeld, Arnold. 1956. *Thermodynamics and Statistical Mechanics*, Academic.

Sommerfeld, Arnold, and Werner Heisenberg. "Die Intensität der Mehrfachlinien und ihrer Zeemankomponenten." *Zeitschrift für Physik* 11 (1922): 131–154.

Sommerfeld, Arnold, and W. Kossel. "Auswahlprinzip und Verschiebungssatz bei Serienspektren." *Verhandlungen der deutschen physikalischen Gesellschaft* 21 (1919): 240–259.

Sommerfeld, Arnold, and Gregor Wentzel. "Über reguläre und irreguläre Dubletts." *Zeitschrift für Physik* 7 (1921): 86–92.

Sponsel, Alistair. "Constructing a 'Revolution in Science': The Campaign to Promote a Favourable Reception for the 1919 Solar Eclipse Experiments." *British Journal for the History of Science* 35 (2002): 439–467.

Stachel, John. *Einstein from "B" to "Z"*. Birkhäuser, 2002.

Staley, Richard. "On the Co-Creation of Classical and Modern Physics." *Isis* 96 (2005): 530–558.

Stanley, Matthew. *Practical Mystic: Religion and Science in the Life of A. S. Eddington.* Dissertation, Harvard University, 2004.

Stanley, Matthew. *Practical Mystic: Religion, Science, and A. S. Eddington.* University of Chicago Press, 2007.

Stark, Johannes. *Die gegenwärtige Krisis in der deutschen Physik.* Johann Ambrosius Barth, 1922.

Stuewer, Roger H. *The Compton Effect: Turning Point in Physics.* Science History Publications, 1975.

Tobies, Renate. "On the Contribution of Mathematical Societies to promoting Applications of Mathematics in Germany." *The History of Modern Mathematics* I (1989): 223–248.

Toulmin, Stephen, ed. *Physical Reality: Philosophical Essays on Twentieth-Century Physics.* Harper and Row, 1970.

Traweek, Sharon. *Beamtimes and Lifetimes: The World of High Energy Physicists.* Harvard University Press, 1988.

Van 't Hoff, J. H. "Die Rolle des osmotischen Druckes in der Analogie zwischen Lösungen und Gasen." *Zeitschrift für physikalische Chemie* 1 (1887): 481–508.

Van 't Hoff, J. H. 1899. *Lectures on Theoretical and Physical Chemistry: Chemical Statics*, volume 2. Edward Arnold.

van der Waerden, B. L. "Exclusion Principle and Spin." In *Theoretical Physics in the Twentieth Century: A Memorial Volume to Wolfgang Pauli*, ed. M. Fierz and V. Weisskopf. Interscience, 1960.

van der Waerden, B. L., ed. *Sources of Quantum Mechanics.* Dover, 1968.

Voigt, Woldemar. "Die anomalen Zeemaneffekte der Spektrallinien vom D-Typus." *Annalen der Physik* 42 (1913): 210–230.

Voigt, Woldemar. "Weiteres zum Ausbau der Koppelungstheorie der Zeemaneffekte." *Annalen der Physik* 41 (1913): 403–440.

"Vorträge und Diskussionen von der 86. Naturforscherversammlung in Nauheim vom 19–25. September 1920." *Physikalische Zeitschrift* 21, no. 23/24 (1920): 649–675.

Warwick, Andrew. *Masters of Theory: Cambridge and the Rise of Mathematical Physics.* University of Chicago Press, 2003.

Warwick, Andrew, and David Kaiser. "Kuhn, Foucault and the Power of Pedagogy." In *Pedagogy and the Practice of Science: Historical and Contemporary Perspectives*, ed. D. Kaiser. MIT Press, 2005.

Wentzel, Gregor. "Fortschritte der Atom- und Spektraltheorie." *Ergebnisse der Exakten Naturwissenschaften* 1 (1922): 299–314.

Wentzel, Gregor. "Zur Systematik der Röntgenspektren." *Zeitschrift für Physik* 6 (1921): 84–99.

Weyl, Hermann. "Die Relativitätstheorie auf der Naturforscherversammlung in Bad Nauheim." *Jahresbericht der Deutschen Mathematiker-Vereinigung* 31 (1922): 51–63.

Wheaton, Bruce R. *The Tiger and the Shark: Empirical Roots of Wave-Particle Dualism*. Cambridge University Press, 1983.

Wien, Wilhelm. *Aus dem Leben und Wirken eines Physikers*. Johann Ambrosius Barth, 1930.

Wien, Wilhelm. "Ueber die Energie der Kathodenstrahlen im Verhältnis zur Energie der Röntgen- und Sekundär-strahlen." In *Festschrift Adolph Wüllner gewidmet zum 70. Geburtstage, 13 Juni 1905*. Teubner, 1905.

Wien, Wilhelm. *Vergangenheit, Gegenwart und Zukunft der Physik: Rede Gehalten beim Stiftungsfest der Universität München am 19. Juni 1926*. Max Hueber, 1926.

Wien, Wilhelm. "Ziele und Methoden der theoretischen Physik." *Jahrbuch der Radioaktivität und Elektronik* 12 (1915): 241–259.

Wien, Wilhelm. "Über die Berechnung der Impulsbreite der Röntgenstrahlen aus ihrer Energie." *Annalen der Physik* 22 (1907): 793–797.

Wien, Wilhelm. "Über eine Berechnung der Wellenlänge der Röntgenstrahlen aus dem Planckschen Energie Element." *Göttingen Nachrichten* (1907): 598–601.

Wise, M. Norton. "How Do Sums Count? On the Cultural Origins of Statistical Causality." In *The Probabilistic Revolution: Ideas in History*, ed. L. Krüger, L. Daston, and M. Heidelberger. MIT Press, 1989.

Wohl, Robert. *The Generation of 1914*. Harvard University Press, 1979.

Wolff, Stefan L. "Physicists in the 'Krieg der Geister': Wilhelm Wien's Proclamation." *Historical Studies in the Physical and Biological Sciences* 33, no. 2 (2003): 337–368.

Yeang, Chen-Pang. "The Study of Long-Distance Radio-Wave Propagation, 1900–1919." *Historical Studies in the Physical and Biological Sciences* 33 (2003): 369–403.

Ziegler, Charles A. "Weapons Development in Context: The Case of the World War I Balloon Bomber." *Technology and Culture* 35 (1994): 750–767.

Index

Abraham, Max, 31, 170
Arrhenius, Svante, 98, 99, 105, 106, 110–112
Atombau und Spektrallinien
　aesthetics and, 202
　as "bible" of quantum spectroscopy, 178
　correspondence principle in, 228, 234–238, 254
　electrodynamics in, 155, 156, 227–230
　Ellipsenverein in, 218
　experimental data in, 173
　models and *Gesetzmässigkeiten* in, 214, 221–224, 238, 254, 257
　pedagogy and, 241
　quantum numbers in, 259, 260
　reception of, 204, 205
　relativity in, 170, 260
　structure of, 227–229
　World War I and, 90
Atomic spectra, 162–165, 168–171, 205–213

Bethe, Hans, 9, 60, 62, 182, 264, 267
Black-body theory
　and electromagnetic worldview, 39–43, 145
　Planck's development of, 8, 95, 112, 114–131, 145, 157
　Sommerfeld and, 14, 30, 35–39
Bohr model, 9, 155, 162–169, 214–223, 251–256
Bohr, Niels
　and Born, 60, 61

and correspondence principle, 233–235, 238–240
and planetary atomic model, 162
and principles, 2, 185, 194–198
second atomic theory of, 251–254, 259
and Sommerfeld, 9, 183, 185, 219, 225–227, 232–236, 239–241
and Sommerfeld School, 143
on Sommerfeld's work, 170, 171
Boltzmann, Ludwig
　at Munich, 49, 50
　and Ostwald, 144
　and Planck's black-body derivation, 35, 36, 116, 119–126, 132, 136
Born, Max
　pedagogy of, 58, 61
　on Planck, 183
　on Sommerfeld, 4, 184, 203–205, 223
　on Sommerfeld School, 3, 48
　as "virtuoso," 186
　and World War I, 73–76, 179

Cahan, David, 5
Cambridge Tripos, 56, 65–68, 191, 193
Cantor, Matthias, 112, 113
"Classical" physics, 1, 139, 155, 157
Compton, Arthur Holly, 236
Compton effect, 236, 237
Constructive theories, 184, 187, 190–193
Correspondence principle, 9, 196–198, 221, 225–227, 232–241, 256

Coster, Dirk, 258, 259
Craft of the quantum, 6, 9, 201–204, 212, 225
Creativity, 7, 66–70
Crisis, 265–269

Darrigol, Olivier, 226, 227
Debye, Peter
  on quantum theory, 141, 217
  and Sommerfeld School, 2, 48, 58, 60, 62
  as "virtuoso," 186
De Sitter, Willem, 191
Dirac, Paul, 193, 194, 265
Disciplining, 65–69
Dynamical method
  and quantum, 8, 142, 143
  of Sommerfeld, 142, 143, 149–156, 161, 162, 165, 171, 173
Dyson, Frank, 184, 190

Eddington, Arthur Stanley, 184, 190–193
Ehrenfest, Paul, 35, 36, 166, 185–187, 196
Einstein, Albert
  and Mach, 188, 189
  and observability, 262, 263
  and politics, 90, 178, 180
  on principles, 2, 184–190
  and relativity, 90, 184, 185, 189–191
  on Sommerfeld, 79, 80
  on Sommerfeld School, 3, 48, 242
Electromagnetic worldview
  and method, 32, 37, 38, 44–46, 145
  and quantum world, 154–156, 229–231, 234, 235
  and relativity, 31
  and resistance to quantum, 30–32, 37–43, 145
  Sommerfeld's adherence to, 3, 6, 14, 30–33, 37, 38, 42, 44, 141, 142, 145, 151, 154–156
  and Sommerfeld School, 33, 44, 156, 242
Electron theory
  and electromagnetic worldview, 31–33, 38–43, 142, 145
  and impulse theory, 146, 147

and quantum theory, 37–43
Sommerfeld and, 17, 31–33, 38, 40, 42, 142, 145, 147, 156, 163, 164
Ellipsenverein, 217, 218, 221
Energeticists, 97, 143–145
Engineering
  in Germany, 20, 21
  and mathematics, 21–25
  and Sommerfeld's physics, 2, 5, 6, 14, 19, 22–25, 201
  and Sommerfeld School, 28, 52
  technische Hochschule and, 5, 19–21
Epstein, Paul, 61, 171
Ewald, Paul
  on Sommerfeld's pedagogy, 17, 47, 49, 62, 63
  on Sommerfeld's physics, 46
  and World War I, 75, 76, 85
Exclusion principle, 247–251, 256, 257

Forman, Paul, 212, 225, 260, 265, 267
Foucault, Michel, 48, 64–69
Fuchs, R., 87–89

Garber, Elizabeth, 145
Gesetzmässigkeiten (lawful regularities)
  and Pauli exclusion principle, 249, 250, 256, 257
  reception of, 223–225
  Sommerfeld's adoption of, 9, 204, 205, 212–214, 219, 223, 263
  Sommerfeld's search for, 4, 207–209
  and spectroscopic laws, 209–211
Gibbs, Josiah Willard, 108
Glitscher, Karl, 170, 242
Graetz, Leo, 50, 51
Gyroscopic motion, 6, 25–27, 79, 84

Haber, Fritz, 73–77
Heilbron, John, 266
Heisenberg, Werner
  on atomic spectra, 250, 263
  experimental incompetence of, 182
  and quantum mechanics, 262–264

and Sommerfeld school, 9, 51, 52, 57–61,
  182, 183, 241, 242, 244, 248
 and *Umdeutung*, 248, 262–264
Helmholtz, Hermann von, 95, 96, 106, 115,
  137
Hermann, Armin, 171
Hertz, Heinrich, 114, 115, 118, 119, 137
Hoerschelmann, Hermann von, 82
Høffding, Harald, 197–200
Hondros, Demetrios, 49
Hopf, Ludwig
 on aircraft design, 85–89
 and hydrodynamics (turbulence), 18, 28, 29,
  45, 49, 85–89
 and Sommerfeld School, 28, 29, 49, 87, 89
 and World War I, 7, 71, 72, 85–89, 90
Hüter, Wilhelm, 75, 83

Ideal processes, 7, 8, 98, 99, 107–113,
  132–136, 173

Jeans, James, 14, 36, 43, 193

Kaiser, David, 64–66
Kaiser's physicists, 7, 72, 73, 84
Kaufmann, Walter, 31
Kirchhoff, Gustav, 95, 115, 137, 145, 146
Klein, Felix, 16, 19, 24–26, 56, 57
Kohlrausch, Friedrich, 5
Kossel, Walter, 211, 214
Kragh, Helge, 237
Kriegsphysik, 77, 86, 90
Kroo, Jan, 217, 218
Kuhn, Thomas
 on black-body theory, 122, 145
 on crises and revolutions, 10, 265–269
 on pedagogy, 64, 65
 on quantum discontinuity, 30, 39, 40, 43

Ladenburg, Rudolf, 75, 76
Landé, Alfred, 221, 224, 239, 244, 249, 250,
  260
Langevin, Paul, 8, 142, 150, 158, 165

Lenard, Phillip, 180
Lenz, Wilhelm, 7, 83, 85
Lorentz, Hendrik Antoon
 and electromagnetic worldview, 31, 38–42
 and electron theory, 17, 31, 37, 38, 39
 and quantum discontinuity, 30, 31, 39–42
 on Zeeman effect, 163, 164
Lummer, Otto, 39

Mach, Ernst, 97, 133, 134, 143, 188, 189, 198
Matrix mechanics, 264, 265
Maxwell, James Clerk, 41, 66, 143, 159, 160
McCormmach, Russell, 32
Mechanical worldview, 32, 97, 134
Modellmässig (model-based) approach
 Bohr and, 162–165, 239, 251–254
 Heisenberg and, 10, 250, 264
 Pauli and, 9, 223, 250, 251, 256, 258, 261
 Sommerfeld and, 9, 163–173, 204–223, 227,
  228, 235, 238–240, 244, 254, 245
"Modern" physics, 1, 139, 155

Natural radiation, 114, 121–126,
Nernst, Walther, 79, 106, 139, 141
Neumann, Franz, 15, 69
Noether, Fritz, 7, 84

Olesko, Kathyn, 69
"On the Unity of the Physical World
  Picture," 133–136
Orr, William McFadden, 133
Ostwald, Wilhelm, 105, 143, 144

Paschen, Friedrich, 163, 164, 170
Pauli, Wolfgang
 and crisis, 265
 and exclusion principle, 247, 249, 251, 256,
  257, 261
 and models, 223, 245, 250, 256
 and "Pauli effect," 181
 and Sommerfeld school, 9, 62, 257
 and *Zweideutigkeit*, 257, 258, 261
Pauling, Linus, 9, 51, 266

Pearson, Karl, 56
Pedagogical economy, 6, 48, 61, 63, 68
Pfrang, Hans, 242, 243
Physics of problems
  and crises and revolutions, 267–269
  historical context of, 5, 7, 19, 72, 91
  and physics of principles, 2, 3, 43
  and Sommerfeld School, 2, 5, 19, 30, 54, 72, 85, 241–243, 249
  as Sommerfeld's style, 3, 44, 45
Planck, Max
  and anthropomorphism, 134–136, 160, 173, 187
  Antrittsrede, 95, 96, 107
  and Arrhenius, 98, 99, 105, 106, 110–112
  and black-body theory, 8, 95, 112, 114–131, 145, 157
  decline of, 183, 184
  on dilute solutions, 101–105
  on experiment, 97, 109, 136, 137
  on ideal processes, 7, 8, 98, 99, 107–113, 132–136, 173
  on irreversibility, 114, 117, 119, 120, 124, 126, 129, 135, 136
  pedagogical practices of, 56
  on principles, 2, 3, 97, 98, 133–138, 173, 187
  on quantum hypothesis, 8, 141, 156–161
  on quantum of action, 158–161
  second theory of, 157, 160, 161
  at Solvay Congress, 157–160
  "statistical" method of, 8, 142, 143, 150, 156–162
  theoretical methodology of, 107–110
  on theory and experiment, 97, 109, 110, 132, 133
  on thermochemistry, 95, 98, 99, 107, 109, 127, 128
  and "thermodynamic method," 97–99, 106, 113, 138, 145, 173
Pockels, Friedrich, 133
Poincaré, Henri, 155
Principles
  Bohr and, 2, 185, 194–198
  in Britain, 194
  crises and revolutions and, 268, 269
  Eddington and, 192, 193
  Einstein and, 184–190
  Høffding and, 198–200
  Jeans and, 193
  Mach and, 134, 188, 189
  physics of, 2, 3, 5, 7, 43, 184, 187, 188, 268, 269
  Planck and, 2, 3, 97, 98, 133–138, 173, 187
  practice of, 2, 7, 97
  Sommerfeld and, 57, 58
Pringsheim, Ernst, 39

Quantum hypothesis
  and electromagnetic worldview, 31, 37–43
  Planck's postulation of, 8, 141, 156–161
  Sommerfeld on, 30, 33–39, 140, 143, 147–149, 151, 153

Rayleigh-Jeans equation, 37–41
Relativity, 31, 42, 170, 189, 190–193
Revolutions, 264–269
Reynolds, Osborne, 28
Riedler, Alois, 19, 21
Roentgen, Wilhelm, 1, 13, 50, 76
Routh, Edward, 56, 57, 67
Rubinowicz, Adalbert, 228, 231, 232
Runge, Carl, 163, 164, 207
Rydberg, Johannes, 211

Scherzer, Otto, 52
Schrödinger, Erwin, 10, 183, 203, 223, 249, 264
Schwarzschild, Karl, 77, 90, 171
Serwer, Daniel, 248
Solvay Conference (1911), 8, 30, 139–143, 149–151, 156–159
Sommerfeld, Arnold
  at Aachen Hochschule, 5, 19, 24
  and aesthetics, 202, 203, 212, 225
  on atomic spectra, 162–165, 168–171, 205–213

and Atom-Mystik, 9, 202, 203, 225, 239, 251, 256
on black-body theory, 14, 30, 34–39
and Bohr, 9, 60, 183, 185, 219, 225–227, 232–236, 239–241
on Bohr's model, 9, 155, 162–169, 214–223, 254
on connection of quantum and electromagnetic, 154–156, 230, 231, 234
on correspondence principle, 9, 228, 233–241
on diffraction, 145–147
dynamical method of, 142, 143, 149–156, 161, 162, 165, 171, 173
and electromagnetic worldview, 3, 6, 14, 30–33, 37, 38, 42, 44, 141, 142, 145, 151, 154–156
on energetics, 144
historiography of, 248
on impulse theory, 146–149
and Klein, 25–27
as mathematician, 16, 24, 25
at Munich, 51
pedagogical practices of, 13, 14, 48, 51–60, 66–68, 172, 242
and quantum hypothesis, 30, 33–39, 42, 140, 143, 147–153, 166
on quantum of action, 148–151, 154, 155
and quantum numbers, 168, 169, 205, 206, 210–212, 231, 232, 259, 260, 266
"refashioning" of, 14, 18
at Solvay Conference (1911), 139, 149–155
on spectral laws, 209–211
and spherical wave theory, 229–232, 236, 237
style of, 3, 13, 44, 45, 142
upbringing of, 15
vision of theoretical physics of, 18, 25, 29, 30, 44, 142
and World War I, 80–82
Sommerfeld School
 and construction of theoretical physics, 6, 14, 28, 44
 and creativity, 66, 68, 70
 and crises and revolutions, 267–269

informal pedagogical economy of, 61–63
practice of pedagogy in, 6, 28, 44, 47, 51–60, 91, 183, 241, 242
and problems, 2, 7, 15, 28, 44, 63, 64, 67, 79, 83–89, 241–244
and quantum physics, 9, 143, 156, 242–244
and "seminary," 59–61
success of, 3, 9, 47, 48
and World War I, 7, 72, 79, 82–87, 91
Stark, Johannes, 147, 180
Statistical method
 of Planck, 142, 143, 150, 156–162
 and quantum, 8, 142, 143
 Sommerfeld's use of, 165–167, 172
Stenström, W., 219, 220
Stokes, George Gabriel, 145
Stoner, E. C., 254–256

Technik, 201–204
Technische Hochschulen, 5, 20–23
Theoretical physics
 construction of, 4, 5, 14, 18, 29, 30
 development of in Munich, 47
 and experiment, 3, 18, 29, 45, 97, 170, 173, 180–183
 status of, 1, 4, 18, 137, 179–182
 World War I and, 72–78, 90, 91, 179, 180
Thermochemistry, 95, 98, 99, 107, 109, 127, 128
Thermodynamic method, 97–99, 106, 113, 138, 145, 173
Traweek, Sharon, 65, 181

Über die Theorie des Kreisels, 25–27, 57, 84
Umdeutung, 248, 262, 263

van 't Hoff, Jacobus, 98, 101, 104, 108, 110, 111
Voigt, Woldemar
 on classical and modern physics, 1
 and phenomenology, 133, 212, 213, 263
 and Sommerfeld, 16, 164, 212, 213, 263
Volkmann, Paul, 15, 17
Von Jolly, Phillip, 137

Vorlesungen über die Theorie der
  Wärmestrahlung, 14, 30, 33, 39, 158, 161

Warwick, Andrew, 6, 44, 64–66, 191
Wave mechanics, 249, 264
Wentzel, Gregor, 237, 238, 245, 258, 259
Wiechert, Emil, 13, 16, 145, 146
Wien, Max, 75
Wien, Wilhelm, 4, 32, 39–42, 73–75, 154, 179, 182, 183, 202, 203
Wien's law, 129–131
Wireless telegraphy, 44, 45, 75, 76, 79–83, 242, 243
World War I
 Hopf in, 7, 71, 72, 85–90
 Sommerfeld and, 80–82
 Sommerfeld School and, 7, 72, 79, 82–87, 91
 and theoretical physics, 72–78, 90, 91, 179, 180

X-rays
 Sommerfeld on, 46, 146–151
 and Sommerfeld School, 85
 spectra of, 214–223, 260
 and World War I, 76, 77, 85

Zeeman effect
 Heisenberg on, 244, 263
 Lorentz on, 163, 164
 Pauli on, 249, 251, 265
 Sommerfeld on, 164, 207–209, 214, 219
Zenneck, Jonathan, 82
Zweideutigkeit, 251, 257, 258, 261

Transformations: Studies in the History of Science and Technology
Jed Z. Buchwald, general editor

Red Prometheus: Engineering and Dictatorship in East Germany, 1945–1990
Dolores L. Augustine

A Nuclear Winter's Tale: Science and Politics in the 1980s
Lawrence Badash

Jesuit Science and the Republic of Letters
Mordechai Feingold, editor

Ships and Science: The Birth of Naval Architecture in the Scientific Revolution, 1600–1800
Larrie D. Ferreiro

H.G. Bronn, Ernst Haeckel, and the Origins of German Darwinism: A Study in Translation and Transformation
Sander Gliboff

Isaac Newton on Mathematical Certainty and Method
Niccolò Guicciardini

Weather by the Numbers: The Genesis of Modern Meteorology
Kristine Harper

Wireless: From Marconi's Black-Box to the Audion
Sungook Hong

The Path Not Taken: French Industrialization in the Age of Revolution, 1750–1830
Jeff Horn

Harmonious Triads: Physicists, Musicians, and Instrument Makers in Nineteenth-Century Germany
Myles W. Jackson

Spectrum of Belief: Joseph von Fraunhofer and the Craft of Precision Optics
Myles W. Jackson

Affinity, that Elusive Dream: A Genealogy of the Chemical Revolution
Mi Gyung Kim

Materials in Eighteenth-Century Science: A Historical Ontology
Ursula Klein and Wolfgang Lefèvre

American Hegemony and the Postwar Reconstruction of Science in Europe
John Krige

Conserving the Enlightenment: French Military Engineering from Vauban to the Revolution
Janis Langins

Picturing Machines 1400–1700
Wolfgang Lefèvre, editor

Heredity Produced: At the Crossroads of Biology, Politics, and Culture, 1500–1870
Staffan Müller-Wille and Hans-Jörg Rheinberger, editors

Secrets of Nature: Astrology and Alchemy in Early Modern Europe
William R. Newman and Anthony Grafton, editors

Historia: Empiricism and Erudition in Early Modern Europe
Gianna Pomata and Nancy G. Siraisi, editors

Nationalizing Science: Adolphe Wurtz and the Battle for French Chemistry
Alan J. Rocke

Islamic Science and the Making of the European Renaissance
George Saliba

Crafting the Quantum: Arnold Sommerfeld and the Practice of Theory, 1890–1926
Suman Seth

The Tropics of Empire: Why Columbus Sailed South to the Indies
Nicolás Wey Gómez